ALLELOPATHY IN CROP PRODUCTION

OTHER RELATED BOOKS

Allelopathy : Field Observation and Methodology *S.S. Narwal*

Allelopathy in Agriculture and Forestry *S.S. Narwal*

Allelopathy in Pests Management for Sustainable Agriculture *S.S. Narwal*

Neem in Sustainable Agriculture *S.S. Narwal*

Research Methods in Plant Sciences : Vol. 1. Soil Analysis *S.S. Narwal*

Research Methods in Plant Sciences : Vol. 2. Plant Protection *S.S. Narwal*

Research Methods in Plant Sciences : Vol. 3. Plant Pathogens *S.S. Narwal*

Research Methods in Plant Sciences : Vol. 4. Plant Analysis *S.S. Narwal*

Research Methods in Plant Sciences : Vol. 5. Plant Physiology *S.S. Narwal*

Plant Science: Research Methdos *S.S. Narwal*

Plant Physiology: Research Methods *S.S. Narwal*

The Influence of One Plant on Another : Allelopathy *S.S. Narwal*

ALLELOPATHY IN CROP PRODUCTION

Shamsher S. Narwal

Department of agronomy
Haryana Agricultural University
Hisar – 125 004 India
E-mail: narwal_1947@yahoo.com

SCIENTIFIC PUBLISHERS (INDIA)
P.O. BOX 91 JODHPUR

Published by:

Scientific Publishers (India)
5 A, New Pali Road, P.O. Box 91
Jodhpur 342 001 (India)

Branch Office
Scientific Publishers (India)
4806/24, Ansari Road, Daryaganj
New Delhi - 110 002 (India)

E-mail: info@scientificpub.com
Website: www.scientificpub.com

Print : 2012

ISBN: 978-81-7233-777-3

Lasertype set: Rajesh Ojha
Printed in India

Dedicated to the
Memory of My Younger Brother
Rajender Singh Narwal
and
Professor Hans Molisch
Father of Allelopathy

FOREWORD

It is a sincere pleasure for me to write a foreword for this book. It is the only scientific monograph that is committed entirely to discussing the roles of allelopathy in crop production. Professor Narwal has also expanded his discussion somewhat outside the field of allelopathy to cover some aspects of chemical interactions between plants and insects and plants and nematodes. Obviously, these are important subjects in crop production also.

This book is quite thorough in its treatment of allelopathy in relation to plants involved in crop production. There are discussions of : terminology, modes of escape of allelochemicals from plants, factors affecting production of allelochemicals, methods of action of allelochemicals, crop□crop interactions, weed-crop interactions, biological weed control, biological disease control, plant-insect interactions, plant-nematode interactions, allelochemicals as herbicides, crop rotations to exploit allelopathy or to eliminate detrimental effects of allelopathy, and elimination of phytotoxins from agricultural lands. There are numerous other discussions on lesser subjects.

Scientists in India have contributed greatly to our knowledge of the field of allelopathy in the past two or three decades. It is thus appropriate that one of them should write this book which has so much to offer to the fields of agriculture and horticulture. All persons connected in any way with food production, including farmers, extension agents, consultants, and scientists should have a copy of this reference readily available.

Elroy L. Rice, Boyd
Professor Emeritus of Botany
University of Oklahoma, Norman
USA

PREFACE

Allelopathy is a new branch of science. Molisch (1937) coined this term, which refers to all stimulatory and inhibitory biochemical interactions between the plants including microbes. It generally refers to the deterimental effects of plants of one species (donor) on the germination growth or development of another species (recipient). The allelopathic effects may have existed in crop production since the start of farming but these could not be isolated due to poor analytical facilities in the past. For example, Theophrastus observed such effects as early as in Ca 300 BC but greater interest in allelopathic research work has been generated since 1960.

My first such experience was in 1975 and 1976 with crop rotation studies in Multiple Cropping Systems under field conditions. In one experiment, mustard (*Brassica juncea*) was sown after the preceding fodder crops of clusterbean, cowpea, maize, sorghum, pearlmillet and fallow. The germination, growth and yield of mustard remained very poor, when sown after pearlmillet crop, compared with other preceding crops even with 90 kg N/ha. In another field trial of 300 per cent crop intensity, greengram was sown after previous crops of pearlmillet, sorghum and greengram. The germination, growth and yield of greengram was very poor after greengram than after sorghum and pearlmillet. I discussed this problem with soil scientists, microbiologists, plant pathologists etc. of my University at Hisar but could not get satisfactory reply. The first book on *Allelopathy* by E.L. Rice came to my Library in 1977 and thus I could know that these effects may be due to allelopathy. Ventura *et. al.* (1984) published results of 8-years study done at IRRI, Philippines and established that in greengram-greengram rotation, the germination, growth and yield of second greengram crop was reduced due to allelopathy.

Most of the allelopathic research on crops has been conducted in the developed countries, where single crop is raised in one year, because of too severe winters to allow raising of the second crop.

Therefore, during fallow period residues of previous crop decompose fully, hence, had no injurious effects on the succeeding crop. However, in the tropical and subtropical irrigated areas, where climate is favourable for round the year cropping, scanty allelopathic research has been done, though allelopathy has greater role to play under these conditions, where an array of crops and weeds exist together. In these areas, when more than three crops are raised annually in the same field (300% cropping intensity) there is a little time gap (10-20 days) between the harvest of previous crop and sowing of next crop. Therefore, the decomposition of residues of previous crop coincides with the seed germination and seedling phase of the next crop, as a consequence the phytoloxic decomposition products may reduce the germination and seedling growth of next crop. Besides, there could be several other mechanisms of allelopathic interactions. The allelopathy has been related to the problems of (a) weed : crop interference, (b) phytotoxicity in stubble mulch farming and (c) soil sickness in certain crop rotations. The past reviews and books on allelopathy have dealt with grasslands, forestry and weeds etc. but none dealt exclusively with crop production. This book, therefore, will acquaint the scientists and students with the current status of the allelopathic research on various aspects of field crops and weeds, which exist together under field conditions and to gainfully utilize these research findings to increase the crop productivity. Only field crops and weeds have been included in this book.

The book consists of six chapters. The first chapter *Introduction* describes the brief history and terminology of allelopathy. The second chapter deals with *Allelochemicals*, their mode of release, origin, factors affecting their production and mode of action. The third chapter *Interactions between plant commu- nities* discusses interactions between crops-crops, crops- weeds and weeds-crops. The fourth chapter *Exploitation of allelopathy for crop production* explains use of allelopathy for controlling weeds, insects, nematodes and diseases and as bioregulators. The fifth chapter *Elimination of phytotoxins* describes the techniques to eliminate allelochemicals from crop fields. The last chapter '*Future prospects*' outlines future lines of research and problems in allelopathic research. The references of all chapters, have been given in the last in alphabetical order.

I am highly indebted to Dr. Ram Singh, Associate Professor (Entomology) for contributing Section 4.2. *Management of Insect Pests* and to Prof. D.S. Bhatti, former Prof. & Head, Department of

Nematology for contributing Section 4.3. *Management of Phyto-Nematodes* for this book.

I am thankful to my wife Mrs. Santosh and sons Rajesh Narwal and Rakesh Narwal who have spared me time for completing this stupendous task and have been constant sources of inspiration. Heart-felt thanks are also to Dr. Gautam Kalloo, Professor of Vegetable Crops, Haryana Agricultural University, Hisar, presently, the Project Director, ICAR, Directorate of Vegetable Crops, Varanasi, Uttar Pradesh, India, for critical comments on the manuscript. I also express my thanks to my Graduate students Mohan Kumar Sarma, Jagdish Chander Tamak and Virender Sardana for helping in various stages of this book. I also thank all the authors who have sent me their reprints for inclusion in this book.

S.S. NARWAL

CONTENTS

CHAPTER - 1

INTRODUCTION

Allelopathy is derived from two Greek words *Allelon* means each other and *Pathos* means to suffer i.e. the injurious effects of one upon another. However, Molisch (1937) coined this term which refers to all biochemical interactions (stimulatory and inhibitory) among plants, including microorganisms. It represents the plant-against-plant aspect of the broader field of chemical ecology. Some authors have used the term in a more restricted sense to describe only the harmful effects of one higher plant upon another. Rice (1984) reported many cases of allelopathy which either directly or indirectly involves microbes. Allelochemicals which inhibited the growth of some species at certain concentrations may stimulate the growth of same or different species at lower concentrations. Therefore, in this review the term `allelopathy' has been used according to Molisch's (1937) definition.

The term allelopathy generally refers to the detrimental effects of higher plants of one species (the donor) on the germination, growth or development of plants of another species (the recipient). It can be separated from other mechanisms of plant interference because the detrimental effect is exerted through release of chemical inhibitors (allelochemicals) by the donor species to the plant environment. It is, therefore, different from competition which involves the removal or reduction of some growth factors from the environment that is required by some other plant sharing the same habitat e.g. water, minerals and sunlight. Confusion has arisen in the literature because some biologists consider allelopathy to be a part of competition. This confusion could be lessened by using the term `allelopathy' to refer to the overall influence of one plant on another (Muller, 1969).

Interference thus includes both allelopathy and competition and in this review the term interference has been used in this sense. Since, there are no known techniques to eliminate the possible allelopathic effects in `competition' experiments, therefore, all such studies have been accurately entitled as `interference' studies.

1.1 PROOF OF ALLELOPATHY

A number of studies have provided excellent evidence for allelopathy but only a few investigators have followed a specific protocol (similar to Koch's postulates for proof of disease) to achieve convincing proof (Fuerst and Putnam, 1983). The proof of allelopathy generally involves the following sequence of studies :

(a) Demonstrate the interference using the suitable controls, describe the symptoms and quantify the growth reduction.

(b) Isolate, characterize and assay the chemical against species that were previously affected. Identification of chemicals that are not artifacts is essential.

(c) Obtain toxicity with similar symptoms when chemicals are added back to the system.

(d) Monitor the release of chemicals from the donor plant and detect them in the environment (soil, air, etc.) around the recipient and ideally, in the recipient plant.

1.2 BRIEF HISTORY OF ALLELOPATHY

Although, the impact of allelopathy on agriculture was recognised by Democritus and Theophrastus in the 5th and the 3rd Century B.C., respectively (Smith and Secoy, 1977) and by DeCandolle in 1832 but most of the progress in this field has occurred in the twentieth century (Rice, 1984). Schreiner and Reed (1907, 1908), Schreiner and Shorey (1909) and Schreiner and Sullivan (1909) reported that plants produced toxic compounds. Molisch (1937) coined the term *Allelopathy* and was pioneer with studies of ethylene on plants. Since the 1960's allelopathy has been increasingly recognised as an important ecological mechanism, which influences the plant dominance, succession, formation of plant communities and climax vegetation and crop productivity. It has been related to the problems with weed: crop interference (Bell and Koeppe, 1972), phytotoxicity in stubble mulch farming (McCalla and Haskins, 1964) and in certain types of crop rotations (Conrad, 1927). Rice (1984) indicated that allelopathy contributed to weed seed longevity problem through two mechanisms, (a) chemical inhibitors in the seed prevented their decay by microbes and (b) the inhibitors kept the seed dormant, although viable for many years.

The past reviews on allelopathy (Altieri and Doll, 1978; Bhandari and Sen, 1983; Lovett, 1983b; Rice 1984, 1985; Einhellig, 1985a,b; Hedin, 1985; Thompson, 1985; Putnam, 1985, 1986; Putnam and Tang, 1986; Waller, 1987; Einhelling and Leather, 1988) have dealt with grasslands, forestry and weeds. Some of these reviews slightly mentioned the crop production but none dealt exclusively with this important aspect. However, the indications of allelopathy had been observed in crops since Theophrastus (*Ca.* 300 B.C.), but a lot of research work has been done in the last few decades. Besides, to meet the food, fuel, fibre and other needs of the ever increasing human population, the cropping intensity is increasing world over (in areas where temperature and water are not limiting for round the year cropping) and new lands are being brought under cultivation. Most of the allelopathic research on crops has been conducted in the developed countries, where monocropping is practised, because of too severe winters to allow raising of the second crop within a calender year. However, in the tropical and subtropical irrigated areas, where the climate is favourable for round-the-year cropping, little research had been done, though presumably, allelopathy has greater role to play under these conditions, where an array of crops and weeds exist together. The present review, therefore, will acquaint the scientists with the current status of allelopathic research on field crops and weeds, which exist together under field conditions and to utilize these findings to increase crop productivity. Only field crops have been included in this review.

1.3. TERMINOLOGY

Since the beginning of research in this field, the various scientists have coined the different terms, which may be confusing to the readers. Therefore to avoid such confusion, all the terms related to allelopathy have been defined in this section.

I. Allelopathy: The various scientists have defined allelopathy as under :

(a) Molisch (1937) coined the term to describe all the chemical interactions among plants (microbes and higher plants), stimulatory as well as inhibitory.

(b) It includes interspecific (antibiotic) as well as intraspecific (autotoxic) chemical co-action (Bonner, 1950).

(c) It refers to the deleterious effects that one higher plant has on another through the production of chemical retardants that escape into the environment (Martin and Redemacher, 1960; Muller, 1966).

(d) Inhibition of germination, growth and metabolism of one plant due to the release of organic chemicals by another (DelMoral and Cates, 1971).

(e) It is interspecific chemical co-action, while auto-allelopathy is intraspecific chemical co-action (Tischler, 1975).

(f) It refers to the detrimental effects of higher plants of one species (the donor) on the germination, growth and development of another (the recipient) plant species (Putnam and Duke, 1978).

(g) It refers to any direct or indirect inhibitory effect by one plant (including microorganisms) on another through the production of chemical compounds that escape into the environment (Rice, 1974). However, later on, Rice (1984) followed the definition of Molisch (1937).

II. Competition : It means that one plant utilizes the necessary environmental growth resources (water, nutrients and sunlight) resulting in a shortage, which is harmful to another plant sharing the same habitat (Muller, 1971). It always implies resource limitation, a specific mechanism of interference.

III. Interference : The adverse effect of a neighbouring plant growing in association is called interference (Muller, 1969). Szczepanski (1977) reported following causes of interference :

(a) **Allelospoly or competition :** It involves depletion of one or more growth resources in association.

(b) **Allelopathy :** The addition of toxins by one or more species in the association.

(c) **Allelomediation :** It involves selective harbouring of an herbivore that might selectively feed on one species, thus lending advantage to another.

(d) **Autotoxicity :** The toxicity within the same species is called autotoxicity.

(e) **Allelochemics or allelochemicals :** Chemicals that cause allelopathic influences on the recipient species are called allelochemics or allelochemicals.

(f) **Allomones :** The allelochemicals which give adaptive advantage to the producer are called allomones e.g. repellents, attractants etc.

(g) **Soil sickness :** The decline in the quality and yield of a crop when grown on the same land year after year (Tukey, 1969).

Besides, Grummer (1955) classified inhibitors based upon their sources of origin and upon the organisms affected by their action (suspect). These terms were not used widely perhaps because they were either non-descriptive (e.g. koline) or non-specific (e.g. antibiotic). Fuerst and Putnam (1983) suggested the terms 'Phytoinhibitins' and 'Saproinhibitins' to describe compounds of plant and microbial origin, respectively, which inhibited growth of higher plants.

CHAPTER -2

ALLELOCHEMICALS

Allelochemicals refer mostly to the secondary metabolites produced by plants and are byproducts of primary metabolic processes (Levin, 1976). They have an allelopathic effect on the growth and development of the same plant or neighbouring plants. The term allelochemicals include, (a) plant biochemicals that exert their physiological/ toxicological action on plants (allelopathy, autotoxicity or phytotoxicity), (b) plant biochemicals that exert their physiological/toxicological action on microorganisms (allelopathy or phytotoxicity) and (c) microbial biochemicals that exert their physiological/toxicological action on plants (allelopathy and phytotoxicity). Secondary compounds are metabolically active in plants and microorganisms, their biosynthesis and biodegradation play an important role in the ecology and physiology of the organism in which they occur (Waller and Nowacki, 1978; Waller and Dermer, 1981). Some of them are accumulated at various stages of growth, while, accumu-lation of some compounds depends upon time of day or season.

Allelochemicals selectively inhibit the growth of soil microorganisms or other plants (or both). They play a vital role in chemical warfare between plants (allelopathic interactions) and include natural herbicides, phytoalexins (microbial inhibitors) and inhibitors of seed germination. Although many allelochemicals are strictly defence substances, others are offensive compounds that act directly in weed aggressiveness, competition and the regulation of plant density. It is expected that in the near future many allelochemicals may be used commercially as herbicides, insecticides, nematicides or as bioregulators (Chapter 4, Section 4.1.2, 4.2.1, 4.3.5 and 4.4). Allelochemicals most often impart plant resistance to insects, nematodes and pathogens (Chapter 4, Section 4.2.2), besides, following their release into environment, some may regulate the distribution and vigour of plants. Most often plants come in contact with the allelochemicals in soil and their effect on crop plants may be modified by soil moisture, soil temperature and other soil factors

(Patrick and Koch, 1958; Patrick *et. al.*, 1964; Wang *et. al.*, 1967b; McCalla and Norstadt, 1974; Bhowmik and Doll, 1983a,b; Einhellig and Eckrich, 1984). Some of the allelochemicals such as terpenoids and polyacetylenes may function in a volatile state, but most of the current research in agroecosystems involves water soluble compounds. The known list of chemicals involved in allelopathy continues to expand and with better isolation and identification techniques, many new substances are being added every year.

Occurrence

The existence of allelochemicals in higher plants and microorganisms, has been documented by several workers. These are produced in above or below ground plants parts or in both, to cause allelopathic effects in a wide range of plant communities. This aspect has been reviewed by many workers, therefore, not discussed in detail. Plant parts known to contain allelochemicals (Rice, 1974) are :

(i) **Roots and rhizomes :** In general, they contain fewer and less potent or smaller amounts of allelochemicals than leaves, but sometimes, it may be the reverse also.

(ii) **Stems :** They contain allelochemicals and are sometimes the principal sources of toxicity.

(iii) **Leaves :** They are the most important sources of allelochemicals. Specific inhibitors in leaves have been identified.

(iv) **Flowers/inflorescence and pollen :** Although studies on flowers or inflorescence are limited, but there is growing evidence that the pollen of corn and *Parthenium* have allelopathic properties.

(v) **Fruits :** Many fruits are known to contain toxins and have been found inhibitory to microbial growth and seed germination.

(vi) **Seeds :** Seeds of many plant families or species have been found to inhibit seed germination and microbial growth.

2.1. MODES OF ALLELOCHEMICALS RELEASE

Since a major requisite of allelopathy is that an organic substance or allelochemical be transferred from a donor plant to a recipient plant, therefore, mode of transfer may play a great role in its toxicity and persistence.

The donor plant which release these chemicals, generally store them in the plant cells in a bound form, such as water-soluble glycosides, polymers including tannins, lignins and salts. It has been suggested that upon cleavage by plant enzyme or environmental stress, the toxic chemicals are released into the environment from special glands on the stems or leaves (Putnam and Duke, 1978; Fisher, 1979; Einhellig, 1985a).

First the terpenoids such as α-pinene, ß-pinene, cineole and camphor are released to the environment through volatilization, which is noticeable under drought conditions. The water-borne phenolics and alkaloids are then moved out by rainfall through leaching. Next, phytotoxic aglycones such as phenolics are produced during the decomposition of plant residues in soil. Finally, many secondary metabolites such as scopoletin and hydroquinones may be released to the surrounding soil through root exudates. Release through leachates and exudates requires water solubility and a broad range of allelochemicals are involved. Whatever may be the mode of formation, it is generally assumed that these chemicals are not toxic to the donor plants. Once the chemicals from the donor plants are released into the environment, they may be either degraded or transformed into other forms. The resultant stages of these chemicals may also be toxic to the host plant. Grodzinsky (1982) reported that the pool of allelochemicals in soil is replenished by the volatization, leachates, root exudates and decompostion of crop residues etc. and at the same time, it is exhausted through absorption by plants, decomposition by microorganisms and carried away by wind and water etc. In higher plants allelochemicals are released from plants through (a) volatilization (b) leaf or stem leachates, (c) root exudates and (d) decomposition of plant residues (Fig. 1) (Rice, 1984; Putnam, 1985).

2.1.1. Volatilization

In few cases, allelochemicals may volatilize and be absorbed directly from the atmoshphere by neighbouring plants, be absorbed from condensate in dew, be absorbed on the soil particles and subsequently taken up by contact with plants or from the soil solution (Muller, 1966). Apparently several terpenoids transfer in these ways, higher plants produce a variety of terpenoids and monoterpenoids as major components of essential oils of plants (Robinson, 1967). From the plants rich in such compounds, these may be released continuously to the atmosphere, this process is enhanced in hot weather, therefore, this phenomenon is observed in the arid regions of the world. The genera which release volatiles are *Artemisia, Salvia, Parthenium* and *Eucalyptus* (Rice, 1984). Camphene, camphor, cineole, dipentene, α-pinene and ß-pinene were identified as volatile inhibitors produced by several shrubs of the southern California chapparal (Muller and Muller, 1964; Muller, 1966; Weaver and Klarich, 1977). More complex terpenoids, such as sesquiterpene lactones may also be involved (McCahon *et. al.*, 1973). Neill and Rice (1971) reported that volatile inhibitors from the leaves of western ragweed may contribute to allelopathic interference in old field succession. Thus, a number of species from several climatic regions may express allelopathy through volatilization (Rice, 1974; Horsley, 1977). Muller (1966) reported that terpenes volatilized by several species inhibited the growth of nearby plants. Kanchan (1975) reported that *Parthenium hysterophorus* also released the allelochemicals through volatiles. Oleszek (1987) found that pulverised leaves of cruciferae species (*Brassica juncea, B. nigra, B. napus, B. rapa* and *B. oeracea*) released volatile substances. The volatiles of *B. juncea* and *B. nigra* were most harmful to germinating seeds of lettuce and wheat, however, the allelochemicals were not identified.

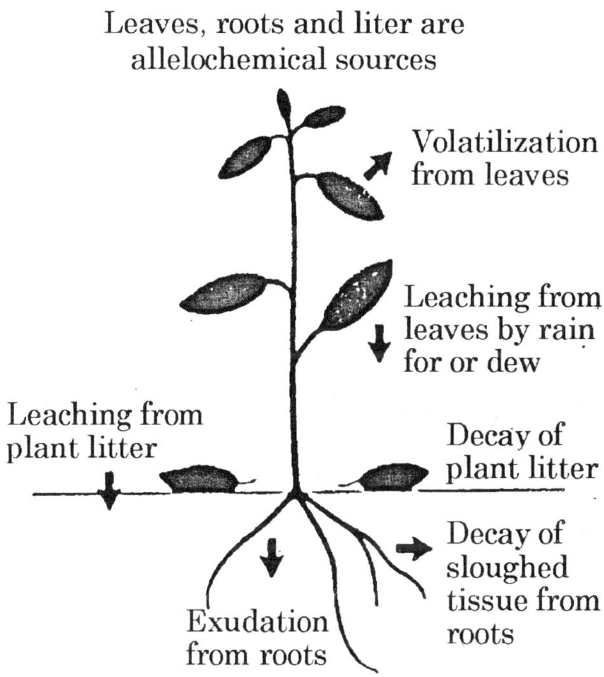

Allelopathic chemicals from plants may be released from living leaves as volatiles or leachates or from roots through exudation or sloughing off of dead tissues. They also may be leached from leaf litter on the soil surface.

Fig. 1. Production of Allelochemicals.

2.1.2. Leaching

Leaching is the removal of water solable substances from plants by the action of aqueous solvents such as rain, dew, mist and fog

(Tukey Jr., 1970), however, earlier authors such as Bonner (1960) have used the term excretion for this phenomenon. All plants seem to be leachable, although the degree depends on type of tissue, stage of maturity and type, amount and duration of precipitation (Tukey Jr., 1970). A large diversity of allelopathic compounds are leached, both organic and inorganic such as phenolic acids, terpenoids and alkaloids (Borner, 1960; Tukey and Mecklenburg, 1964; Tukey Jr., 1970). The leaching of mineral nutrients, carbohydrates and phytohormones, may be beneficial for the growth of associated species, however, mainly toxic effects have been studied. Although seed leachates may contribute toxins (Gressel and Holm, 1964; Tukey and Mecklenburg, 1964), but foliage leachates have been investigated the most. Toxin-bearing leachates are important in weed-crop associations (Grummer and Beyer, 1960; Kanchan and Jayachandra, 1980b) and grasslands (Wilson and Rice, 1968; Rice, 1971, Lodhi and Rice, 1971; Neill and Rice, 1971).

Colton and Einhellig (1980) suggested that leaf leachates of velvet leaf were inhibitory to soybean. The rainfall leachates from the leaves of *Camelina* weed adversely affected the growth of linseed plants (Grummer and Beyer, 1960). The leaf washings of *Camelina* contained a number of potentially toxic chemicals including benzylamine, which retarded the seedling growth of linseed (Lovett, 1987). The extract of broccoli plants not soaked in rain was more inhibitory to growth of barley, rye grass, vetch and radish as compared to that from plants soaked in rain, because, the rainfall leached inhibitors from its plant tissues (Jimenez- Osorino and Gliessmann, 1987). Under field conditions, the occurrence of the allelopathic phenomenon depends on the amount of rainfall during late spring or early summer, presumably at a critical stage for crop growth and development. Substances leached from plants influence not only the leached plant but also the adjacent plants and thus play a role in allelopathy. Leached material may fall to the soil beneath, from where they can be absorbed directly through roots by either (a) the same plant from which they were leached or (b) adjacent plants of the same or different species. Besides, leached materials may also be intercepted before they reach the soil by stems, branches and foliage of the same or adjacent plants. Mecklenburg and Tukey (1964) showed that both inorganic cations and anions as well as organic materials can cycle from plant to plant in this manner.

2.1.3. Root exudates

Root exudates are those substances released from intact live plant roots into the surrounding medium (Rovira, 1969). Although their volume is small i.e. 2-12% of the total gross phytosynthates

(Grodzinsky, 1974), but they play significant role in allelopathy (Woods, 1960; Rovira, 1969; Whittaker, 1971; Rice, 1974). The root exudates are mediators in the interrelationships between higher plants and microorganisms. In some cases, they (a) provide plants with immunity against phytopathogens, (b) sustain the life activity of microflora in the rhizosphere and (c) sustain the life of mycorrhiza to improve mineral nutrition in the plants. There are known instances when higher plants cannot absorb mineral nutrients without associated heterotrophs. These root exudates are also important for the elimination of waste metabolic products from the plants (Grodzinsky, 1982). The root exudates of wheat, barley, oat, carrot and radish contains inhibitors (Evanari, 1949).

A variety of compounds are exuded from the roots, which may influence the growth of microorganisms and associated higher plants. The identification of allelochemicals in root exudates is very troublesome, because under many experimental conditions, microorganism activity may alter the initial exudate. Certainly in a soil environment transformations by microorganisms of the rhizosphere must occur, in some cases inactivating the original exudation compounds and in other cases creating active allelochemicals. Exudates vary according to plant species and age, temperature, light, plant nutrition, microorganism activity around the roots and the nature of the medium supporting the roots (Rovira, 1959, 1969). Roots of peas and oats excreted amino acids, while those of wheat excreted sugars, amino acids, organic acids, nucleotides, flavones and enzymes. The oat roots excreted fluorescent compounds and their principal toxin was identified as scopoletin (Martin 1957). Oat varieties vary in their ability to exude scopoletin and four accessions exuded upto three times as much as the standard cultivar 'Garry' oat (Fay and Duke, 1977). The roots of seedlings in *Lolium* spp., *Nigella* spp. and *Papaver* spp. also excreted scopoletin (Eberhardt and Martin, 1957). Grodzinsky (1969) studied root exudates of maize, yellow lupine and horse bean using C_{14} and found that maize root excretions were mainly carbohydrates and that of legume roots were amino acids. The root excretions may be absorbed by plant roots of the same or other neighbouring species.

Root exudates of the most species were found inhibitory to other plant species except a few such as *Sorghum bicolor* which stimulated the germination of witch weed (Netzy *et. al.*, 1988). Fletcher (1912) found that root exudates of *Sorghum vulgare* Pers. and maize inhibited the growth of sesame. The root exudates of sorghum x sudan grass hybrid `De Kalb' proved inhibitory to the growth of alfalfa, Italian rye grass and curly cress (Forney *et. al.*, 1983). Elkan (1961) reported that when a non-nodulating near isogenic strain of soybean was planted in pots with normal nodulating soybean and ladino clover, its root excretions significantly decreased the nodulation of both plant species. The root exudates of buckwheat inhibited the growth of *Brassica campestris, Digitaria adscendens*, purple nutsedge and buckwheat (Tsuzuki, 1980). In further studies in sand culture, Tsuzuki *et. al.* (1984) applied root exudates of barley, white clover, lupine and spurrey to all these species and found that all the exudates decreased growth of all the test species. Tsuzuki and Kawagoe (1984) studied the effect of root exudates of soybean, alfalfa and white clover on the growth of soybean, alfalfa, rice, maize, barley, radish and turnip. They found that root exudates of soybean reduced the dry weight of radish and turnip and had no inhibitory effect on maize and soybean. Root exudates of white clover decreased the dry weight of radish only, while that of alfalfa reduced the dry weight of alfalfa, barley and radish; however, the maximum reduction occurred in alfalfa showing its autotoxicity. Sadhu and Dass (1971a) found that root exudates from selected species of rice affected seedling growth of each other but were non-toxic. Rye grown with one year-old grapes (*Vitis vinifera*) reduced their growth even under high moisture and nutrients owing to its root exudates (Cubbon, 1925). The root exudates of maize grown in sterile culture inhibited the growth of maize plants grown in the same solution (Rempe, 1967; Rempe and Gryunberg 1970) perhaps because, organic molecules of root exudates moved between herbaceous neighbouring plants (Ivanov, 1962). The root exudates of cucumber exerted allelopathic inhibitions on weeds and crops (Putnam and Duke, 1974). Kim and Kil (1987) found that the root exudates of tomato inhibited the growth of lettuce and egg plant. El-Habbasha and Behairy (1975) studied the effect of root exudates of cabbage `Badaly', bean (*P. vulgaris*) `Cemenal', onion `Behairy', tomato `Pritchard' and squash `Eskanderani' on seed germination and seedling development of each of these crops. Germination of bean was increased by root exudates from squash seedlings and decreased by root exudates from other seedlings. The fresh weight of bean seedlings was reduced by exudates from cabbage, squash, onion, bean and tomato in descending order. Seedling dry weight was slightly reduced by cabbage, bean and onion exudates and increased by others. Das and Ghosh (1978) reported that soaking of wheat seeds in root exudates of jute decreased number of roots per plant, root length, shoot length and fresh weight of wheat seedlings. The root exudates of Chinese cabbage inhibited the growth of mustard as well as its own radicle growth, fresh and dry matter (Akram and Hussain, 1987). The application of root exudates of asparagus to its three varieties (`Mary Washington', `California 309', `California 711') reduced plant height and number of stalks, thus showing autotoxicity (Young, 1984).

The scopoletin liberated from root exudates of oat inhibited the growth of *Sinapis arvensis* and *Papaver rhoeas* but not that of *Matricaria maritima* which was more affected by rye than by wheat root exudates (Martin and Rademacher, 1960). Root secretions of white lupine and maize inhibited the growth of lambsquarter and redroot pigweed (Dzyubenko and Petrenko, 1971). Application of root exudates of soybean cultivars to velvet leaf and foxtail millet reduced the dry weight of velvet leaf plants by 15% and had no inhibitory effect on foxtail millet (Rose *et al.*, 1984). Deperalta and Estioko (1923) showed both stimulatory and inhibitory effects of root excretions of weeds on rice production. The root leachates of hedge bindweed reduced seedling development of wheat and proved autotoxic (Quinn, 1974).

2.1.4. Decomposition of residues

The decomposition of plant residues potentially provides the largest quantity of allelochemicals that may be added to the rhizosphere. At plant death, materials compartmentalized in cells are released and recycled. Important variables in this process with regard to allelopathy are : the nature of the plant residues, the soil type and the conditions of decomposition. Patrick *et al.* (1964) reported that depending on the decomposing conditions, substances highly toxic, non-toxic or stimulatory to plants may be formed during the decomposition of similar plant residues. In general, more severe and persistent toxicity has been reported from cold and wet soils (Patrick and Koch, 1958; McCalla and Haskins, 1964; Patrick *et al.,* 1964; Wang *et. al.*, 1967b).

Since the decomposing plant materials are never equally distributed throughout the soil, the soil adjacent to the decomposing debris may contain more decomposition products than other areas (Patrick *et. al.*, 1963, 1964; Patrick, 1971; Drost and Doll, 1980). Therefore, as the roots grow through the soil, at some points they may come in contact with decomposing plant residue and are affected by allelochemicals, while at other locations there may be no such influence.

The soil is a dynamic system, where activity of substances released can be quite transitory, as they are subjected to destruction, soil absorption and inactivation and transformation by soil microflora. However, this does not preclude their having effects on plant growth during decomposition and often sustained toxicity may occur as new toxic products are formed in some of the transformations (Patrick *et. al.*, 1964). Besides, microorganisms active in decomposition may themselves produce inhibitory allelochemicals i.e. microbial toxins (McCalla and Haskins, 1964;

Norstadt and McCalla, 1968, 1969, 1971; McCalla and Norstadt, 1974). In decaying crop residues, different types of allelochemicals differing in their activity are produced depending upon the organisms that digest them. For example, some *Penicillium* spp. synthesize patulin and pecolinic acid; *Aspergillus* spp. synthesize malphermine etc. All these substances produced from the crop residues inhibited the growth of crops (McCalla, 1971). Turner (1971) listed many fungal compounds which influenced the growth of both higher plants and microorganisms. Microbes in the rhizosphere can produce toxic compounds by enzymic degradation of conjugates or polymers present in the plant tissue e.g., microbe action on cyanogenic glycosides of Johnson grass produces the toxin HCN and the corresponding benzaldehydes (Conn, 1980). According to Rice (1979), investigations in Russia by Kudhrir have demonstrated several allelopathic situations where bacteria, actinomycetes and fungi of the rhizosphere, synthesize allelopathic compounds.

Some of the toxic effects of decomposition products on plants are inhibition of seed germination, stunted growth, inhibition of the primary root system and increase in secondary roots, inadequate nutrient absorption, chlorosis, slow maturation and delay or failure of reproduction (Patrick and Koch, 1958; Patrick *et. al.*, 1964). Debris of wheat, oat, barley, rye, corn, sorghum, rice and sugarcane (Patrick and Koch, 1958; Borner, 1960; Patrick *et. al.*, 1963; Toussoun and Patrick, 1963; Guenzi *et. al.*, 1967; Toussoun *et. al.*, 1968; Kimber, 1973a,b; Wu *et. al.*, 1976a) and weeds such as quackgrass, Canada thistle, Johnson grass, yellow nutsedge and giant foxtail (Schreiber and Williams, 1967; Bell and Koeppe, 1972; Kovacs, 1972; Drost and Doll, 1980; Wilson, 1981) have been implicated in allelopathy during decomposition (Rice, 1974).

Investigations using aqueous plant extracts have shown that water-soluble inhibitors may be present in crop plants (McCalla and Duley, 1948; Borner, 1960; Guenzi and McCalla, 1962; Guenzi *et. al.*, 1964 and Overland, 1966) and in weeds of crop fields (Einhellig and Rasmussen, 1973; Rasmussen and Einhellig, 1975, 1979; Bendall, 1975; Ashraf and Sen, 1978; Colton and Einhellig, 1980). These water-soluble inhibitors are quickly released during decomposition. Further, Patrick *et. al.* (1964) estimated that upto 30% of the plant's dry matter entering the decay cycle was lignin which is formed from alcohols and acids with the carbon skeleton of phenylpropane (Brown, 1969; Robinson, 1967), which are known inhibitors. Thus, it is reasonable that upon decay some of these compounds may become active; indeed, phenolics are released on decomposition of plants.

The inhibition resulting from plant residues, could not be explained by changes in nutrient availability (especially nitrogen) or due to microbial activity because, definitive experiments on this aspect are rare. Kimber (1973a) reported that both nitrogen immobilization and toxins affect the wheat yields, when grown in excessive straw residues and addition of nitrogen fertilizer did not overcome the harmful effects of straw. In contrast, Elliott *et. al.* (1981) reported that higher rates of nitrogenous fertilizers reversed the negative effects from winter wheat straw. Chandramohan *et. al.* (1973) also found that nitrogen applied to rice `Co. 13' planted in the residues of previous rice crop increased its growth. They also monitored the phenolic acids levels in the soil following fertilizer application and found that after 35 days the phenolic content decreased in inverse proportion to the fertilizer application. Therefore, nitrogen played a crucial role of increasing nutrient availability and in reducing the phenol toxicity. Chou and Lin (1976) concluded that the decline in productivity of rice subsequent to the first crop was primarily due to allelochemicals from the decaying residues in paddy.

The toxicity arising from plant residues provides some of the more challending problems and opportunities for agronomists and weed scientists. Toxins from the decomposing crop residues could affect young crop plants sown between the mature plants e.g., in relay cropping or into the stubbles of the preceding crop in multiple sequential cropping (Trenbath, 1976). In areas where stubble mulch farming is practised for soil and water conservation, toxins from the stubbles have proved toxic to certain crops in rotations (McCalla and Haskins, 1964). In modern agriculture, there is a movement to use conservation tillage (including no tillage) practices, which preserve surface plant residues. These residues influence not only the crop emergence, growth and productivity but may also influence such aspects of weed growth. Putnam and De Frank (1983) have shown that management of selected crop residues greatly reduced weed germination and growth.

2.2. ORIGIN AND NATURE OF ALLELOCHEMICALS

According to Whittaker (1970) and Whittaker and Feeny (1971), most allelochemicals are secondary substances, which do not occur in all living matter but appear sporadically. These substances are biosynthesized from the metabolism of carbohydrates, fats and amino acids and arise from acetate or the shikimic acid pathway (Robinson, 1963). Many of these compounds have been implicated as allelopathic agents and have been classified into various classes (Horsley, 1971; Rice, 1974; 1979; Mandava, 1985; Putnam, 1985).

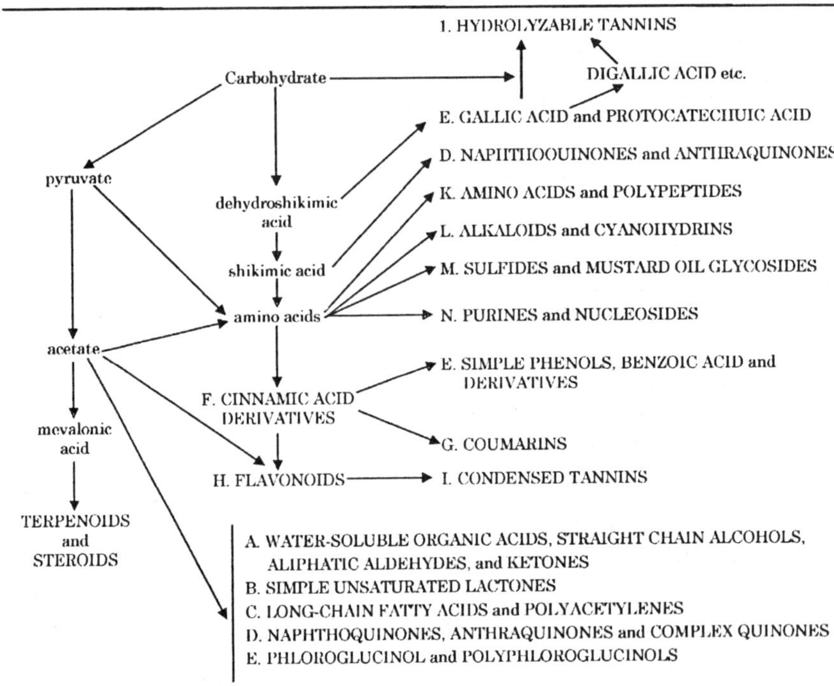

Fig. 2. Probable major biosynthetic pathways leading to production of the various categories of allelopathic agents.

Rice (1974) has divided these compounds into fourteen chemical categories; (a) cinnamic acid derivatives, (b) coumarins, (c) simple phenols, benzoic acid derivatives, gallic acid and protocatechuic acid, (d) flavonoids, (e) condensed and hydrolyzable tannins, (f) terpenoids and steroids, (g) water soluble organic acids, straight chain alcohols, aliphatic aldehydes and ketones, (h) simple unsaturated lactones, (i) long chain fatty acids, (j) naphthoquinones, anthraquinones and complex quinones, (k) amino acids and polypeptides, (l) alkaloids and cyanohydrins, (m) sulfides and mustard oil glycosides and (n) purines and nucleotides as depicted in Fig. 2. However, Putnam (1985) grouped these chemicals into eleven classes : (a) toxic gases, (b) organic acids and aldehydes, (c) aromatic acids, (d) simple unsaturated lactones, (e) coumarins, (f) quinones, (g) flavonoids, (h) tannins, (i) alkaloids, (j) terpenoids and steroids and (k) miscellaneous and unknown. All the naturally occurring compounds of the above categories are not allelochemicals. Derivatives of cinnamic acid, benzoic acid, coumarin and terpenoids are the most often reported allelochemicals. The terpenoids are of limited distribution and are produced in small quantities, whereas, phenolic compounds (cinnamic acid, benzoic acid and coumarin) are present in fair abundance (Mandava, 1979; Robinson, 1967).

2.3. FACTORS AFFECTING PRODUCTION OF ALLELOCHEMICALS

Factors affecting the amounts of allelochemicals produced by the plants are extremely important in allelopathy but little research has been done on this aspect. However, knowledge of these factors may enhance the crop production through better plant stand, improve the insect-pest resistance of crop plants and improve the weed control by exploiting smothering ability of field crops. Lovett (1982a) reviewed the factors affecting production of phytotoxins from decomposing residues, while Rice (1984) outlined following factors which affect the amount of allelochemicals produced : (a) radiation, (b) mineral deficiences, (c) water stress, (d) temperature, (e) allelopathic agents, (f) age of plant organs, (g) genetics, and (h) pathogens and predators. All these factors except radiation and temperature could be exploited under field conditions to improve crop productivity but it needs more research. It may be accomplished either by selecting the plant types that are more allelopathic or by enhancing inhibitors production by exerting the proper stress on plants.

2.4. MODE OF ACTION OF ALLELOCHEMICALS

A frequent approach followed to assess the mode of action of specific allelochemicals had been to monitor their effects on major plant functions. While we know that diverse secondary plant products exhibit allelopathic activity, it is difficult to understand their mechanism of action (Mandava, 1985) partly because of (a) complications in separating the secondary effects from primary causes, (b) the uncertainty in translating the observed effects in isolated enzymes and other biochemical systems to intact plant systems and (c) the lack of understanding of the effects of the allelopathic agents on all photosynthetic processes (changes in stomatal opening, membrane permeability, water content) and many other processes that affect the overall photosynthesis process.

Rice (1984) indicated that allelopathic agents influence the plant growth through following physiological processes : (a) cell division and cell elongation, (b) phytohormone induced growth, (c) membrane permeability, (d) mineral uptake, (e) availability of soil phosphorus and potash, (f) stomatal opening and photosynthesis, (g) respiration, (h) protein synthesis and changes in lipid and organic acid metabolism, (i) inhibition of porphyrin synthesis, (j) inhibition or stimulation of specific enzymes, (k) corking and clogging of xylem elements, stem conductance of water and internal water relations and (l) miscellaneous mechanisms. These processes that contribute significantly to mode and mechanism of action have been also reviewed by Mandava (1985) and Einhellig (1986).

CHAPTER - 3

INTERACTIONS BETWEEN PLANT COMMUNITIES

In the crop fields, at any given time there are at least more than one plant species growing together. During fallow period these species may be weeds, while in sole cropping these may be crops and weeds; because complete weed control is not possible for longer periods. In crop mixtures or intercropping systems, the major plant species are crops, besides, some weeds may also be present. When the two plant species grow together they interact with each other either inhibiting or stimulating their growth or yield through direct or indirect allelopathic interactions. Allelopathy is generally associated with interactions between living plants and have been observed in agricultural lands from last many centuries. In the past, competition for light, space, water and nutrients was considered the major factor for the plant interactions. Now detailed studies have shown that allelopathic activity gives selective advantage to many plants, particularly under stress conditions, such as nutrients deficiency, drought, waterlogging and chilling etc., which may occur in combinations under field conditions. Such combinations of stress factors could synergistically increase the allelopathic activity of inhibitory species.

In many plant ecosystems, plants tend to pattern themselves as pure stands or as individuals, spaced in rather specific densities or configurations. For example, many desert species show obvious zones of inhibition around them, in which few, if any, alien species are able to invade. These patterns often cannot be adequately explained by competition alone and are probably caused by a combination of factors including allelopathy. This phenomenon happens in herbaceous plants as well as in woody shrubs and trees, perhaps, to not share their deficient resources of water and nutrients with other neighbouring plants.

In general, allelopathy has been related to the problems with crop production in (a) crop monocultures or sole crops (Patrick *et al.*,

1963; Kozel and Tukey, 1968; Patrick, 1971), (b) certain crop rotations (Schreiner and Reed, 1907b; Patrick *et al.*, 1963; Patrick, 1971) and (c) crop mixtures or intercropping systems (Rakhteenko *et al.*, 1973b; Tsuzuki, 1980; Nielson *et al.*, 1981; Cruz *et al.*, 1988). Stubble mulch farming (McCalla and Duley, 1948, 1949; Guenzi and McCalla, 1962; Patrick *et al.*, 1963; McCalla and Haskins, 1964; Guenzi *et al.*, 1967) is practised over large areas in the world and stubbles or crop residues affect all types of cropping systems viz., monocultures, crop rotations and intercropping or crop mixtures. Therefore, crop residues (stubbles) have been dealt separately in detail (Section 3.1.1.1). Besides, weeds use allelopathy as a part of their interference with crops. The increased interest in various agricultural systems such as intercropping, multiple cropping, monoculture and no-till planting etc., where plant interactions are critical, the knowledge of allelopathy has become necessary. It plays a significant role in plant-plant and plant-microbial interactions, however, in this book only plant-plant interactions have been discussed.

3.1. CROPS

Theophrastus (Ca, 300 B.C.) stated that chickpea did not reinvigorate the soil but rather, exhausted it compared with other legumes and destroyed weeds particularly caltrop. Plinius Secundus (1 A.D.) reported that chickpea, barley, fenugreek and bitter vetch all "scorch up" corn land, owing to blend of injurious scents or juices. Further, the radish was harmful to the grape vine, for the grape vine could be inferred to possess a sense of smell and was affected by odours to a great degree. He quoted Virgil as stating that cornland was also "scorched" by flax, oat and poppy. Culepreper (1633) stated that there was such an antipathy between grape vine and cabbage plants, that one died where the other grows. Gorobets and Serdyuk (1981) found accumulation of n-hydroxybenzoic and n-coumaric acids, phenol glucosides and coumarins after cabbage crop in perlite substrate. Young (1804) reported that clover failed in those districts, where it had been cultivated constantly, because the soil became sick of clover. T.A. (1845) pointed out that clover sickness can be prevented by allowing an interval of 7 to 8 years between clover crops. DeCandolle (1832) suggested that soil sickness problem in agriculture might be due to root exudates of crop plants and crop rotation could help to alleviate the problem. He observed that thistles in field injured oats, while *Euphorbia* and *Scabiosa* injured flax, rye and wheat plants. Voichenko *et al.* (1982) reported highest accumulation of inhibitors in onion, tomato, radish, buckwheat and salad mustard out of 25 crops. The crops included in this book have been listed in Table-1, Chapter 7.

3.1.1 Crop-crop interactions

The field crops generally add phytotoxins or allelochemicals to the soils mainly through crop residues and partially through root exudates, therefore, their allelopathic effects have been studied most.

3.1.1.1. Crop residues

In monocropping, the crop and weed residues do not pose any management problem. Because residues are incorporated into the soil sufficiently ahead of planting time to allow their complete decomposition and thus toxins released during the decay become harmless to the succeeding crop. However, since 1960's multiple cropping systems have been introduced (owing to availability of short duration and high yielding varieties of crops) in areas where climate and irrigation facilities are not limiting for crop production, throughout the year. The adoption of multiple cropping systems in subtropical and tropical countries under irrigated conditions have firstly, led to a greater production of crop residues over mono-cropping. Secondly, where more than two crops are raised per calender year, there is little time gap between the harvest of previous crop and sowing of next crop and for the decomposition of crop residues. Thereby, the succeeding crops are sown in the crop residues of previous crops. In these systems, the crop residues are incorporated into the soil to facilitate the sowing of subsequent crops.

While in areas where soil erosion and conservation of soil moisture are major problems for crop production, reduced or no-tillage agriculture along with stubble mulch is being increasingly adopted. In such cropping practices, decomposing plant residues have shown allelopathic effects. In reduced tillage and stubble mulch agriculture, plant residues are maintained on the soil surface to protect the soil from erosion and to conserve soil moisture. It is accomplished, either by growing cover crops and dessicating them with herbicides or leaving stubbles of crops after their harvest (stubble mulch) on soil surface. The crops are sown with least displacement of crop residues on the soil surface. Stubble mulch farming is practised over large areas in various parts of the world. While in areas where soil erosion is not a serious problem and seedbed is prepared with conventional tillage, all the crop residues or stubbles are incorporated into the soil. Generally land is ploughed immediately after the crop harvest to incorporate crop residues in the soil and to hasten their early decomposition. The crop residues add a variety of organic and inorganic compounds to the soil system. These compounds may be liberated during the decomposition of stubble residues or synthesized by microbes using the residues as a nutrient source, besides in stubble mulch agriculture these may be also leached directly from stubble residues. The production, accumulation, transformation and destruction of such chemicals is influenced by a large number of factors and the dynamic nature of soil system ensures that their concentration varies over space and time.

I. Factors affecting production of phytotoxins

Skinner (1918), Takijama (1963), Patrick *et al.* (1964), Patrick and Toussoun (1965), Lynch, (1977) Chou and Chiou (1979), Lynch *et al.* (1981a), Ventura *et al.* (1984), Waller *et al.* (1987) and Hicks *et al.* (1988) indicated that soil toxicity from organic constituents was associated with soil aeration and tillage practices. Besides, some other factors affecting phytotoxins production are: crop type (Cochran *et al.*, 1977), age of plant (Kimber, 1973a), soil type (Patrick, 1971) and microorganisms (Krasilnikova, 1958; McCalla and Haskins, 1964), but these are of little importance, therefore, not discussed in detail.

Table 1. Persistence of phytotoxicity from decomposing soil incorporated crop and weed residues

Plant spp.	Duration of phytotoxicity	References
A. Crops		
Wheat	8 weeks	Guenzi *et al.* (1967), Kimber (1973b)
Barley	7 weeks	Toussoun *et al.* (1968)
Oats	8 weeks	Guenzi *et al.* (1967)
Rice	16 weeks	Chou and Lin (1976), Chou *et al.* (1977), Chou and Chiou (1979)
Corn	22 weeks	Guenzi *et al.* (1967)
Sorghum	30 days	Breazeale (1924)
Rye	30 days	Patrick *et al.* (1963) Chou and Patrick (1976)
Broccoli	30 days	Patrick *et al.* (1963)
Sweet potato	12 weeks	Harrison Jr. and Peterson (1986)
B. Weeds		
Datura stramonium L.	20*, 56** weeks	Levitt and Lovett (1984)
Sorghum halepense (L.) Pers.	16 weeks	Friedman & Horowitz (1971)
Amaranthus palmeri (L.) Wats.	11-16 weeks	Bradow and Connick Jr. (1987)
Cirsium arvense (L.) Scop	9 weeks	Wilson Jr. (1981)
Agropyron repens (L.) Beauv.	7 weeks	Lynch and Penn (1980)
Parthenium hysterophorus L.	4 weeks	Kanchan and Jayachandra (1980)
Argemone mexicana L.	10 days	Leela (1981).

* Podzol soil, ** Black soil

(i) Soil aeration : There are various routes for breakdown of organic matter in the soil. The soil oxygen status is most important factor determining the type of products formed, their concentration and persistence. During decomposition of complex organic materials in soil, both under aerobic and anaerobic conditions simple organic compounds are formed, which are often further metabolized (Greenwood, 1961; Currie, 1962; Whitehead, 1963). Under aerobic conditions, the organic compounds disappear rapidly and microbial material is synthesized. While under oxygen deficient conditions, volatile fatty acids and other organic acids accumulate and synthesis of microbial material is suppressed. Under such conditions, methane, hydrogen sulfide, ethylene, acetic, lactic, butyric, formic and other organic acids, phenolic compounds (syringaldehyde, vanillin, p-hydroxy benzoic and benzoic acids), various amino acids and many intermediate breakdown products are formed, many of them are toxic in laboratory studies (Skinner, 1918; Martin, 1957; Borner, 1960; Whitehead, 1964; Wang *et al.*, 1967a,c; Wu *et al.* 1976 a,b; Chou *et al.*, 1977; Chou and Chiou, 1979). The oxygen availability, not only affects the products formed during plant residue decomposition but also affects their activity, concentration and persistance in soil (Greenwood and Lee, 1960; Parr and Renszer, 1962). Patrick and Koch (1958) and Patrick *et al.* (1963) found that aqueous extracts of soils in which residues from rye, barley, timothy, corn, wheat, tobacco and other plants were allowed to decompose under waterlogged conditions, were considerably more toxic than when similar residues were decomposed at optimum moisture levels. Greatest phytoxicity from decomposing residues, occurred during early stages of decomposition. Because phytotoxins appeared five days after decomposition, reached peak after three weeks and declined after three to five weeks. However, Guenzi *et al.* (1967). Toussoun *et al.* (1968), Kimber (1973a), Chou and Lin (1976), Chou and Chiou (1979) and Chou *et al.* (1977) reported that phytotoxicity from decomposing residues of rye, sorghum, barley, wheat, oat, sweet potato, rice and corn persisted in soil for 4, 4, 7, 8, 12, 16 and 22 weeks, respectively (Table 1). Waterlogging caused oxygen deficiency in the soil and oxygen level in paddy soils of Taiwan reached almost zero at 25 cm soil depth. This was even more pronounced during the second paddy crop season, when the soil incorporated crop residues of previous crop started decomposing and producing significant amount of phytotoxic substances (Patrick and Mikkelson, 1971). The production of these compounds reached peak in the first month after their soil incorporation, resulting in suppression of root growth, panicle initiation and yield in succeeding rice crop (Chou and Lin, 1976; Chou *et al.*, 1977; Chou and Chiou, 1979).

(ii) Tillage practices : In cultivation of crops, two types of tillage practices are generally practised, i.e., conventional tillage and minimum or no-tillage (direct seeding). The former is adopted in areas where soil erosion is not a problem, therefore, all crop residues are incorporated into the soil. The later is practised in areas where soil erosion is a major problem, thus all crop residues are left or maintained on the soil surface as stubble mulch. It is known that the decomposition of crop residues occur early, when incorporated into the soil, than when lying on the soil surface. Harper and Lynch (1979) found that oat straw decomposed rapidly when buried into the soil, than lying on the soil surface. Within six weeks of burial into the soil, the ability to produce toxins was reduced to 15% of that of fresh straw and 50% of cellulose and hemicellulose had been degraded; while surface straw reached this stage, after 3 months. Oat straw present on soil surface decreased the yields of subsequent wheat crop, compared to its burning or soil incorporation (Table 2) (Lynch *et al.*, 1981b). Straw residues injured the crop plants during the early stages of growth, because the concentration of phytotoxins reached maximum, minimum and nil at 10-25, 30 and 60 days, respectively, after its soil incorporation (Patrick and Toussoun, 1965). McCalla and Duley (1948) reported poor performance of corn, in stubble mulch of white sweet clover and wheat straw. Wheat, barley, pea, lentil and blue grass straw layered on soil surface in August did not produce phytotoxicity till end September. Thereafter, pea and lentil straw extracts inhibited root elongation of test wheat seedlings upto 90%; wheat and barley straw produced phytotoxins during fall, winter and spring and blue grass in the fall. In all cases, the phytotoxins inhibited the root growth of test seedlings of wheat and toxins were produced from wet straw at air temperatures above 15°C (Cochran *et al.*, 1977).

Table 2. Effect of oat straw and method of its disposal on wheat yield.

Straw treatment	Furrow opener	Yield (t/ha)
Burnt	Triple disc	8.51
	Single disc	8.14
Chopped and spread		
(a) left *in situ*	Triple disc	3.77
(b) disced	Single disc	5.37
(c) rotovated	Single disc	4.50
LSD (P = 0.05)		0.35

Source : Lynch *et al.* (1981 a)

II. Effect of residue phytotoxins on crop plants

Every growing crop produces compounds, which may be toxic to other plants or to itself. The toxic effects vary with nature of the soil, the plant affected and the vigour of the plant, producing toxin. In crop fields, the toxins are mainly produced from the decomposing crop residues incorporated in the soil. The soil heating produces toxins from the soil organic matter and their quantity is grater than that produced during crop growth. In both cases, the toxins after oxidation increased the soil fertility (Pickering, 1914). Residues left on or near the soil surface, reduces growth and yield of the next crop compared with their incorporation into the soil. Reduced yields have been attributed to the toxic compounds leached from crop residues and microbial toxins produced during residue decomposition (Krasilnikova, 1958; McCalla and Haskins, 1964; Kimber, 1967; Whittaker, 1970; Rice, 1974; Cochran *et al.*, 1977; Lynch *et al.*, 1981a; Kehr *et al.*, 1983; Ventura *et al.*, 1984). Phytotoxicity and its possible relationships with crop residues on the soil surface and with heavy residues layered immediately below the plough layer have been reviewed (Krasilnikova, 1958; McCalla and Haskins, 1964; Patrick *et al.*, 1964; Whittaker, 1970; Patrick, 1971; McCalla and Norstadt, 1974; Rice, 1974, 1984; Hedin, 1985; Thompson, 1985; Putnam and Tang, 1986; Waller *et al.*, 1987). The phytotoxins from crop residues have mostly nagative effects on crop plants such as (a) delayed or complete inhibition of germination, (b) reduced plant population, (c) stunted and deformed roots and shoots, (d) deranged nutrient absorption, (e) lack of seedling vigour, (f) reduced tillering, (g) chlorosis, (h) wilting, (i) predisposition to root rots and (j) seedling death (Norman, 1959; Patrick *et al.*, 1963; Guenzi *et al.*, 1967; Norstadt and McCalla, 1968; Toussoun *et al.*, 1968; Horricks, 1969; Kimber, 1973b; Cochran *et al.*, 1977; Lynch, 1977; Kuo *et. al.*, 1981; Walker and Jenkins, 1986; Waller *et al.*, 1987; Oleszek and Jurzysta, 1987; Hicks *et al.*, 1988). However, the major effects of phytotoxins on crop plants are : (i) inhibition of nitrification and biological nitrogen fixation, (ii) predisposing the plants to diseases and (iii) inhibition or stimulation of germination, growth and yield.

(i) Inhibition of nitrification and biological nitrogen fixation : The phytotoxins produced during the decomposition of crop residues, inhibit the nitrification process in the soil and biological nitrogen fixation in legumes. The maintenance of corn residues on the soil surface, increased the concentration of nitrification inhibitors (ferulic and p-coumaric acids) in the soil, which decreased the population of nitrosomonas bacteria and thus increased the concentration of NH_{4+} over NO_3^- compared with soil without corn residues (Lodhi, 1981). In south Taiwan, soybean following rice yielded higher, when the rice residues were burnt than when decomposed in the field (Table 3) (Asian Vegetable Research and Development Centre, 1978). Because phenolics produced from decomposing rice residues inhibited the growth of N fixing bacteria (*Rhizobium japonicum*), reduced nodule number and thus decreased biological nitrogen fixation in soybean crop (Rice *et al.*, 1981).

Similarly, soil incorporation of vines and storage root residues of sweet potato, reduced the nodulation and nitrogen fixation in cowpea plants (Walker and Jenkins, 1986).

Table 3. Effect of mulching and burning of rice straw on soybean grain yield and fresh weight of weeds

Treatment	Mulch application rate (t/ha)	Soybean grain yield (kg/ha)	Fresh weight of weeds at harvest (g/m²)		
			Broad leaf	Others	Total
No mulching (check)	—	1165	98	574	672
Rice straw mulch	0.2	1049	71	689	760
Rice straw mulch	4.0	1243	119	611	730
Rice straw mulch	6.0	1344	80	548	628
Rice straw burnt	4.0	1740	7	222	229
Rice straw burnt	6.0	1752	4	243	247
Rice straw burnt	8.0	1833	7	133	140

Source : Asian Vegetables Research and Development Centre (1978). Preceding rice crop produced 4.0 tonnes rice straw/ha.

Major weeds : *Echinochloa colonum, Echinochloa crusgalli, Alternanthera sessilis*; Minor weeds : *Cyperus* spp; *Ammania bacoifera, Ludwigia octovalis* and *Chenopodium album*.

(ii) Predisposing the plants to diseases : Generally organic amendments incorporated in the soil in adequate quantity and at the right time, suppress the root diseases but they may also cause diseases. The residues may be colonized by pathogens, which use them as food base and may produce phytotoxins which predispose the plants to diseases. The phytotoxic products produced during decomposition, may predispose the roots to infections. Such phytotoxins may increase susceptibility to root rot of tobacco (Patrick and Koch, 1963), cotton (Linderman and Toussoun, 1968), beans (Toussoun and Patrick, 1963) and sugarcane (Rands and Dopp, 1938). Plant roots injury from these phytotoxins opens the way for secondary root decay (Cochrane, 1948; Carley and Watson, 1967). Volatile compounds from decomposing alfalfa residues, stimulated the germination of sclerotia of *Sclerotinia rolfsii* and microsclerotia of *Verticillium dahliae* followed by lysis, which reduced their population. Sulphur containing volatiles from decomposing *Crucifer* residues, were found inhibitory to *Aphanomyces euteiches* and thus reduced root rot in peas (Lewis and Papavizas, 1975).

(iii) Inhibition or stimulation of germination, growth and yields : One crop may inhibit or stimulate the germination, growth and yield of associated crop growing with it (crop mixture or intercropping) or of following crop (monoculture or crop rotations), through release of leachates or washings from germinating seeds or from decomposing crop residues. Very scanty information is available, on the inhibitory or stimulatory effects of seed or seedling leachates or germinating seeds of one crop, upon the seed germination or seedling growth of another crop. Germinating seeds of barley, inhibited seed germination of white mustard (*Sinapis arvensis*), while the seed leachates of wheat, oat, maize, vetch and prosomillet increased its seed germination (Prutenskaya, 1972). Seed coat extracts of kidney bean seeds inhibited the root growth of flax seedlings (Siegel, 1950), while gibberellins in the leachates from the pericarps of cuba bean stimulated the growth of rice seedlings (Ota, 1983). Seed and root leachates of barley, inhibited seed germination and seedling growth of wheat and tobacco but leachates of living barley plants were more inhibitory than dead ones (Overland, 1966). The seedling sap from maize seedlings, increased the growth and yield of mungbean (Peswani, 1981) and ascorbic acid in the root extracts of radish stimulated the seedling growth of pearlmillet (Sharma and Singhvi, 1981). Many phytotoxic compounds from crop residues or their extracts, have been described and identified both before and during the decomposition. These include (a) toxins indigenous to the plant residues, (b) those released from the residues during decomposition and (c) their byproducts produced in microbial decomposition. But toxins produced during residue decomposition, were phytotoxic to associated plants.

Collison and Conn (1925) concluded that two mechanisms are associated, with phytotoxicity of plant residues: (a) toxic compounds which act quickly and are usually inactivated through their adsorption by colloidal matter in the soil and (b) stimulation of microbial population, which immobilizes much of nitrogen making it unavailable to the plants. This review is confined to the first mechanism i.e. inhibitory or stimulatory effects of compounds released from crop residues. Such effects have been observed with residues of cereals, oilseeds and grain legumes etc. (Lovett and Jessop, 1982). However, maximum allelopathic studies have been conducted on cereals (wheat and sorghum) and in legumes, the alfalfa has been investigated most (Tables 4, 5).

3.1.1.1.1. Cereals

Allelopathic effects of crop residues of cereals, such as wheat (McCalla and associates; Kimber, 1967,1973a,b; Lynch, 1977; Lynch *et al.*,1981a; Hairston *et al.*, 1987; Lodhi *et al.*, 1987; Santos *et al.*, 1983; Hicks *et al.*, 1988), corn (McCalla and Army, 1961; Guenzi *et al.*, 1967; Chou and Patrick, 1976; Lodhi, 1981), rice (Chandramohan *et al.*, 1973; Wu *et al.*, 1976; Chou and Lin, 1976; Chou *et al.*, 1977; Asian Vegetables Research and Development Centre, 1978; Chou and Chiou, 1979; Rice *et al.*, 1980, 1981), rye (Patrick and Koch, 1963; Patrick *et al.*, 1963; Barnes and Putnam, 1983b, 1986), barley

Table 4. Inhibitory effects of crop residues/plant extracts/leachates on seed germination and growth of test crops

Donor crop			Receipient crop		Reference
Crop	Crop residue/plant extract/plant leachate	Phytotoxin	Crop	Inhibitory effect	
1	2	3	4	5	6
Crop residues (straw/stubble) or extracts					
Wheat	Crop residue decomposition	ND	Wheat	Inhibition of seed germi-nation, root and shoot growth, shortened internodes, low grain yield	McCalla and Haskin (1964), Kimber (1967 1973a b), Guenzi *et al.* (1967), McCalla and Norstadt (1974), McCalla (1971 1976)
	Crop residue decom-position (Anaerobic)	ND	Wheat	Inhibited root growth of seedlings	Lynch (1977)
	Straw extract	ND	Wheat	Reduced root and shoot growth, roots become discoloured and slimy with enlarged tips	Collison (1925)
	Straw extract	ND	Barley, rice, rye	Suppressed growth	Hozumi *et. al.* (1974)
	Crop residue decomposition	Patulin*	Corn	Reduced seed germination, inhibited and stunted growth, chlorosis	McCalla and Duley (1948), Guenzi and McCalla (1962), Norstadt and McCalla (1963)
	Crop residue decomposition	Ferulic p-coum-aric, syringic. p-benzoic, vanillic acids	Cotton	Reduced seed germination, seedling emergence, growth and lint yield	Lodhi *et al.* (1987), Hicks *et al.* (1988 1989)
	Straw leachate	ND	Soybean	Inhibited radicle growth at higher concentration. Decreased plant stand and grain yields	Hairston *et al.* (1987), Sanford (1982)

Contd.....

1	2	3	4	5	6
	Crop residue decomposition	ND	Sugarbeet	Reduced plant population and yields	Elliott *et. al.* (1979)
Rice	-do-	p-hydroxybenzoic, p-coumaric, syringic, vanillic, ferulic, propionic, o-hydrophenylacetic and butyric acids, H_2S* gas	Rice	Suppressed root growth, tillering, panicle initiation and decreased yields	Chou and Lin (1976), Chou *et. al.* (1977), Chou and Chiou (1979)
	-do-	ND	Soybean	Inhibited growth, yield, nodules and nitrogen fixation	Asian Vegetable Research and Development Centre (1978), Rice *et al.* (1981)
Rye	Crop residue decomposition	ND	Rye	Inhibited seed germination and radicle elongation of seedling	Barnes and Putnam (1983a)
	Crop residue decomposit ion	ND	Lettuce	Inhibited seedling emergence and growth	Chou and Patrick (1976), Barnes and Putnam (1986)
Corn	-do-	ND	Corn	Stunted seedling growth	Garcia (1983)
Sorghum	-do-	ND	Wheat	Reduced seedling emergence, killed the seedlings	Breazeale (1924), Lovett (1983a)
	-do-	ND	Wheat, oat	Reduced grain yields	Vinall and Getty (1921), Cole and Halstead (1922)
	-do-	ND	Corn	grain yields	Ten Eych and Shoesmith (1906)
	-do-	ND	Corn, snapbean	Inhibited crop growth	Ruiz-Sifre and Reis (1983)

Contd.....

1	2	3	4	5	6
	-do-	ND	Sorghum	Inhibited crop growth	Burgeos-Leon *et. al.* (1980)
	Stem extract	p-hydroxybenzoic acid, p-hydroxybenzaldehyde	Tomato	Inhibited root growth	Weston *et. al.* (1989)
	Stem extract	ND	Vetch	Inhibited crop growth	Hawkins (1925)
	Stubble extract	ND	Oat, barley	Inhibited seed germination	Narwal and Gupta (1991d)
	Stubble extract	ND	Rape, Indian rape, Indian colza, Indian mustard, Brown *sarson*	Inhibited seed germination and root growth of seedlings	Narwal *et. al.* (1990)
Pearlmillet	Crop residue decomposition	ND	Indian mustard	Inhibited seed germination, growth, branching and grain yield	Singh *et. al.* (1983)
	Stubble extract	ND	Lentil	Inhibited seed germination	Narwal *et. al.* (1989)
	-do-	ND	Wheat, barley, chickpea, *Eruca sativa*	Inhibited seedling growth	Narwal *et. al.* (1989a)
	-do-	ND	Rape, Indian rape, Indian mustard, Brown *sarson*	Inhibited seed germination and seedling growth	Narwal and Gupta (1991b)
	Root stem, stubble extract	ND	Indian colza	Inhibited seed germination and seedling growth	Narwal *et. al.* (1989b)

Contd.....

1	2	3	4	5	6
Soybean	Crop residue extract	ND	Soybean, alfalfa, radish, turnip	Inhibited seed germination	Tsuzuki and Kawagoe (1984)
	Crop residue decomposition	ND	Wheat	Delayed seedling emergence, reduced coleoptile length, poor plant stand and decreased grain yields	Lovett (1983b), Huber and Abney (1986)
Pea	-do-	ND	Wheat	Reduced seed germination, seedling emergence, root length, 1000- grain weight	Lovett (1983 b)
Pigeonpea	-do-	ND	Pigeonpea	Inhibited seed germination and seedling growth	Badillo-Feliciano and Lugo-Lopez (1983)
White sweet clover	Crop residue decomposition	Coumarin	Corn	Abnormal and inhibited seed germination, reduced root and seedling growth	McCalla and Duley (1948, 1949) Guenzi and McCalla (1962)
Crimson clover, Hairy vetch	-do-	ND	Cotton	Decreased seedling emergence and growth	White *et. al.* (1989)
Alfalfa	-do-	Saponin	falfa	Inhibited seed germination and seedling growth	Stepanova and Prutenskaya (1975)
	-do-	Saponin	Cotton	Inhibited seed germination, seedling growth and yield	Mishushtin and Naumova (1955), Pederson (1965), Megie *et. al.* (1967), Stepanova and Prutenskaya (1975)
	Crop residue extract	ND	Alfalfa, turnip, radish	Inhibited seed germination	Tsuzuki and Kawagoe (1984)
	Soil extract	ND	Alfalfa, wheat, radish	Inhibited root and shoot growth	Read and Jensen (1989)

Contd.....

1	2	3	4	5	6
White lupine	Crop residue decomposition	ND	Cotton	Decreased seed germination, seedling growth	Lehle *et al.* (1983)
Sunflower	-do-	ND	Sorghum	Decreased seed germination, seedling growth	Schon and Einhellig (1982)
Rape	-do-	ND	Soybean	Decreased yields	Santos *et al.* (1983)
	-do-	ND	Wheat, barley oats	Stunted growth and decreased dry weight	Horricks (1969)
Oilseed rape	-do-	ND	Soybean	Reduced seedling elongation and plant population	Vilhordo *et al.* (1985)
Yellcw mustard, Mustard, Indian mustard, Black mustard, rapeseed	Crop residue decomposition	ND	Wheat	Reduced tillers, plant height, dry weight and grain yield	Mason-Sedun *et al.* (1986)
Chinese cabbage	-do-	ND	Soybean	Reduced leaf area, stunted plants and lower dry weight	Song *et al.* (1985)
Chinese cabbage	Crop residue decomposition and extract	ND	Mungbean	Inhibited seed germination and seedling growth	Kuo *et al.* (1981)
Asparagus	Crop residue decomposition	ND	Tomato	Decreased plant growth	Putnam *et al.* (1983)
	-do-	ND	Asparagus	Predisposed the plant to *Fusarium* disease	Hartung *et al.* (1989)
	Root extract	ND	Asparagus, tomato, lettuce, cucumber, wheat, barley	Reduced germination (tomato, lettuce) and seedling growth (Asparagus, lettuce, barley)	Hazebroek *et al.* (1989)

Contd.....

1	2	3	4	5	6
Sugarcane	Crop residue decomposition	p-hydroxy benzoic, p-coumaric, syringic, ferulic, vanillic, formic, acetic, oxalic, melonic, tartaric, malic and fusaric* acids	Sugarcane	Decreased cane yields	Wang *et al.* (1967a,c), Wu *et al.* (176a,b)
Sugarbeet	-do-	ND	Soybean	Decreased growth and yield	Matsui and Nishiiri (1989)
Sweet potato	Crop residue decomposition	ND	Sweet potato	Inhibited plant growth, dry matter, reduced uptake of P, K, Ca, Mg, S, Zn, B, Cu	Walker and Jenkins (1986)
	-do-	ND	Cowpea	Inhibited plant growth and nodulation	Walker *et. al.* (1989)
	-do-	ND	Alfalfa	Inhibited plant growth	Harrison and Peterson (1986)
	Crop residue extract	ND	Alfalfa	Inhibited seed germination	Harrison and Peterson (1986)
Taro	-do-	ND	Mungbean, corn, okra, egg plant	Inhibited germination (mungbean, okra) and growth (mungbean)	Pardales Jr. and Dingal (1988)
Seed/Plant extracts or leachates					
Kidneybean	Seed coat extract	ND	Flax	Inhibited root growth of seedling	Siegel (1950)
Rye	Seed and root leachates	ND	Rye	Decreased root growth	Barnes and Putnam (1983a, 1986)
Barley	Seed and root leachates	ND	Wheat, tobacco	Inhibited seed germination and seedling growth	Overland (1966)
	Seed leachates	ND	White mustard	Inhibited seed germination	Prutenskaya (1972)

Contd.....

1	2	3	4	5	6
Pearlmillet	Stem extract	ND	Rape, Indian rape, Indian mustard, Brown *sarson*	Inhibited seed germination and seedling growth	Narwal and Gupta, (1991a)
	Root extract	ND	Rape, Brown *sarson*	Inhibited seed germination and seedling growth	Narwal and Gupta (1991d)
Soybean	Root extract	ND	Rape, Indian mustard	Reduced and delayed seed germination	Gupta *et. al.* (1987)
Pigeonpea	Leaf extract	ND	Maize, Tomato, beans, soybean, cowpea	Inhibited seed germination	Badillo-Feliciano and Lugo-Lopez (1983)
Alfalfa	Root extract	Medicogenic acid, glucosides	Wheat	Inhibited seed germination and root growth of seedlings	Oleszek and Jurzysta (1987)
White sweet clover	Shoot extract	Coumarin	Wheat, maize, sunflower	Reduced growth of seedlings	Moroz and Leshenko (1972)
White sweet clover	Flower extract	ND	Tomato, radish	Inhibited root growth of seedlings	Nicollier and Thompson (1982)
Sunflower	Leaf extract	ND	Sorghum	Reduced leaf water potential, increased leaf diffusive resistance and seedling growth	Schon and Einhellig (1982)
Yellow mustard, Mustard, Indian mustard, Black mustard, rapeseed	Plant extract	ND	Wheat	Inhibited coleoptile and radicle growth	Mason-Sedun *et. al.* (1986)

ND = Not determined, *Microbial origin

Table 5. Stimulatory effects of crop residues/extracts/leachates on growth and yield of test crops

Donor crop			Recipient crop		Reference
Crop	Crop residue/plant extract/ plant leachate	Phytotoxin	Crop	Stimulatory effect	
Crop residues/extracts					
Wheat	Crop residue decomposition (aerobic)	Acetic acid	Barley	Stimulated root elongation of seedling	Lynch (1977)
	Crop residue decomposition	ND	Soybean	Increased yields	Santos *et al.* (1983)
	Straw leachate	ND	Soybean	Stimulated radicle growth at low concentration	Hairston *et al.* (1987), Sanford (1982)
Rice	Crop residue (burnt)	ND	Soybean	Increased yields	Asian Vegetable Research & Development Centre (1978)
Sorghum	Crop residue decomposition	ND	Soybean, corn	Stimulated crop growth	Ruiz Sifre and Ries (1983)
	Crop residue decomposition	ND	Pea	Stimulated crop growth	Hawkins (1925)
Soybean	Crop residue extract	ND	Corn	Increased growth dry matter	Kalantari (1982), Mosca *et al.* (1988)
Crimson clover, hairy vetch	Crop residue decomposition	ND	Corn	Increased growth	White *et al.* (1989)
Alfalfa	Alfalfa meal decomposition	ND	Tomato, lettuce, cucumber	Stimulated growth	Ries *et al.* (1977)
Seed/plant extract leachates					
Maize	Seedling sap	ND	Mungbean	Increased growth and yield	Peswani (1981)
Cuba bean	Pericarp or cotyledon leachates	Gibberalin	Rice	Stimulated growth of seedling	Ota (1983)
Radish	Root extract	Ascorbic acid	Pearlmillet	Stimulated growth of seedling	Sharma and Singhvi (1981)
Soybean	Root extract	ND	Rape, Indian mustard	Stimulated growth of seedling	Gupta *et al.* (1987)

(Toussoun *et al.*, 1968), sorghum (Behmer and McCalla, 1963; Burgeos-Leon *et al.*, 1980; Narwal *et al.*, 1990; Narwal and Gupta, 1991d) and pearlmillet (Singh *et al.*, 1983; Narwal *et al.*, 1989a,b; Narwal and Gupta, 1991a, b,c,d) have been studied on various crops.

I. Wheat : The phytotoxic effects of wheat residues have been studied on the germination, growth and development of wheat (McCalla and Haskins, 1964; Kimber, 1967, 1973a, b; Guenzi *et al.*, 1967; McCalla and Norstadt, 1974; McCalla, 1971,1976), barley (Collison, 1925; Hozumi *et al.*, 1974; Lynch, 1977), corn (McCalla and Duley, 1948; Guenzi and McCalla 1962; Norstadt and McCalla, 1963), cotton (Lodhi *et al.*, 1987; Hicks *et al.*, 1988), soybean (Santos *et al.*, 1987, Hairston *et al.*, 1987) and sugarbeet (Elliot *et al.*, 1979) (Table 4). The decomposing wheat residues contains primary organic acids (acetic, propionic, butyric, pentanoic and traces of isobutyric and isopentanoic acids) (Tang and Waiss Jr, 1978) and a number of phenolic acids (p-coumaric, syringic, vanillic and hydroxybenzoic acids). The phenolic acids were dominant, therefore, wheat residues added 1.5 tonnes phenolic acids per acre under no-till conditions (McCalla and Haskins, 1964; McCalla, 1971, 1976; McCalla and Norstadt, 1974; Lodhi *et al.*, 1987). The wheat residues are not phytotoxic until the start of decomposition. The production of phytotoxins from decomposing wheat residues increases upto 4 weeks, declines thereafter and stops after 8 weeks, therefore, the phytotoxicity of decomposing residues persisted for 8 weeks (Guenzi *et al.*, 1967; Kimber, 1967, 1973a,b; Cochran *et al.*, 1977; Tang and Waiss Jr, 1978). The extracts from decomposing green straw were more toxic than from ripe straw (Kimber 1973a).

Wheat straw inhibited seed germination (LeTourneau *et al.*, 1956; Guenzi *et al.*, 1967; Kimber, 1973b), decreased root and shoot growth (McCalla and Norstadt, 1974; Guenzi *et al.*, 1967; Kimber, 1967, 1973a,b; Cochran *et al.*, 1977), shortened internodes (Ellis and McCalla, 1973) and reduced grain yields of wheat (Kimber 1967, 1973b; Ellis and McCalla, 1973) and had no stimulatory effect. McCalla and Norstadt (1974) concluded that wheat sown in stubble mulch make poor growth due to phytotoxicity from microbial toxin patulin produced by *Pencillium urticae* Bainer. In greenhouse studies, its concentration reached 50 ppm after 11 days of incubation and reduced the shoot growth by 50%. while, its concentration was 20,20 and 75 ppm in solution, sand and soil culture, respectively, It reduced seed germination and shoot and root growth of wheat seedlings. In vicinity of roots, it shortens the plant internodes and reduces seed yields (Ellis and McCalla, 1973). Wheat rhizosphere soil extract showed phyto-toxicity to wheat seeds in bioassays. More alleopathic materials were found in the wheat-rice rotation soils than in the rice soil. Extracts obtained under basic conditions (pH 8.0), proved more inhibitory than those obtained under acidic conditions (pH 5.0), because the former contains more phytotoxic compounds. Wheat straw also contained certain phytotoxins (Young *et al.*, 1989). In Australia, Kimber (1967) observed poorest yields of wheat, when the fall was dry and the first rains coincide with the sowing of subsequent wheat crop into the residues of previous wheat crop. Further studies (Kimber, 1973b) found that the yields of wheat, grown in excessive straw were reduced due to phytotoxic effects of decomposing crop residues, which had more effect on root growth than on shoot growth. The different varieties of wheat responded differently to these phytotoxins. Guenzi *et al.* (1967) also tested ripe straw of nine wheat varieties (Nebred, Warrior, Cheyenne, Ponca, Yogo, Wichita, Pawnee, Bison and Omaha) against wheat. the extract of `Ponca' wheat most depressed the germination, while 'Nebred' wheat was least toxic to root and shoot growth and 'Omaha' wheat was most toxic to shoot growth. The aqueous extracts of soil surface residues of wheat and barley, blue grass and pea and lentil inhibits the root elongation in wheat seedlings by 30, 60 and 90%, respectively (Cochran *et al.*, 1977). The crop residues of wheat, barley, oat, rye, corn, sorghum, lucerne, pea, lentil, rape, tomato and pumpkin inhibits seed germination and decreases radicle growth of wheat seedlings (Becker *et al.*, 1950; Winter and Schonbeck 1953a, b; LeTourneau *et al.*, 1956; Guenzi *et al.*, 1967; Kimber, 1973a; Cochran *et al.*, 1977). The straw or root extract of wheat, barley, oat and rye inhibited root growth of wheat seedlings. The concentration of phytotoxins was found higher under the stubble rows than between the rows, confirming that phytotoxins were released from cereal remains (Becker *et al.*, 1950; Winter and Schonbeck; 1953a,b). The crop residues of wheat, maize, oat, pea, tomato and pumpkin inhibited the seed germination of wheat, however, tomato residues were most toxic (LeTourneau *et al.*, 1956). Wheat, oat, corn and sorghum residues were toxic to the growth of wheat seedlings in order wheat > oat > corn > sorghum. The toxicity of wheat and oat residues persisted for 8 weeks and that of corn and sorghum residues for 22-28 weeks (Guenzi *et al.*, 1967). The aqueous extracts of wheat, barley, oat, lucerne, pea and rye grass residues, which has been rotted for 21 days, inhibited the growth of wheat and the order of toxicity was wheat > pea > oat > barley > lucerne > rye grass (Kimber, 1973a). Wheat and sorghum residues reduced the seed germination of wheat, both at lower (8°C) and higher temperatures (24°C), but pea and rape residues were toxic only at higher temperature and had no effect on seedling growth of wheat (Jessop and Stewart, 1983).

Wheat straw mulch present on soil surface decreased the tillering, number of earheads and straw and grain yield of oat compared with burning of straw (Table 5) (Lynch *et al.*, 1981a).

Presence of decomposing wheat straw in soil, inhibited seed germination (McCalla and Duley, 1948; Norstadt and McCalla, 1963) and stunted the growth of corn (Guenzi and McCalla, 1962; Norstadt and McCalla, 1963). Wheat straw mulch in wet and cool soils, reduced the corn seed germination (44%) and growth compared with unmulched plots due to combined inhibitory effects of phytotoxins present in the plant material and those produced (including patulin) by microorganisms (McCalla and Duley, 1948; Norstadt and McCalla, 1963). Crop residues of wheat and oat straw, maize and sorghum stalks, soybean, sweet clover and brome grass hay contained water soluble substances that depressed the growth of corn, sorghum and wheat plants. The order of increasing toxicity toward test crop of wheat was: sweet clover stems > wheat straw > soybean hay > brome grass hay > oat straw > maize stalks > sorghum stalks > sweet clover hay. In another study, the aqueous extracts of 2.3 g wheat straw, completely inhibited the growth of one corn seed, this was equivalent to about 101 lb of straw per acre assuming a plant population of 20,000 plants per acre (Guenzi and McCalla, 1962). Putnam and DeFrank (1983) studied the allelopathic effects of cover crop residues of wheat, barley, corn, oat, rye, sorghum and sorghum x sudangrass on the plant stand of cucumber, lettuce, pea, radish, snap beans, tomato and weeds. All crop residues reduced the plant stand of lettuce, radish and tomato but increased that of cucumber, pea and beans. However, rye residues were inhibitory only to lettuce and tomato and stimulatory to other vegetables.

Decomposing wheat straw may inhibit or stimulate the root growth of barley seedlings, depending upon the decomposition conditions (Table 4 and 6). The aerobic decomposition, causes stimulation but anaerobic decomposition causes inhibition, due to higher concentration of acetic acid (Lynch, 1977). The aqueous extracts of wheat straw, reduced root and shoot growth and caused abnormal root growth of barley seedlings (Collison, 1925). Rice sown in wheat stubbles immediately after its harvest, gave poor grain yields than that sown after one month. The production of phytotoxins was high during the early stages of stubble decomposition and declined with time. These phytotoxins decreased seed germination and plant stand of rice (Hozumi *et. al.*, 1974). Wheat straw inhibited the growth and yield of soybean under field conditions. Wheat straw left on soil surface or incorporated in soil, reduced plant population, plant height and grain yield of following soybean crop compared with burning of straw (Sanford, 1982; Hairston *et al.*, 1987). The leachates of wheat straw, stimulated the radicle growth of soybean at low concentrations and inhibited at higher ones. Soybean following wheat, barley, rape, flax and lupine gave higher yields after wheat and barley and lowest after rape, because the substances from rape residues were inhibitory to soybean (Santos *et al.*, 1983). Soil incorporation of wheat and oat cover crops, reduced the plant population of sugarbeet due to phytotoxity of their crop residue toxins to sugarbeet seedlings (Elliot *et al.*, 1979). Wheat straw and its mulch, reduced the seed germination and seedling growth of cotton and wheat due to presence of ferulic, p-coumaric, p-hydroxybenzoic, syringic and vanillic acids. Quantitatively, ferulic acid concentration was higher in soil than p-coumaric acid. As little as 0.5% wheat straw by weight of the soil in seedbed, decreased seedling emergence and development of cotton (Table 7). Cotton varieties differed in resistance to inhibition by wheat straw extracts. The straw of wheat varieties, also differed in their ability to inhibit the development of cotton seedlings (Table 8) (Lodhi *et al.*, 1987; Hicks *et al.*, 1988, 1989).

Table 6. Effect of wheat straw burnt or present on soil surface on growth and yield of oats

Parameter	Straw present	Straw burnt	LSD (P=0.05)
Tillers/plant	1.2	2.7	0.9
Yield/m of crop row			
Straw (g)	94	115	17
Grain (g)	77	117	31
Number of ears	65	89	5
Grains per ear	37	38	NS
1000-grain weight (g)	32	35	NS

Source : Lynch *et al.* (1981 a)

Table 7. Effect of wheat straw soil incorporated or maintained on soil surface on seedling emergence and dry matter of cotton varieties (14 days after planting)

Wheat straw in soil (%)	Cotton Variety			
	% emergence	'Acala 246' Dry weight (g)	% emergence	'Pay Master 404' Dry weight (g)
0.0 Control	60a	0.32a	78a	0.39a
0.5*	47ab	0.23ab	46b	0.27ab
0.5**	28bc	0.23ab	38b	0.29ab
1.0*	20c	0.22ab	44b	0.20b
1.0**	16c	0.13bc	36b	0.25ab
2.5*	8c	0.08c	20b	0.14b
2.5**	18c	0.16bc	40b	0.17b

Means within a variety not followed by the same letters are significantly different LSD P = 0.05. * Incorporated, ** Surface

Source : Hicks *et al.* (1988)

Table 8. Effect of water extracts of wheat straw varieties on the root growth of cotton varieties.

Wheat variety	Cotton root fresh weight (% of control) Cotton Variety	
	'Pay Master 404'	'Acala A 246'
Scout 66	96.3 a	57.0ab
Vona	82.0 ab	59.0 ab
Hawk	76.0 ab	52.0 ab
Centurk 78	69.0 ab	72.0 a
TAM 105	67.0 ab	57.0 ab
Newton	65.0 ab	52.0 ab
Sturdy	58.0 ab	72.0 a
Sioux land	55.0 ab	78.0 a
Mustang	54.0 ab	57.0 ab
TAM 107	50.0 ab	34.0 b
TAM 108	37.0 b	42.0 b
TAM 101	35.0 b	52.0 ab

Mean within a variety followed by the same letter are significantly different at the 5% level of probability

Source : Hicks *et al.* (1988).

II. Corn : Decomposing corn residues decreased the respiration of tobacco seedlings (Patrick and Koch, 1958), caused chlorosis and stunted the growth of maize (McCalla and Army, 1961) and inhibited the nitrification process in soil (Lodhi 1981) due to production of allelochemicals (Table 4) (Chou and Patrick, 1976). Corn grown in stubble mulch farming systems of humid areas, become chlorotic and stunted during seedling stage and do not recover fully, therefore, gave poor grain yields (McCalla and Army, 1961). Soil previously sown with maize, stunted the maize seedlings compared with fallow soil. The soil from the maize crop field also inhibited seedling growth of maize owing to its root exudates or rain leachates. The phytotoxins in root exudates and rain leachates also inhibited germination of garden cress (Garcia, 1983). Because, firstly, the phytotoxicity of decomposing corn residues remained high till 22 weeks of decomposition, i.e. throughout maize crop duration (Guenzi *et al.*, 1967). Secondly, the decomposing corn residues, perhaps reduced availability of NO_3^- nitrogen to the growing corn plants, due to inhibition of nitrification process from high concentrations of ferulic and p-coumaric acids (Lodhi, 1981). Chou and Patrick (1976) indentified eighteen compounds in the decomposing corn residues; four were volatiles (butyric, salicyladehyde, phenylacetic and 4-phenylbutyric acids), eleven were non- volatiles (p-hydroxy-benzoic, vanillic, ferulic, o-coumaric, o-hydroxybenzoic, salicylic, syringic, p-coumaric, trans-cinnamic and caffeic acids) and others (resorcinol, p-hydroxybenzaldehyde and phloroglucinol).

III. Rice : In multiple cropping systems of tropical and subtropical regions, many crops follow rice but allelopathic studies have been done only on rice and soybean. In Taiwan, rice is planted twice a year in continuous monoculture system. The yields of second rice crop are generally 25% less than first crop, particularly in areas of poor water drainage. After crop harvest, rice stubbles are left in the field and are submerged with water for decomposition during fallow period (3 weeks between first and second crop; 10 weeks between second and first crop). The decomposing rice straw contained p-hydroxybenzoic, p-coumaric, syringic, vanillic, ferulic, o-hydroxyphenylacetic, propionic and butyric acids (Chandramohan *et al.*, 1973; Chou and Lin, 1976; Wu *et al.*, 1976b). The o-hydroxyphenylacetic and cinnamic acids inhibited the growth of rice seedlings, even at low concentrations. The phytotoxicity of decomposing rice straw persisted upto four months, therefore, the tillering and panicle formation of rice are reduced significantly, which decreased the rice yield (Chou and Lin, 1976; Chou *et al.*, 1977; Chou and Chiou, 1979). Besides, application of ammonium sulphate to rice residues increased the formation of H_2S gas which is toxic to rice growth (Wu *et al.*, 1976a; Chou *et al.*, 1981). Incorporation of rice residues in soil, reduced the grain yield of succeeding soybean crop compared with burning of rice crop residues (Asian Vegetables Research and Development Centre, 1978). This reduction in grain yield of soybean was partly due to its decreased biological nitrogen fixation. Because phenolic compounds produced from decomposing rice residues, inhibited the growth of *Rhizobium japonicum*, reduced nodulation and their haemoglobin content and thus decreased biological nitrogen fixation of soybean (Rice *et. al.*, 1981).

IV. Rye : Rye residues became most phytotoxic, 10-30 days after the start of decomposition (Patrick *et al.*, 1963; Chou and Patrick, 1976) and contained vanillic, ferulic, phenylacetic, 4- phenylbutyric, o-coumaric, p-coumaric, p-hydroxybenzoic, benzoic, butyric, hydrocinnamic and salicylic acids and salicylaldehyde (Patrick 1971; Chou and Patrick, 1976). Phenylacetic, 4- phenylbutyric and salicylic acids were most inhibitory to growth of lettuce seedlings. These compounds were concentrated around their loci of production and injure the plant parts which encounter them (Patrick *et al.*, 1963; Patrick and Koch, 1963). Decomposing rye residues, inhibited the seed germination, seedling emergence and radicle elongation of rye seedling. Its dry residues, proved more inhibitory than green ones. Under no-till conditions, rye `wheeler' residues reduced the seedling emergence of lettuce `Ithaca' by 58% over wood shaving control mulch. Its shoot tissues caused 52% more inhibition of lettuce seed germination than root tissue (Barnes and Putnam, 1983b, 1986).

V. Oat : Harper and Lynch (1979) studied the effect of aging on phytotoxin production of oat straw, when incorporated in soil or maintained on soil surface. Soil buried residues decompose very rapidly than that lying on soil surface. Within six weeks of soil incorporation, main sources of phytotoxins (cellulose and hemicellulose) were degraded upto 50% and phytotoxin production was decreased by 85%. Oat straw was most toxic at maturity and its toxicity persisted upto 8 weeks after the start of decomposition (Guenzi *et al.*, 1967). The aqueous extracts of decomposing oat residues, were more inhibitory to seedling growth of test species (oat, corn, timothy, soybean, alfalfa, peas) compared with corn stover, timothy hay and alfalfa hay (Nielson *et al.*, 1960).

VI. Sorghum : The toxic effects of sorghum on following crops have been reported, since the beginning of 20th century. It has been found harmful to succeeding crops of maize (Ten Eych and Shoesmith, 1906), oat and wheat (Vinall and Getty, 1921; Cole and Halstead, 1922; Breazeale, 1924; Sewell, 1933; Myers and Hallsted, 1942) and cotton (Burhan and Mansi, 1967), but improved the yields of vetch (Hawkins, 1925). Aqueous extracts of decomposing sorghum stubbles, killed the wheat seedlings and toxicity persisted for 30 days (Breazeale, 1924). Decomposing sorghum and corn residues reduced the height of wheat seedlings (Behmer and McCalla, 1963). Ruiz-Sifre and Ries (1983) studied effects of young residues of several cultivars of sorghum (*S. vulgare* Pers) on the growth of snap beans and maize in the greenhouse and field. Shoot residues of sorghum stimulated the growth of maize seedlings, but it was not true for root residues in glass house studies. While roots and whole plant residues of sorghum `Bird-a-Boo' decreased the maize growth in field. Sorghum residues increased or decreased the growth and yield of snap beans in field, depending on the sorghum plant part, quantity, cultivar and soil environment. In Senegal, West Africa, the growth of succeeding sorghum was poor following sorghum in sandy soils than in montmorillonite rich soils. The microflora of sandy soils was unable to detoxify the soil toxins (p-coumaric, o-hydroxybenzoic and proto-catechuic acids) at faster rate to prevent inhibition of subsequent crop (Burgeos - Leon *et al.*, 1980). Aqueous extracts of 4-weeks old sorghum herbage was most inhibitory to seed germination of cress and inhibition declined sharply as the plants matured (Lehle and Putnam, 1982). Weston *et al.* (1989) studied allelopathic potential of Sudex (Sorghum x sudan grass) measuring radicle elongation of indicator species (tomato, foxtail millet and burpee curlygarden cress). The allelopathic potential decreased with increasing sudex age and was found to be maximum in 7- days old shoots tissues. It proved more inhibitory to small-seeded, broad leaf species than grass species. Two major phytoinhibitors (p-hydroxybenzoic acid and p-hydroxybenzaldehyde) were isolated from the aqueous extracts of sudex shoot, these were breakdown products of cynogenic glucoside dhurrin. Young (7-days old) sudex tissues possessed greater percen-tage of these phytoinhibitors than older tissues. With increasing age of sudex tissue, the percentage of p-hydroxybenzaldehyde increased, while, that of p-hydroxybenzoic acid decreased. The aqueous extracts of sorghum stubble, inhibited the seed germination and root growth of Indian rape, Indian mustard, oilseed rape and brown *sarson*. But maximum inhibition occurred in seed germination of Indian rape and brown *sarson* and root growth of Indian rape and Indian mustard (Narwal *et al.*, 1990). Amongst test spp. of oat, barley, chickpea, lentil and *Eruca sativa* the aqueous extract of sorghum stubble inhibited the seed germination of oat and barley and had no effect on root or shoot growth of test crops (Narwal and Gupta, 1991d).

VII. Pearlmillet : In field studies, Indian mustard sown after pearlmillet showed poor stand, stunted growth, very few branches and pods and poor seed yield than that sown after maize, sorghum, cowpea and clusterbean. Nitrogen application upto 90 kg/ha did not overcome the injurious effects of previous pearlmillet crop (Singh *et al.*, 1983). Subsequently, Narwal and co-workers pursued further bioassay studies to determine the effect of aqueous extracts of pearlmillet (stem, root and stubble) on the seed germination and seedling growth of cereals (wheat, barley), pulses (chickpea, lentil), oilseeds (oilseed rape, Indian rape, Indian mustard, brown *sarson, Eruca sativa*) and fodders (oats, sweet clover, Egyptian clover) (Narwal *et al.*, 1989 a,b, 1990; Narwal and Gupta, 1991 a,b, 1992 a,b). In all these studies, the inhibitory effects of aqueous extracts increased with their increasing concentration. The inhibitory effect of aqueous extract of pearlmillet stem, root and stubble on the seed germination and seedling growth of test species, (oilseed rape, Indian rape, Indian mustard and brown *sarson*) followed the order stem > root > stubble. The stem extract reduced the seed germination of Indian rape and Indian mustard, shoot growth of Indian rape and oilseed rape and root growth of all test crops except mustard. The root extract inhibited the seed germination and shoot growth of Indian rape and root growth of oilseed rape and brown *sarson*. The stubble extract inhibited the seed germination of Indian rape, the shoot growth of Indian rape and Indian mustard and the root gowth of all crops except oilseed rape. All these extracts inhibited the seed germination and seedling growth of Indian colza beyond 4% concentration (v/v), however, root extract proved more inhibitory compared with stem and stubble extracts (Narwal *et al.,* 1989 b; Narwal and Gupta, 1991 a,b, 1992 b.). Stubble extract drastically inhibited the seed germination of lentil and fodder oat compared with wheat, barley, chickpea, rocket salad, sweet clover and Egyptian clover, (Narwal *et al.,* 1989 a; Narwal and Gupta, 1992a).

VIII. Barley and Buckwheat : Decomposing barley residues became phytotoxic 7-10 days after soil incorporation, phytotoxicity reached peak at 20 days and declined after 45-50 days. These residues contained benzoic, phenylacetic, 3-phenylpropionic and 4-phenly-butyric acids. Decomposing residues of barley, cotton, cowpea and soybean inhibited the seed germination and root elongation of tobacco seedlings due to the presence of benzoic and phenylacetic acids (Toussoun *et al.,* 1968). Tsuzuki *et al.* (1987) and Tsuzuki and Yamamoto (1987) identified four fatty acids (palmitic, stearic, arachidic and behenic) and three phenolic acids (ferulic, caffeic and chlorogenic) in buckwheat plants. The roots were rich in caffeic (>10 ppm) and ferulic acids than stem. Caffeic or ferulic and fatty acids inhibited the growth of rice seedlings at 10 and 250 ppm, respectively.

3.1.1.1.2. Legumes

Allelopathic effects of crop residues of alfalfa (Mishushtin and Naumova, 1955; Pedersen, 1965; Megie *et al.*, 1967; Stepanova and Prutenskaya, 1975), clovers (Cochrane, 1948; McCalla and Duley, 1948, 1949; Guenzi and McCalla, 1962), soybean (Kalantari, 1982; Lovett, 1983 a; Tsuzuki and Kawagoe, 1984; Huber and Abney, 1986; Gupta *et al.,* 1987), lupine (Lehle *et al.,* 1983), pigeonpea (Badillo-Feliciano and Lugo-Lopez, 1983) and pea (Lovett, 1983a) have been studied on the germination, growth and yield of other crops.

I. Alfalfa

Cotton following a 3-4 years old alfalfa crops gave poor yields, because alfalfa roots excrete saponins during vegetative growth (Mishushtin and Naumova, 1955). Besides its decomposing residues, reduced seed germination and radicle growth of cotton seedlings (Pedersen, 1965; Megie *et al.*, 1967). Saponins inhibited the cotton seed germination by blocking the oxygen diffusion to embryo (Marchaim *et al.*, 1972). Its root and stubble extracts were inhibitory to test species (alfalfa, cotton, wheat) at high concentrations and stimulatory at low concentrations. The compounds from its shoots were most active than from roots and those from `Azskh. I-262' alfalfa were more active than from `Azskh I-1' alfalfa. Cotton was most susceptible and wheat most resistant to the decomposition products of alfalfa (Stepanova and Prutenskaya, 1975). According to Oleszek and Jurzysta (1987), the root extracts of alfalfa suppressed the seed germination and seedling growth of wheat due to presence of inhibitory medicagenic acid glucoside. The powdered roots of alfalfa completely inhibited root growth of wheat seedlings on sand compared with heavy soils due to sorption of more inhibitors by heavy soils. The phytotoxicity decresed quickly on heavy soils than on sand, owing to hydrolysis of glucosides by microorganisms (Table 9). Read and Jensen (1989) studied problems associated with continuous plantings of alfalfa, using extracts of alfalfa soil, fallow soil or soil after a cereal crop. In germination chamber experiments, water soluble compounds present in methanol extracts of soil cropped to alfalfa or barley decreased seedling root lenghts of test crops viz., alfalfa `L-720', wheat `Nugaines', radish, `Crimson giant'. Five days after germination, seedling dry weight of alfalfa and radish in alfalfa soil extract was lower than wheat or redclover `Kenland'. The soil extracts from cropped soil had no harmful effect on growth of red clover. The extracts of crop residues screened from soil cropped to alfalfa or barley reduced root length of seedlings, however, the extracts of alfalfa residues proved more inhibitory than barley. Aqueous extracts of alfalfa inhibited seed germination of alfalfa, radish and turnip but not that of rice and soybean, yet it reduced seedling growth of rice and barley (Fig. 3) (Collison, 1925; Tsuzuki and Kawagoe, 1984). However, soil incorporation of alfalfa hay stimulated the growth and yield of tomato, cucumber, lettuce, rice and corn due to presence of the compound triacontanol. Its spray increased the growth of rice, corn and barley (Ries, *et al.*, 1977).

Table 9. Effects of alfalfa and red clover roots incorporated into soils with different texture on wheat seedling growth.

Soil type	Wheat seedling growth inhibition (%)			
	Alfafa		Red clover	
	Shoot	Root	Shoot	Root
Loose sand	52 a	86 c	26 b	33 b
Coarse sand	41 c	53 b	25 b	30 b
Loamy sand	41 c	47 b	24 b	31 b
Clay loam	27 b	17 a	15 a	7 a

Value within a column followed by different letters differe (P = 0.05)

Source : Oleszek and Jurzysta (1987).

II. Clovers : The extracts from residues of ladino clover and rye grass caused browning of radish root compared with extracts from corn and soybean (Cochrane, 1948). Aqueous extracts of sweet clover residues, caused poor and abnormal seed germination and depressed seedling growth of corn than did extracts of alfalfa hay, wheat straw or oat straw due to presence of coumarins in sweet clover residues (McCalla and Duley, 1948, 1949; Guenzi and McCalla, 1962). Aqueous extracts of white clover shoot, reduced the seed germination in radish and decreased seedling growth in wheat, maize and sunflower; while MeOH extracts of its flowers inhibited root growth in radish and tomato seedlings (Moroz and Leshenko, 1972; Nicollier and Thompson, 1982). The soil incorporated debris (0.8-6.7 mg/g soil) of crimson clover and hairy vetch, decreased the seedling emergence and dry weight of cotton but had no adverse effect on seedling emergence in corn and increased its dry weight (20-75%). The root biomass and root exudates of these species, had no inhibitory effect on the emergence and growth of test crops seedlings. However, the aqueous extracts (8.3-33.3 g debris/l) of crimson clover and hairy vetch, decreased germination and seedling growth of corn and cotton. The extracts of hairy vetch were found more phytotoxic than that of crimson clover (White *et al.*, 1989).

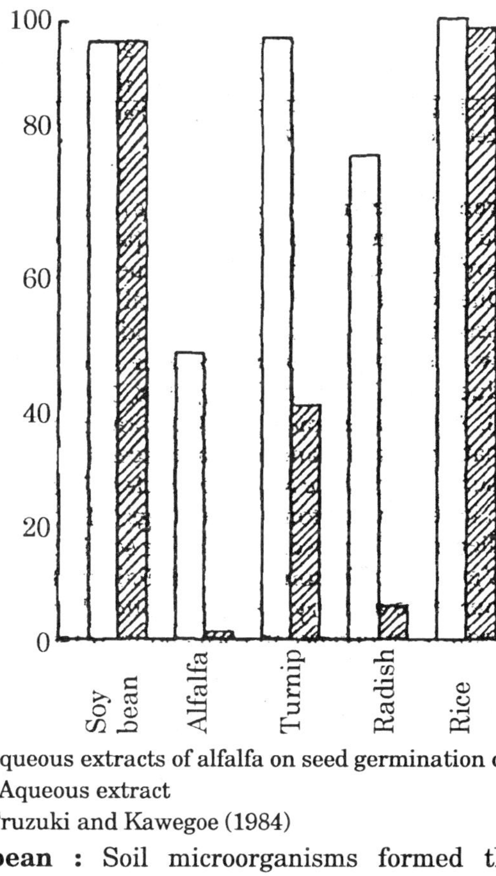

Fig. 3. Effect of aqueous extracts of alfalfa on seed germination of crops.
Controls, Aqueous extract
Source : Truzuki and Kawegoe (1984)

III. Soybean : Soil microorganisms formed the inhibitory substances from the decomposing soybean residues, such as amino acids (valine, leucine, phenylamine), organic acids and polyphenolic compounds (Prutenskaya *et al.*, 1970). Planting of wheat in soybean residues, delayed its seedling emergence and decreased the growth and grain yield (19-29%). Shallow incorporation of residues in soil, resulted in delayed seedling emergence, shortened coleoptile length and thin plant stands. The phytotoxicity increased with increasing rates of soybean residues (Huber and Abney, 1986). Aqueous extracts of soybean, inhibited the seed germination of soybean, alfalfa, turnip and radish and not that of rice (Fig. 4) (Tsuzuki and Kawagoe, 1984). Its root extracts delayed and inhibited the seed germination of oilseed rape and Indian mustard, inhibited their seedling growth at high concentrations but caused stimulation at lower concentrations (except roots of oilseed rape). Extract at high concentrations promoted branching of roots but inhibited accumulation of dry weight in seedlings (Table 10) (Gupta *et al.*, 1987). Chloroform extracts of soybean, increased root weight of maize seedlings upto 20 percent (Kalantari, 1982). In corn-soybean rotation, the corn was benefited from soybean effect, owing to release of 30-60 kg N/ha and allelopathy (Mosca *et al.*, 1988).

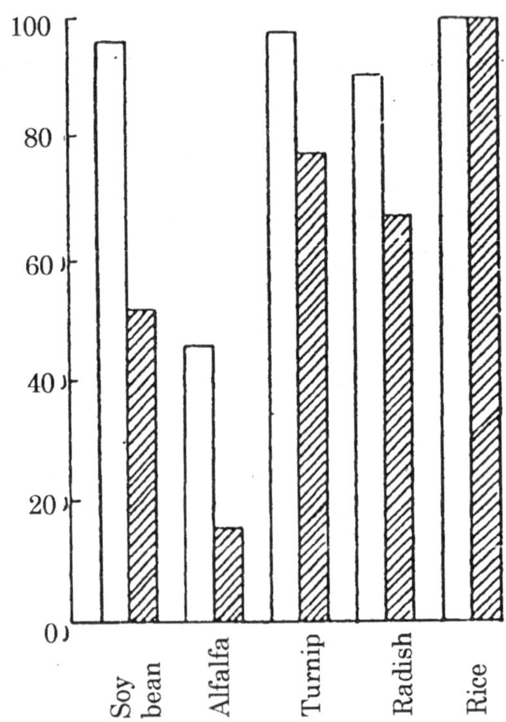

Fig. 4. Effect of aqueous extracts of soybean on seed germination of crops.
▯ Controls, ▯ Aqueous extract
Source : Truzuki and Kawegoe (1984)

Table 10. Effects of aqueous root extract of soybean on the per cent inhibition (-) or stimulation (+) of shoot and root growth of Indian mustard and oilseed rape compared with control

Aqueous extract concentration	Shoot			Root		
	Hours	after	Sowing	Hours	after	sowing
	96	120	168	96	120	168
Indian mustard						
Control						
1.0 g/20 ml	+44.0	+74.2	+132.3	+33.7	+46.6	+4.0
1.0 g/10 ml	+31.8	+58.2	+101.9	+42.3	-48.1	-41.4
LSD (P=0.05)	11.2	9.5	22.2	8.6	1.1	19.9
Oilseed rape						
Control						
1.0 g/20 ml	+32.0	+43.9	+59.9	-18.5	-2.9	-13.7
1.0 g/10 ml	+2.4	+5.3	+34.2	-57.4	-48.5	-53.9
LSD (P=0.05)	17.6	18.4	12.1	21.4	23.1	17.5

Source : Gupta *et al.* (1987)

IV. Other Legumes : Rhizosphere soils of mungbean contained high concentrations of p-hydroxybenzoic and salicylic acids than non-rhizosphere soils (Pareek and Gaur, 1973). Soil incorporation of white lupine, decreased the seed germination of cotton and seedling growth of cotton and soybean (Lehle *et al.*, 1983). Soil application of pigeonpea dry leaves (2.7-11% w/w), decreased the seed germination (54-93%), height and fresh weight of pigeonpea seedlings (Table 11) and its leaf extract decreased the seed germination of maize, tomato, beans, soybean, cowpea and pigeonpea (Badillo-Feliciano and Lugo-Lopez, 1983). According to Lovett and Jessop (1982), the residues of pea, field bean, soybean, lupine, chickpea, safflower, sunflower, rapeseed, sorghum, oats, barley and wheat inhibited seed germination, seedling emergence, height of coleoptile and length of seminal roots of wheat. Soil incorporation of crop residues, increased the phytotoxicity than left on soil surface. Rapeseed, soybean and lupine exerted the greatest phytotoxic effects. Pea residues drastically reduced seed germination of wheat, while lupine residues were most inhibitory and oat residues the least inhibitory to seedling growth of wheat.

Table 11. Effects of soil incorporation of dry pigeonpea leaves on seedling emergence of beans

Treatment	Seedling emergence (%)
Soil + 2.75% of dry leaves	46
Soil + 5.50% of dry leaves	40
Soil + 11.0% of dry leaves	7

Source : Badillo - Feliciano and Lugo - Lopez (1983)

3.1.1.1.3. Oil Seeds

Allelopathic effects of residues of sunflower (Schon and Einhellig, 1982), rapeseed (Horricks, 1969: Santos *et al.,* 1983), oilseed rape (Vilhordo *et al.,* 1985.), Mustard and broccoli (Patrick *et al.,* 1963; Jimenez-Osornio and Gliessman, 1987) and chinese cabbage (Kuo *et al.,* 1981; Song *et al.,* 1985) have been studied on the germination, growth and yield of other crops.

I. Sunflower : Aqueous extracts of sunflower leaf reduced the growth of sorghum seedlings through decreased leaf water potential and increased diffusive resistance (Table 12). Soil incorporation of its residues @ 2 g/80 g soil, inhibited seed germintion and seedling growth of sorghum (Schon and Einhellig, 1982). Germacranolide isolated from sunflower inhibited the growth of oat coleoptiles (Spring *et al.,* 1982).

Table 12. Effect of aqueous leaf extracts and leaf residues of sunflower on the water status and growth of grain sorghum

Sunflower treatment	Dry weight (mg ± SE)			Water* potential (bars ± SE)	Osmotic* potential (bars ae SE)	Turgor* pressure (bars ± SE)	Abaxial* leaf resistance (s/cm ± SE)
	Shoot	Root	Plant				
Aqueous extract (Leaf weight : medium)							
Control	124.1±3.2a	53.3±2.5a	177.2±4.5a	-5.9±0.3a	10.2±0.4a	4.3±0.5a	19.3±0.8a
1:120	98.2±2.9b	32.0±1.4b	130.1±2.8b	-5.9±0.3a	-9.5±0.3a	3.7±0.5a	28.6±3.7b
1:60	88.5±4.2c	29.5±2.0b	118.0±5.7b	-7.0±0.4a	11.1±0.5a	4.0±0.7a	33.3±5.5b
1:20	40.1±2.7d	17.7±1.4c	57.7±3.8c	14.6±1.2b	14.1±0.5b	-0.2±0.8b	63.9±12.5c
Dried leaf residue							
Control	42.0±1.5a	25.3±0.8a	67.3±2.1a				
0.5 g/80 gsoil	39.4±1.2a	21.7±1.0bc	61.1±2.0b				
1.0 g/80 g soil	35.6±1.3b	22.8±0.6b	58.5±1.9b				
2.0 g/80 g soil	26.2±0.9c	19.4±0.9c	45.6±1.8c				

*Mean of 6 days, For extracts and residues means in a column followed by the same letter are not significant different (P=0.05). Source : Schon and Einhellig (1982).

II. Brassica oilseed crops : Decomposing residues of rapeseed, stunted the plants and reduced dry weight of oat, wheat and barley and the inhibition was maximum in oat (Horricks, 1969). Soil incorporated roots of rapeseed, initially inhibited the growth of wheat seedlings but caused stimulation after four days (Martin and Rademacher, 1960). Application of brassinolide steroid (a growth regulator from rapeseed pollens) @ 1.0 ng increased the growth of beans (Maugh, 1981). Extracts of oilseed rape, reduced seed germination and radicle growth of tomato and lettuce seedlings but soil incorporated residues had no effect on soybean (Peters *et al.*, 1982: Vilhordo *et al.*, 1985). Decomposing plants of broccoli, inhibited the seedling growth of lettuce, rye and barley and phytotoxicity persisted upto 30 days after soil incorporation (Patrick *et al.*, 1963). Jimenez-Osornio and Gliessman (1987) studied inhibitory effects of broccoli and mustard extracts on root growth of monocots (barley, rye grass) and dicots (broccoli, lettuce, vetch, radish, wild radish) seedlings and found greater inhibition in monocots than in dicots. Mustard extracts were more inhibitory compared to broccoli and toxicity was greater at flowering stage. Rainfall leached the inhibitors from the dry residues of broccoli. In green mustard plants, leaf extract was most inhibitory to root growth of barley but in dry plants extracts, the order of inhibition was root > leaf > stem > flower. The water soluble inhibitors present in the leaf litter of black mustard inhibited seed germination and seedling growth of grassland species (Went *et al.*, 1952: Scherger, 1954). Decomposing residues of chinese cabbage, inhibited seed germination of mungbean and decreased seedling growth of mungbean, soybean, corn, cabbage, chinese cabbage, tomato and sweet potato, while; extracts from oat and its root exudates reduced the root growth and dry matter of chinese cabbage and mustard. However, extracts from fresh roots were more toxic than from dry ones (Kuo *et al.*, 1981; Song *et al.*, 1985; Akram and Hussain, 1987).

Table 13. The effective of volatiles from *Cruciferae* species on lettuce barnyard grass and wheat

Cruciferae species	Germination (% control)			Coleoptile length (% control)			Root length (% control)		
	Lettuce	Barnyard grass	Wheat	Lettuce	Barnyard grass	Wheat	Lettuce	Barnyard grass	Wheat
Brassica juncea '603'	11	81	100	5	63	89	0	0	85
B. juncea	0	84	100	0	65	83	0	14	80
B. nigra	0	81	100	0	77	80	0	34	87
B. napus 'Bronowski'	95	82	100	121	89	96	83	84	95
B. napus 'Gorczanski'	100	90	100	129	89	90	183	21	90
B. rapa var. *rapifera*	100	91	100	100	84	90	142	52	95
B. oleracea var. acephala	99	87	100	86	95	90	83	20	96
Sinapis alba	100	92	100	122	81	95	175	73	98

Source : Oleszek (1987)

Besides crop residues, volatile substances from the pulverised leaves of rapeseed, Indian mustard, black mustard, white mustard, turnip and broccoli delayed the seed germination and reduced seedling growth of wheat, lettuce and barnyard grass. The volatiles of black mustard and Indian mustard were most inhibitory (Table 13). The inhibition was due to degradation products of glucosinolates (Oleszek, 1987).

3.1.1.1.4. Other Crops

This section includes least studied crops such as sugar crops (sugarcane, sugarbeet,) tuber crops (sweet potato, cassava, taro) and crops like asparagus, radish and tobacco.

I. Sugar Crops : Sugarcane yield declined in continuous monocropping, partly due to phytotoxicity of decomposing residues or microbial toxins. Decomposition of sugarcane leaves under waterlogged conditions produced p-hydroxybenzoic, p-coumaric, syringic, ferulic, vanillic, formic, acetic, oxalic, malonic and malic acids; which at 50 ppm in water culture inhibited the growth of sugarcane roots (Wang *et al.,* 1967b). Besides, rhizosphere soils of poor sugarcane ratoon contained a large population of *Fusarium oxysporum* fungi compared with rhizosphere soils of good ratoon or newly planted sugarcane. This fungus produced fusaric acid, which inhibited the growth of young sugarcane plants (Wu *et al.,* 1976b). Sugarbeet decreased the stem and leaves growth of subsequent soybean crop in dry type volcanic ash soil in District Hokkaido, Japan. It was speculated to be caused by some allelopathic compounds (Matsui and Nishiiri, 1989).

II. Tuber Crops : Soil from sweet potato field or soil incorporated residues or its residue extracts, inhibited the seed germination and reduced the dry matter accumulation in seedlings of alfalfa. The phytotoxicity from the decomposing residues persisted upto 12 weeks (Fig. 5) (Harrison and Peterson, 1986). Soil incorportaion of vines and storage roots of sweet potato reduced the shoot dry weight of sweet potato and cowpea and also inhibited the nodulation in cowpea plants compared with that grown in sand (Table 14) (Walker and Jenkins, 1986). Sweet potato vines or roots @ 1.67% (w/w) added to coarse silica sand or silt loam soil decreased the shoot dry weight and uptake of P,K, Ca, Mg, S, Zn, B and Cu in indicator plants of sweet potato 'Centennial' compared with control (Walker *et al.*, 1989). Pardales Jr. and Dingal (1988) studied allelopathic effects of taro residues on the germination and seedling growth of test crops viz., mungbean, corn, okra and egg plant. Its residues extract inhibited germination and seedling growth of mungbean and okra but had no effect in corn and egg plant. Increasing levels of taro residues incorporated in a sand growing medium, stimulated production of more but shorter roots in taro plants. Among residue extracts of sweet potato, cassava and taro, only that of taro decreased germination and hypocotyl elongation in mungbean. The allelopathic effects of taro residues were species-specific and level dependent.

Table 14. Influence of sweet potato vines and storage root residues on growth of sweet potato vines and cowpea plants

Plant part	Shoot weight (g)		Root weight (g)	
	Fresh	Dry	Fresh	Dry
	Sweet potato `Jewel'			
Vines	22.33 a	3.22 a	13.73 a	2.30 a
Roots	6.59 b	1.21 b	4.64 b	0.97 b
Control	33.51 c	4.74 a	19.19 c	2.95 a
	Sweet potato `Centennial'			
Vines	22.35 a	3.90 a	15.37 a	2.37 a
Roots	8.86 b	1.30 b	4.48 b	0.79 b
Control	25.81 a	4.77 a	20.58 c	3.07 a
	Cowpea `Brown crowder'			
Vines	4.18 a	0.69 a	0.80 a	—
Roots	1.61 b	0.30 a	0.67 a	—
Control	16.21 c	3.21 b	3.35 b	—

Means within a variety not followed by the same letters are significantly different P = 0.05.

Source : Walker and Jenkins (1986).

III. Asparagus : Rhizome and root tissues of asparagus were less inhibitory to the growth of tomato and barnyard grass compared with roots. Root tissues reduced root and shoot growth of one year old asparagus crowns by 51 and 66%, respectively, because the root extracts killed the isolated cells of asparagus (Putnam *et al.,* 1983). Aqueous extracts of asparagus roots, inhibited seed germination in tomato and lettuce but not in cucumber. The extracts reduced hypocotyl growth in lettuce, shoot growth in asparagus and inhibited radicle elongation in barley, lettuce and asparagus. However, seedling growth was not affected in tomato and two cultivars of wheat. The inhibition was concentration dependent. Radicle growth of lettuce 'Grand Rapids' was sensitive to an extract concentration as low as 0.05 g dry root tissue/100 ml water. Asparagus radicles were more sensitive than its shoots. Aqueous root extracts of *Asparagus racemosis* willd. inhibited germination and radicle growth of lettuce 'Grand Rapids". A compound from crude extract inhibited growth of lettuce and asparagus roots (Hazebroek *et al.,* 1989). Hartung *et al.*

(1989). reported that allelochemicals from asparagus have direct physiologiclal and biochemical effects on its plants, that predisposes them to fusarium diseases. Therefore, the yields of asparagus declined in continuous monocultures owing to autotoxicity (Takatori and Souther, 1978; Hartung and Stephens, 1983).

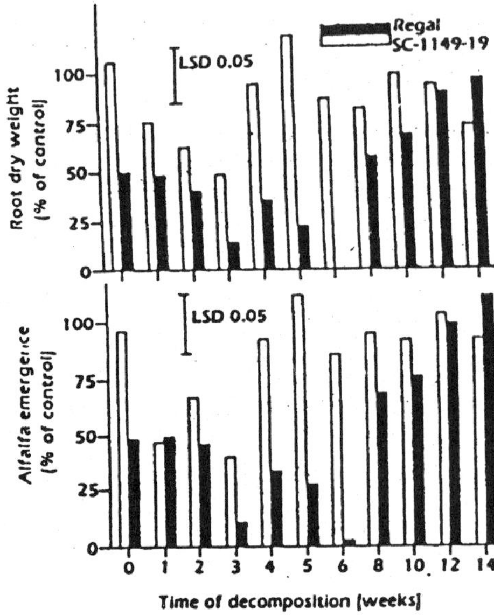

Fig. 5. Emergence and root dry weight of alfalfa plants in potting medium at various incubation periods following the incorporation of Regal and SC 1149-19 sweet potato plants (Source : Harrison and Peterson (1986))

IV. Radish : The juices of radish completely inhibited the germination of tobacco, *Phaseolus* spp., *Trifolium* spp, lettuce spp., *Capitata* and *Allium porrum* (Uygur *et al.,* 1990).

V. Tobacco : Nicotine adversely affected the germination, root and shoot length and seedling vigour of rice but increased height, specific leaf weight and chlorophyll content of maize (Rizvi *et al.,* 1989a), because it increased amylase activity in later crop. The positive effect on maize was partially due to increase in solubilization of stored starch through the activity of amylase (Rizvi *et al.,* 1989b).

3.1.1.2. Root exudates

Root exudates of crops affect the germination, growth and yield of other crop plants, therefore, play a major role in crop mixtures or intercropping systems (Table 15). DeCandolle (1832) was the first scientist, to report harmful effects of root exudates of one plant on the growth of other plants. Sorghum and maize root exudates inhibited the growth of sesame plants, therefore its plants could not be grown closer than 60 cm to live sorghum plants, which released natural toxins in the growing medium (Fletcher, 1912 : Breazeale, 1924 : Hawkins, 1925 : Conrad, 1927 : McKinley, 1931). Of buckwheat, alfalfa, red clover, pea, soybean, rye, vetch and blue grass root exudates, only that of buckwheat reduced the yield of tomato (Alderman and Middleton, 1925).

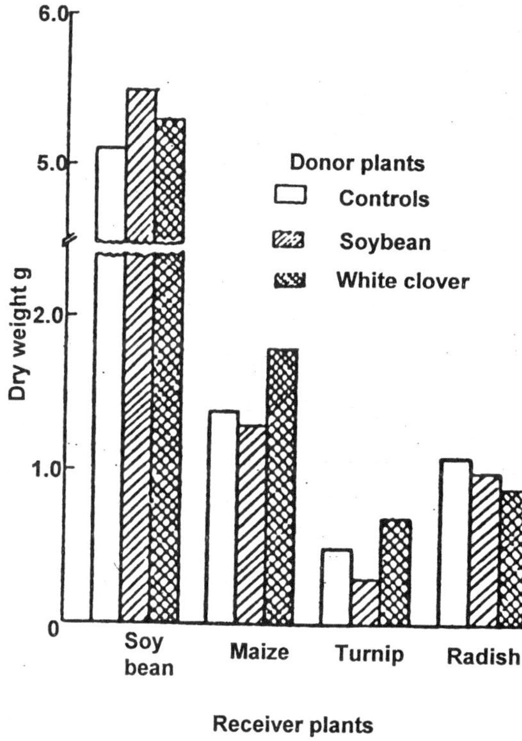

Fig. 6. Effects of root exudates from soybean and white clover on dry weight of crops (*Source* : Tsuzuki and Kawagoe, 1984).

According to Overland (1966), barley is excellent smother crop due to its extensive root growth and root exudates which inhibited the germination and growth of tobacco, chick weed and shepherd's purse. However, its root exudates had no inhibitory effect on wheat plants. The root exudates from living plants contained the alkaloid 'gramine' and were more inhibitory than aqueous leachates of dead roots, this proved active metabolic secretion of allelopathic substances. Root exudates of rice varieties 'CB-1' and 'Rupsail' inhibited the root and shoot growth of test seedlings of both these varieties owing to presence of phenolic compounds such as abscisic acid. The maximum release of inhibitors in root exudates occurred under favourable climatic conditions for rice growth (Sadhu and Das, 1971a,b; Sadhu, 1975). Tobacco root exudates inhibited seed germination and seedling growth of maize, mustard and foxtail millet (Haq and Hussain, 1979), while, that of chinese cabbage reduced radicle growth and drymatter of chinese cabbage and mustard (Akram and Hussain, 1987).

Table 15. Effect of crop root exudates on the germination and growth of crops

Donor crop	Phytotoxin	Recipient crop		Reference
		Crop	Inhibitory/stimulatory effects	
Inhibitory effects				
Rice	Abscisic acid	Rice	Inhibited root and shoot growth of seedling	Sadhu and Das (1971 a,b)
Barley	Gramine	Chickpea, tobacco	Inhibited germination and growth	Overland (1966)
Sorghum	ND	Sesame	Inhibited growth	Fletcher (1912)
Buck-wheat	ND	Tomato	Reduced yields	Alderman & Middleton (1925)
Tomato	ND	Lettuce, egg plant	Decreased seed germination and seedling growth	Kim and Kil (1987)
Tobacco	ND	Maize, mustard, Italian millet	Decreased seed germination and seedling growth	Haq and Hussain (1979)
Chinese, Cabbage	ND	Chinese cabbage, mustard	Reduced radicle growth and dry matter	Akram and Hussain (1987)
White clover	ND	Radish	Reduced dry weight	Tsuzuki and Kawagoe (1984)
Alfalfa	ND	Alfalfa, radish barley	Reduced dry weight	Tsuzuki and Kawagoe (1984)
Soybean	ND	Radish, turnip	Reduced dry weight	Tsuzuki and Kawagoe (1984)
Stimulatory effects				
Jute	ND	Wheat	Stimulated root and shoot growth	Das and Ghosh (1978)
Squash	ND	Bean (*Phaseolus vulgaris*)	Increased germination and seedling dry matter	El-Habbasha and Behairy (1975)

ND = Not determined

Tsuzuki and Kawagoe (1984) studied the inhibitory effects of root exudates of soybean and white clover on the dry weight of soybean, maize, radish and turnip; and that of alfalfa root exudates on alfalfa, barley and radish. The root exudates of white clover and soybean, decreased the dry weight of radish, but that of soybean also reduced dry weight of radish, but that of soybean also reduced dry weight in turnip (Fig. 6). While root exudates of alfalfa reduced the dry weight in all test species (Fig. 7). In another study, tomato root exudates depressed the seedling growth of lettuce and egg plant and its volatiles reduced the dry weight of neighbouring lettuce and grapevine (Kim and Kil, 1987). El-Habbasha and Behairy (1975) studied inhibitory effects of root exudates from cabbage, bean, onion and tomato on their own seed germination and seedling growth. The root exudates of all crops, except squash were inhibitory to other crops, squash root exudates stimulated the germination and dry matter of bean plants. The root exudates of jute also stimulated the root and shoot growth of wheat seedlings (Das and Ghosh, 1978).

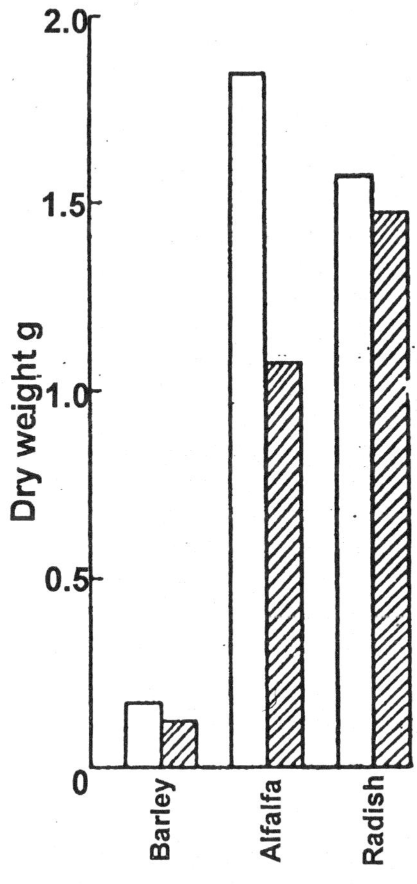

Fig. 8. Effect of root exudates from alfalfa on dry weight of crops.
Ž Controls Ž Root exudate (Source : Tsuzuki and Kawagoe (1984))

The root exudates of wheat, onion and rye grass reduced the rate of nitrification. The presence of nitrification inhibitors in the root exudates of rye grass reduced it upto 80% (Moore and Waid, 1971).

3.1.1.3. Cropping systems

Generally crops are grown in three types of cropping systems viz., monocultures, crop rotations and crop mixtures or intercropping. Allelopathy plays a significant role in all these cropping systems. It decreases the crop growth and yields in continuous monocultures, due to build up of phytotoxins or harmful microbes or both in the growing medium and gives rise to autotoxicity and soil sickness. Crop rotations are practised to overcome such harmful effects of monocultures, the preceding crops may be inhibitory or stimulatory to the succedings crops. Allelopathy also plays a great role in crop mixtures or intercropping systems, due to inhibitory or stimulatory effects of one plant on another.

3.1.1.3.1. Monocultures

According to Francis (1989), "*Monoculture is growing of same sole crop on the same land*" and this definition has been used in text. Under optimal conditions of growth, monocultures of annual crops do not show inhibitory effects on yields but it may not be true in their continuous monocultures or in perennial crops. This is due to accumulation of phytotoxins or build up of harmful microbes or both, which leads to autotoxicity and/or soil sickness.

I. Autotoxicity : Autotoxicity limits the yields of perennial crops and of annual crops in continuous monocultures. It results from the accumulation of phytotoxins from decomposing plant residues, root exudates and from multiplication of harmful pathogens in the soil. The autotoxicity has been observed in alfalfa (Miller, 1983; Read and Jensen, 1989), sugarcane (Wu *et al.,* 1976b), rice (Chou and Chiou 1979), asparagus (Hartung and Stephens, 1983; Hartung *et al.,* 1989; Hazebroek *et al.,* 1989), pigeonpea (Hepperly and Diaz, 1983), sunflower (Curtis and Cottam 1950,) and summer squash (Itulya, 1987) (Table 16). Similar problems of yield decline occur in wheat and corn, in areas of stubble mulch farming.

Preceding alfalfa crop released phytotoxins in soil, which adversely affected the re-establishment of alfalfa crop, however, saponins had no such effect (Miller, 1983). In germination chamber experiments, water soluble compounds present in methanol extracts of soil, cropped to alfalfa or barley decreased seedling root lengths of test crops viz., alfalfa `L-720', wheat `Nugaines', radish, 'Crimson giant'. Five days after germination, seedling dry weights of alfalfa and radish in alfalfa soil extract were lower than wheat or redclover, 'kenland'. The soil extracts from cropped soil had no harmful effect on growth of red clover. The extracts of crop residues from soil cropped to alfalfa or barley reduced root lengths of seedlings, however, the extracts of alfalfa residues proved more inhibitory than barley (Read and Jensen, 1989). Sugarcane yields were declined in subsequent ratoons, partly due to phytotoxicity of decomposing residues (under waterlogged conditions) or microbial toxins (Wang *et al.,* 1967a; Wu *et al.,* 1976b). Autotoxicity is one of the major reason of low yields of second or third crop of rice, in paddy growing areas of the world and it has been investigated by Chou and co-workers in Taiwan. The phytotoxins produced from the decomposing residues of preceding rice crop, under waterlogged or submerged or saturated soil conditions suppressed root growth, panicle initiation and grain yields of rice (Chou and Lin, 1976; Wu *et al.,* 1976a: Chou and Chiou, 1979: Chou *et al.,* 1981). For details on autotoxicity of rice and sugarcane see Chapter 3, Section 3.1.1.1.1.III, pp. 42, 3.1.1.1.4.I. pp. 54-55. Asparagus may give high yields upto 25-30 years after planting, but its autotoxicity markedly declined the yields after few years of successive growth in the same field. The phytotoxic compounds produced from decomposing residues of asparagus and from pathogenic fungi *Fusarium* spp. were toxic to its own growth (Kitahara *et al.,* 1972; Lauffer and Garrison, 1977; Takatori and Souther, 1978; Hartung and Stephens, 1983). The extracts reduced hypocotyl growth in lettuce, shoot growth in asparagus and inhibited radicle elongation in barley, lettuce and asparagus (Hazebroek *et al.,* 1989). Hartung *et al.* (1989) reported that allelochemicals from asparagus have direct physiological and biochemical effects on its plants, that predisposes them to *Fusarium* diseases. Sorghum autotoxicity was observed only in sandy soils of Senegal, West Africa, due to low absorption capacity of the soil and inability of the soil microflora to detoxify the soil phytotoxins, before the sowing of next sorghum crop (Burgeos Leon *et al.,* 1980). Continuous monocultures of pigeonpea, declined the growth and yield of subsequent crops, owing to phytotoxicity from terpenoids, polyphenols and other allelochemicals (Hepperly and Diaz, 1983). The autotoxins produced from the decay of prairie sunflower *(Helianthus rigidus)* residues, reduced the plant population and inflorescence numbers and size in the centre of clone (Curtis and Cottam, 1950). Itulya (1987) reported that the leachates of summer squash reduced its own growth due to autotoxicity.

Table 16. Crops exhibiting autotoxicity and soil sickness in monocultures

Donor crop	Source of phytotoxicity		Recipient crop		Reference
	Compounds	Soil microbes	Name	Inhibitory effect	
1	2	3	4	5	6
Autotoxicity					
Sorghum	ND	—	Sorghum	Inhibited crop growth	Burgeos-Leon *et al.* (1980)
Rice	p-hydroxybenzoic, syringic, vanillic, ferullc, propionic, o-hydrophenylacetic and butyric acids H_2S gas	*putida*	Rice	suppressed root growth, tillering, panicle initiation and low yield	Chou and Lin (1976), Wu *et al.* (1976ab), Chou *et al.* (1977), Chou and Chiou (1979), Chou *et al.* (1981)
Surgarcane	p-hydroxybenzoic, p-coumaric, syringic, ferulic, vanillic, formic, acetic, oxalic, melonic, tartaric, malic and fusaric acids	*Fusarium oxysporum*	Sugarcane	Poor germination and growth	Wu *et al.* 1976ab), Wang *et. al.* (1967abc)
Asparagus	ND	*Fusarium* spp.	Asparagus	Poor growth and yield	Kitahara *et al.* (1972), Hartung & Stephens (1983)
Alfalfa	ND	—	Alfalfa	Poor re-establishment	Miller (1983)
Sunflower	ND	—	Sunflower	Reduced plant stand and yield attributes	Curtis and Cottam (1950)
Summer squash	ND	—	Summer squash	Reduced seedling growth	Itulya (1987)
Pigeonpea	Terpenoids, polyphenols	—	Pigeonpea	Declined growth and yield	Hepperly and Diaz (1983)

Contd......

1	2	3	4	5	6
Soil sickness					
Alfalfa	Saponin	—	Alfalfa	Reduced seedling vigour, retarded root development and nodules, poor plant stand of short spindly yellowish plants and lower herbage yields	McElgunn and Heinrichs (1970), Webster *et. al.* (1967), Klein and Miller (1980), Kehr *et. al.* (1983)
Alfalfa	Saponin	—	Cotton	Decreased yields	Mishushtin and Naumova (1955)
Wheat	ND	33 fungal special from *Penicillum, Fusarium, Mortierella* genera. *Gaeumannomyces graminis* fungus and nematodes (*pratylenchus pratensis* and *Tylenchohynchus dobius*)	Wheat	Decreased shoot and root growth and tillering	Golovko *et al.* (1981), Grodzinsky and Golovko (1983), Waller *et al.* (1987), Wildermuth and Rovira (1977), Sigareva (1982)
Sugarbeet	ND	*Haterodera schachtii* and *Pratylenchus napus* nematodes increased, *Azotobacter* and nitrifying bacteria decreased	Sugarbeet	Decreased germination, attack of black root rot	Geller *et al.* (1977), Sigareva (1982)
Corn	ND	*Pratylenchus pratensis*, and *Helicotylenchus dihystera*	Corn	Increased soil toxicity and reduced crop growth	Kalmykova (1973)
Lupine, potato	ND	*Penicillium, Aspergillus* fungi	Lupine, potato		Nadkernichnyi (1974)
Cowpea	ND		Cowpea	Inhibited crop growth	Schreiner and Sullivan (1909)
Rice	ND	*Fusarium moniliforme*	Rice	*Reduced growth and yield*	*Ventura et al.* (1984)
Mungbean	ND	*Rotylenchulus, Meloidogyne*	Mungbean	Reduced growth and yield	Ventura *et al.* (1984)
Cowpea	ND	Meloidogyne	Cowpea	Reduced growth and yield	Ventura *et al.* (1984)

II. Soil sickness : In tropical and subtropical regions of the world, crops may be grown round the year, if irrigation water is not limiting. But when a crop is grown continuously in the same soil for long periods, subsequent plantings often grow poorly, compared with similar planting in soil never planted to the species concerned. The condition is species specific and other crops grow well in the same soil (Table 16). In sick soils phytotoxic materials liberated from the plants or plant residues may accumulate in soil through physical sorption (Seredyuk 1982). Clover crop failed in the field, where it had been grown constantly owing to soil sickness (Young, 1804) caused by root exudates and may be corrected by crop rotations having an interval of 7-8 years between the clover crops (DeCandolle 1832; T.A. 1845). According to Schreiner and associates, soil sickness is caused by single cropping for long periods (Schreiner and Reed, 1907a,b, 1908; Schreiner and Shorey, 1909; Schreiner and Sullivan, 1909). Extracts of cowpea sick soil, inhibited cowpea growth but after the extraction of inhibitors from the soil it did not inhibit cowpea growth (Schreiner and Sullivan, 1909). Continuous cropping of oat, wheat and barley caused soil sickness, owing to accumulation of toxins (Periturin, 1913). Soil sickness may be caused by (a) unbalanced use of nutrients, micronutrient deficiencies and excess of fertilization, (b) destruction of soil structure and physical properties of soil, (c) evolution of phytopathogenic microflora, (d) disproportionate development of several groups of microflora, (e) increased production of pests and weeds, (f) change in soil pH and (g) accumulation of phytotoxins in the soil (Krasilnikova and Garkina, 1946; Grummer, 1955; Patrick, 1955; Grodzinsky, 1974; Grodzinsky *et al.,* 1979; Grodzinsky and Golovko, 1983; Kehr *et al.,* 1983; Ventura *et al.,* 1984; Waller *et al.,* 1987). However, in this review only the reasons having phytotoxic effects (c,d and g) have been discussed. In the early phases of soil sickness, most of the toxins are native metabolites of higher or lower plants but in the later phases, humic substances (phenolic acids and closely related aromatic compounds) are dominant.

(i) Alfalfa : Areas in Alberta, Canada which once produced good crops of alfalfa, now grow consistently poor crops having short, spindly and yellowish plants, which do not respond to fertilization and irrigation. However, these soils are suitable for growing of alsike clover and red clover (Webster *et al.,* 1967). Soils under monocultures of alfalfa, suppressed the production of herbage roots and nodules (McElgunn and Heinrichs, 1970). Alfalfa sown after corn or sorghum, produced higher plant population and yield compared with continuous alfalfa (Table 17). Autophytotoxins were found in continuously alfalfa cropped soil, which inhibited the seedling establishment and growth and not in the root exudates of alfalfa (Klein and Miller, 1980). According to Kehr *et al.* 1983), the relative yields of alfalfa for land with no known history of alfalfa cropping, were 26, 40 and 35% higher in first, second and third year after sowing, respectively, compared with the same land area resown to alfalfa (Table 18). The stand of alfalfa, became poor in subsequent years of continuous alfalfa crop, owing to phytophthora and anthracnose diseases which caused soil sickness. Plant stand and yield declined rapidly, under continuous alfalfa than in alfalfa-soybean rotation. Soil fatigue caused premature death of red clover plants, owing to choking of their root xylem vessels with brown mass (Bogdan *et al.,* 1982).

Table 17. Plant population and yield of alfalfa (after 6 years under various cropping sequences) Illinois, Urbana

Cropping sequence	Plants/ft^2	Dry matter yield (t/ha)
Corn-alfalfa	15.0	3.78
Corn-soybean-alfalfa	12.5	3.48
Continuous alfalfa	6.5	1.90

Source : Klein and Miller (1980)

Table 18. Average plant stand and forage yield of alfalfa in the year of seeding and in subsequent years at Mead, Nebraska

Year seeded	Plant stand (%)			Forage yield (t/ha)		
	Year of seeding	Subsequent I	Years II	Year of seedling**	Subsequent I	Years* II
1970	95	95	90	8.25	17.15	16.46
1976	96	96	86	****	12.42	12.22
1977	97	82	—	6.59	11.97	—
1978	98	—	—	6.52	—	—

*Three cuttings per year, **Four cuttings per year, ***comparable data not available.

Source : Kehr *et al.* (1983).

In germination chamber experiments, water soluble compounds present in methanol extracts of soil, cropped to alfalfa or barley decreased seedling root lengths of test crops viz., alfalfa 'L- 720', wheat 'Nugaines', radish, 'Crimson giant'. Five days after germination, seedling dry weights of alfalfa and radish in alfalfa soil extract were lower than wheat or red clover 'Kenland'. The soil extracts from cropped soil had no harmful effect on growth of red clover. The extracts of crop residues from soil cropped to alfalfa or barley, reduced root lengths of seedlings, however, the extracts of alfalfa residues proved more inhibitory than barley (Read and Jensen, 1989). The soil sickness from alfalfa, may also decrease yields of other crops which follow alfalfa. Cotton yields were decreased, when it was grown after 3-4 years old lucerne crop owing to excretion of saponins in the soils (Mishushtin and Naumova, 1955).

Table 19. Weat bioassay of soxhlet extract of wheat soil collected at monthly intervals

Experimental soil	Root length (mm)	Shoot length (mm)	Inhibition (%)		Amount of crude organic extract (mg/seed)	Amount of soil extracted (g/seed)
			Root	Shoot		
April 9, 1985						
Control, distilled water	23.1±0.9	6.6 ± 0.2	—	—	—	—
No-Till, aqueous extract	15.5±2.9b	5.3 ± 0.7a	32	19	0.76	3.5
Conventional Till, aqueous extract	20.7 ±4.2c	6.56±0.8c	10	2	0.75	3.6
June 10, 1985						
Control, distilled water	18.0±1.7	5.7 ± 0.1	—	—	—	—
No-Till, aqueous extract	4.9±1.3a	3.55 ± 0.5a	83	39	0.7	2.8
Conventional Till, aqueous extract	3.3±0.5a	2.99 ± 0.3a	92	49	1.4	2.5
July 9, 1985						
Control, distilled water	23.8±4.6	6.5±0.7	—	—	—	—
No-Till, aqueous extract	16.6±1.9b	5.2±0.3b	30	20	0.45	1.0
Conventional Till, aqueous extract	18.7±2.5b	5.2±0.6b	22	22	0.58	2.0

(ii) Wheat : Inhibitory compounds in the aqueous extracts of soil, under wheat monoculture with conventional and no tillage inhibited the root growth of wheat seedlings by 32 and 10% and shoot growth by 19 and 21%, respectively (Table 19) (Waller *et al.*, 1987). Continuous monocultures of wheat, decreased tillering and crop yields of wheat (Grodzinsky and Golovko, 1983). owing to accumulation of allelochemicals in the soil (Berestetskiy and Kalmykova, 1974; Krupa, 1981; Krupa and Dzyubenko, 1981; Grodzinsky *et al.*, 1982). The allelochemicals may change the hydrophillic nature of the soil colloids and thus directly inhibit the germination of crops. Maximum soil toxicity occurred during the kernel formation stage of wheat. Continuous monocultures of wheat accumulated 1.4-1.7 times more allelochemicals (n-coumarin, n-oxybenzoic, ferulic and vanillic acids), resulted in build up of harmful microbes in the soil, than that under crop rotations (Berestetskiy and Kalmykova, 1974; Krupa, 1981; Krupa and Dzyubenko, 1981; Grodzinsky *et al.,* 1982). The extracts of these phytotoxic microbes were found to decrease the seed germination of radish by 50-80%. Golovko *et al.* (1981) isolated and identified 33 species of fungi mainly from *Penicillium, Fusarium* and *Mortierella* genera which made the soil sick and their toxicity ranged from 20 to 100%. Continuous monocultures of wheat also enhanced the accumulation of fungus *Gaeumannonyces araminis* responsible for root rot (Wildermuth and Rovira, 1977) and parasitic nematodes such as *Pratylenchus pratensis* and *Tylencohynchus dobina* (Sigareva, 1982).

(iii) Other crops : Continuous monocultures of corn on soddy podzolic soil, increased the soil toxicity owing to build up of toxic forms of bacteria, fungi and parasitic nematodes such as *Pratylenchus pratensis* and *Helicotylenchus dihysteria* (Kalmykova, 1973; Sigareva, 1982). However, Ventura *et al.* (1984) did not observe soil sickness from 8 years of corn and sorghum monocultures compared with rice, mungbean and cowpea. Growth inhibition in monocultures of rice, mungbean or cowpea began with second crops and increased with succeeding crops. Soil sickness was not due to the nutrient status of the soil, but was owing to the toxins produced from the decomposing crop residues and from build up of fungi in rice and nematodes in legumes. Soil sickness was observed in continuous monoculture of rice and mungbean for 8 years but not in rice-mungbean rotation (Table 20). Continuous cultivation of sugarbeet, lowered its seed germination from 77 to 33%, decreased the availability of nitrogen and phosphorus to crop plants owing to reduction in the population of N and P compounds transforming bacteria and that of *Azotobacter* by 5-10 times (Geller *et al.,* 1977). It also lead to the build up of nematodes *(Heterodera schachti* and *Pratylenchus napus),* which increased crop susceptibility to black root rot (Sigareva, 1982). Continuous monocultures of tobacco, increased the incidence of root diseases caused by soil borne pathogen *Thielaviopsis basicola* compared to rye grass - corn - tobacco rotation (Patrick and Koch, 1963). Because, fungitoxic substances produced from decomposing residues of corn or rye grass, inhibited the germination of conidia or chlaymydospores of *T. basicola* and thus reduced its population and root rot diseases (Chou and Patrick, 1976). Continuous cultivation of lupines or potato increased the population of toxic fungi *(Penicillium* and *Aspergillus*) compared to rye-lupine or rye-potato rotations (Nadkernichnyi, 1974). The continuous monocultures of cabbage for eight crop seasons, accumulated amino acids in growing substrate of perlite (Gorobets, 1981).

3.1.1.3.2. Crop rotations

In crop rotations, one crop may harm another one through depletion of plant nutrients, phytotoxins released from decomposing residues may injure the plant roots in contact with them and toxins in root exudates may be harmful to subsequent crops (Miller, 1931). Since all the plants are not equally sensitive to allelochemicals, therefore, ideal crop rotations avoids negative effects of allelochemicals and capitalize on their stimulatory effects.

Table 20. Effect of previous cropping condition on yield of upland rice

Crop	Previous crop treatment	Grain yield (t/ha)
Rice `IR 2061-464-2-4'	3 continuous rice crops	0.36 b
	15 month continuous fallow	1.75 a
Rice `IR 52'	16 continuous rice crops	0.85 b
	2 year continuous fallow	2.35
Rice `IR 747-B2-6-3'	4 continuous rice crops	0.95 b
	4 continuous sorghum crops	1.55 a
Rice `IR 5'	6 continuous rice crops	0.87 b
	7 continuous cowpea crops	3.02 a
Mungbean `MG 50-10A'	16 continuous mungbean crops	0.15 b
	8 continuous mungbean crops	0.53 a
	6 continuous sorghum crops	1.21 a
	2 Mungbean + 1 year fallow	0.60 a
Cowpea `EG green pod'	5 continuous cowpea crops	0.60 b
	5 continuous maize crops	1.64 a
Mungbean `MG 50-10A'	16 continuous mungbean crops	0.42 a
	16 continuous cowpea crops	0.47 a
Sorghum `Cosor 2'	4 continuous sorghum crops	3.45 a
	4 continuous rice crops	3.71 a
	16 continuous sorghum crops	2.95 a
	16 continuous cowpea crops	3.30 a
Maize `DMR 2'	5 continuous maize crops	3.63 b
	5 continuous cowpea crops	4.47 a

In crops the same cropping period means followed by a common letter are not significantly different at 5% level.

Source : Ventura *et al.* (1984).

Good crop rotations maintain the soil productivity, control the insects and diseases and overcome the soil sickness or autotoxicity in crops. Crop rotations remove the ill effects of soil sickness arising from continuous cropping of tobacco (Patrick and Koch, 1963), lupine

(Nadkernichnyi, 1974), potato (Nadkernichnyi, 1974), alfalfa (Klein and Miller, 1980; Kehr *et al.*, 1983), rice (Ventura *et al.*, 1984) and mung bean (Ventura *et al.*, 1984) and autotoxicity caused by pangola grass (Chou, 1986). Crop rotations overcome the major causes of soil sickness such as (i) accumulation of phytotoxins in the soil (Periturin, 1913; McCalla and Haskins, 1964; Krupa 1981; Grodzinsky *et al.*, 1982; Ventura *et al.*, 1984; Chou, 1986), (ii) multiplication of phytopathogeinc soil microflora (Kalmykova, 1973; Nadkernichnyi, 1974; Golovko *et al.*, 1981), (iii) increased incidence of pests (Sigareva, 1982) and diseases (Patrick and Koch, 1963; Kehr *et al.*, 1983) and (iv) destruction of soil physical properties (Grodzinsky and Golokvo, 1983). Numerous studies have been done on the crop rotations but very few dealt with allelopathic aspects and their causative agents. Cereals (wheat, oat, barley, rice, corn, sorghum) based rotations have been studied most compared with rotations of other crops. Inhibition or stimulation of crop yields have been observed in rotations of cereals-cereals, cereals-legumes and other crops (Table 21, 22).

Table 21. Inhibitory allelopathic effects in crop rotations

Crop rotation	Donor crop	Phyto-toxin	Recipient crop		Reference
			Crop	Inhibitory effects	
Sorghum-maize	Sorghum	ND	Maize	Decreased yield	Ten Eych and Shoesmith (1906)
Sorghum-sorghum	Sorghum	ND	Sorghum	Decreased growth	Burgeos-Leon *et. al.* (1980)
Sorghum-sorghum/ pearlmillet/ maize-foxtailmillet	Sorghum	ND	Sorghum, pearlmillet, maize, foxtailmillet	Reduced seed germination, radicle growth, dry matter and water content	Hussain and Gaddon (1981)
Sorghum-wheat-oat	Sorghum	ND	Wheat, oat	Decreased yield	Vinall and Getty (1921)
Sorghum/ maize-oat	Sorghum/ maize	ND	Oat	Reduced rate of seedling emergence	Wozniak *et. al.* (1981)
Sorghum/millet-wheat/rape	Sorghum/ millet	ND	Wheat/rape	Decreased grain yield	Su and Fang (1981)
Tobacco-rice	Tobacco	Nicotine	Rice	Inhibited growth	Rizvi and Rizvi (1987)
Barley-barley	Barley	Phenols	Barley	Decreased grain yields	Muller-Wilmes and Zoschke (1980)

Contd.....

Crop rotation	Donor crop	Phyto-toxin	Recipient crop		Reference
			Crop	Inhibitory effects	
Rice-soybean	Rice (crop resid-ues buried in soil)	ND	Soybean	Decreased grain yields	Asian Vegetables Research & Development Centre (1978)
Cowpea-mungbean	Cowpea	ND	Mungbean	Decreased grain yields	Ventura *et. al.* (1984)
Mungbean-mungbean	Mungbean	ND	Mungbean	Decreased grain yield	Ventura *et. al.* (1984)
Alfalfa-alfalfa	Alfalfa	ND	Alfalfa	Decreased fodder yield	Klein and Miller (1980)
Pea-rye	Pea	ND	Rye	Decreased germination and seedling growth	Neustrueva and Terekhina (1975)
Sesame-cabbage	Sesame	ND	Cabbage	Decreased yields	Skinner (1913)
Sunflower-sorghum	Sunflower	ND	Sorghum	Interfere with water-balance, reduced grain yields	Schon and Einhellig (1982)
Surgarbeet-soybean	Sugarbeet	ND	Soybean	Reduced growth and yield	Matsui and Nishiiri (1989)

ND = Not determined.

Table 22. Stimulatory allelopathic effects in crop rotations

Crop rotation	Donor crop	Phyto-toxin	Recipient crop		References
			Crop	Stimulatory effect	
Soybean-oat	Soybean	ND	Oat	Produced bold seeds	Wozniak *et al.* (1981)
Soybean-corn	Soybean	ND	Corn	Increased grain yields	Mosca *et al.* (1988)
Sorghum-barley/pea	Soybean	ND	Barley-pea	Increased grain yields	Su and Fang (1981)
Maize-groundnut/wheat	Maize/groundnut	ND	Wheat	Increased grain yields	Su and Fang (1981)
Wheat/rape/pea-maize	Wheat/rape/pea	ND	Maize	Increased grain yields	Su and Fang (1981)

Contd.....

Crop rotation	Donor crop	Phyto-toxin	Recipient crop		References
			Crop	Stimulatory effect	
Wheat/ barley/ oat-sorghum	Wheat/ barley/oat	ND	Sorghum	Increased grain yields	Su and Fang (1981)
Corn-alfalfa	Corn	ND	Alfalfa	Increased grain yields	Klein and Miller (1980)
Tobacco-potato	Tobacco	ND	Potato	Incseased yields	Garner *et al.* (1925)
Tobacco-maize	Tobacco	Nicotine	Maize	Stimulated radicle and plumule growth	Rizvi and Rizvi (1987)
Rice-soybean	Rice (crop residue burnt)	ND	Soybean	Increased grain yields	Asian Vegetable Research & Development Centre (1978)
Watermelon-pangola grass	Water-melon	ND	Pangola grass	Increased growth and yield	Chou (1986)

ND = Not determined.

Allelopathy has been implicated in yield decline in crops when sown after sesame (Skinner, 1923), sorghum (Ten Eych and Shoesmith, 1906; Vinall and Getty, 1921; Cole and Halstead, 1922; Su and Fang, 1981; Wozniak *et al.,* 1981),corn (Garner *et al.,* 1925; Wozniak *et al.,* 1981), rice (Asian Vegetables Research and Development Centre, 1978; Ventura *et al.,* 1984), pea (Neustrueva and Terekhina, 1975), mungbean (Ventura *et al.,* 1984) and sunflower (Schon and Einhellig, 1982) (Table 21). Besides, allelopathic effects also reduced crop yields in continuous alfalfa (Miller, 1983) and pangola grass (Chou. 1986) and continuous monocultures of sorghum (Burgeos-Leon *et al.,* 1980), barley (Muller-Wilmes and Zoschke, 1980), rice (Ventura *et al.,* 1984), tobacco (Patrick and Koch, 1963) and mungbean (Ventura *et al.,* 1984). Sorghum was one of the first crops to be recognised for having deterimental effects on the succeeding crops of corn, wheat and oat (Ten Eych and Shoesmith, 1906; Schreiner and Reed, 1907b; Vinall and Getty, 1921; Cole and Halstead, 1922). Sorghum decreased the growth and yield of following sorghum crop in sandy soils of Senegal, West Africa due to low absorption capacity of the soil microflora to detoxify the soil toxins before the sowing of next sorghum crop (Burgeos-Leon *et al.,* 1980). Aqueous extract of sorghum field soils, decaying mulch and its residues reduced seed germination, radicle growth and dry matter of sorghum, pearlmillet, maize and foxtail millet by creating water deficit in these crops (Hussain and Gaddon, 1981). In sorghum/corn/ soybean-oat rotations, heavy residues of sorghum and corn inhibited growth of oat plants, therefore, oat following soybean produced bold and heavy seeds compared with those after corn or sorghum (Wozniak *et al.,* 1981). Su and Fang (1981) studied sorghum/corn/ rape/barley/ peas, sorghum/millet-wheat and corn/groundnut-wheat rotations. Corn and groundnut proved better preceding crops of wheat than sorghum or millets. While previous crop of sorghum was favourable to barley and pea but harmful to rape and reverse hold true for preceding corn. Wheat, peas and rape were good previous crops for maize as were wheat, barley and oat for sorghum. In corn-soybean rotation, the corn was benefited from soybean effect, owing to release of 30-60 Kg N/ha and to allelopathy (Mosca *et al.,* 1988) (Table 22).

In barley/potato-barley rotations, barley sown after barley gave 5-18% less yield than that after potato. Because phenols from the decomposing residues of preceding barley reduced the root growth of succeeding barley crop at heading stage (Muller-Wilmes and Zoschke, 1980.)

Table 23. Effect of previous crops on plant population and yield of mungbean `MG 50-10A' in field

Crop history	Plant population (x10^3/ha)			Grain yield
	4 week	6 week	8 week	(kg/ha)
15 Continuous mungbean crops	221a	98c	64c	61.5d
2 year fallow	263a	185b	158b	419.6a
1 rice crop-1 mungbean crop-1 rice crop	284a	263a	228a	473.1a
1 year mungbean and 1 season fallow	271a	210b	190ab	514.7

In a column means followed by a common letter are not significantly different LSD P = 0.05.

Source : Ventura *et al.* (1984).

In rice-soybean rotations, soil incorporation of rice straw decreased the grain yields of soybean compared with burning of rice straw, owing to inhibition of biological nitrogen fixation in soybean by the phytotoxins from the decomposing rice straw (Asian Vegetables Research and Development Centre, 1978; Rice *et al.,* 1980, 1981). In 8-years study, rice-mungbean rotation maintained high yields over their monocultures (Table 23), because, soil, fungi and root infecting nematodes caused soil sickness in monocultures of rice and mungbean, respectively. While, rice-mungbean rotation mitigated the soil sickness problem by suppressing the population of these microbes

(Table 24) (Ventura *et al.*, 1984). Similarly, corn- rye grass- tobacco rotation controlled the root rot disease of tobacco monocultures caused by *Thielaviopsis basicola* (Patrick and Koch, 1963). Alfalfa grown in corn-alfalfa rotation gave 50-60% higher yields than its monocul-tures (Klein and Miller, 1980).

Sesame reduced the yield of following cabbage due to the injurious effects of its residues (Skinner, 1913). Preceding pea crop inhibited the germination and growth of succeeding rye crop, because of release of inhibitory free radicles from the decomposing pea stubbles (Neustrueva and Terekhina, 1975). Allelochemicals released from the sunflower plants and residues decreased yields of subsequent sorghum, through interference in water balance (Schon and Einhellig, 1982). Sugarbeet decreased the stem and leaves growth of subsequent soybean crop in dry type volcanic ash soil in District Hokkaido, Japan. It was speculated to be caused by some allelopathic compounds (Matsui and Nishiiri, 1989). Potatoes gave better yields when grown after tobacco than after corn, while small grains did better following potatoes than following tobacco (Garner *et al.*, 1925) (Table 22). In tobacco/corn-rice rotations, tobacco leaf litter added 5.32 kg nicotine/ha which stimulated the radicle and plumule growth of corn but inhibited that of rice (Rizvi and Rizvi, 1987). Pangola grass-water melon rotation increased the growth of pangola grass by 40% over its monocultures, because water melon eliminated the phytotoxins of pangola grass residues (Chou, 1986).

3.1.1.3.3 Crop mixtures

Productivity of crop mixtures or intercropping systems may be decreased or increased depending on the inhibitory or stimulatory effects of component crops on each other, provided growth resources (light, water, nutrients and space) are not limiting (Table 25). Much of the research on agrophytocenosis has been conducted in the USSR and major part of this section is based on their english summaries. In crop mixtures or intercropping systems, the growth and yield of component crops increases due to greater nutrient absorption, better weed control etc. than in pure crops, yet the mechanism have not been fully understood. Root exudates play a major role in the productivity of crop mixtures, as they may improve growth (Lykhvar and Nazarova, 1970; Rakhteenko *et al.*, 1973a; Zabyalyendzick, 1973) and yield (Kaurov, 1970; Pronin and Yakovlev, 1970; Pronin *et al.*, 1972; Zabyalyendzick, 1973) of component crops, through improved ion exchange (Rakhteenko *et al.*, 1973a), greater nutrient uptake (Kaurov, 1970; Lastuvka, 1970; Pronin and Yakovlv, 1970; Tomashevskaya and Lugovskaya, 1970; Rakhteenko *et al.*, 1973b) and partial weed control (Jimenez-Osorino and Schultz, 1981) compared with pure crops.

Table 24. Plant parasitic nematodes in soil from continuous upland rice `IR 2061-464-2-4' and mungbean `MG 50-10A' field plots

Crop history	Present crop	Nematode density (Mean no./300 mg soil + 1g roots)								Rice grain yield (t/ha)
		Rotylenchulus		Meloidogyne		Nelicotylenchus		Total		
		1 month*	Flowering	1 month*	Flowering	1 month*	Flowering	1month*	Flowering	
14 continuous mungbean crops	Mungbean	263a	907a	1a	45a	4a	5a	272a	960a	
1 year fallow	Rice	5a	Ob	Oa	Ob	44a	64a	49b	67b	1.9a
1 rice crop + mungbean crop	Rice	24b	1b	Oa	Ob	35a	2a	59b	3b	2.0a
10 continuous rice crops	Rice	Ob	1b	Oa	Ob	9a	19a	9b	20b	1.0b

For each growth stage, means in a cloumn followed by a common letter are not significantly different P = 0.05.

* 1 month after sowing, **Total of *Rotylenchulus, Helicotylenchus, Meloidogyne, Tylenchorhynchus, Hemicycliophora, Hemicriconemioides* and *Criconemoides*.

Source : Ventura *et al.* (1984).

Table 25. Allelopathic effects in crop mixtures

Crop mixture	Donor crop	Source of allelopathy	Recipient Crop			Reference
			Crop	Inhibitory effect	Stimulatory effect	
Alfalfa + oat	Alfalfa	Root exudate	Oat	Reduced growth and yield	—	Lambert (1959) Nielsen *et al.* (1981)
	Oat	Root exudate	Alfalfa	–	Improved growth	Lambert (1959) Nielsen *et al.* (1981)
White mustard + pea	White mustard	—	Pea	—	Increased dry matter and P absorption	Sokolova and Mikryukov (1970), Sokolova (1973)
Buckwheat + mustard	Buckwheat	—	Mustard	Inhibited growth	—	Tsuzuki (1980)
Buckwheat + oat	Oat	Root exudate	Buckwheat	Inhibited growth and yield	—	Tsuzuki (1980)
	Buckwheat	Root exudate	Oat	—	Increased yields	Tsuzuki (1980)
Buckwheat + lupine/mustard	Lupine/ mustard	Root exudate	Buckwheat	—	timulated growth and development	Zabyalyendzick (1973)
	Buckwheat	Root exudate	Lupine	Inhibited growth and yield	—	Zabyalyendzick (1973)

Contd....

Crop mixture	Donor crop	Source of allelopathy	Recipient Crop			Reference
			Crop	Inhibitory effect	Stimulatory effect	
Lupine+bird'sfoot trefoil	lupine	Root exudate	Bird'sfoot trefoil	—	improved mineral nutrition	Kaurov (1970)
Barley/oat + pea/vetch	Pea/vetch	Root exudate	Barley/oats	—	stimulated photosynthesis and N,P,K,Ca uptake	Rakhteenko *et al.* (1973b)
	Barley/oat	Root exudate	Pea/vetch	Inhibited photosynthesis & N,P,K,Ca uptake	—	Rakhteenko *et al.* (1973b)
Soybean/chickpea bean+corn	Soybean/bean/ chickpea	Shoot volatile	Corn	Reduced 32P uptake	—	Petrova (1977)
	Soybean/bean/ chickpea	Root-shoot volatilves	Corn	—	stimulated 32p uptake, yield	Petrova (1977)
Corn+water melon	Corn	Pollens	Water melon	Reduced respiration and growth	—	Cruz *et al.* (1988)

Several legumes have been shown to increase maize growth in mixtures (Lykhvar and Nazarova, 1970; Pronin and Yakovlev, 1970: Dzyubenko and Petrenko, 1971). To improve the productivity of mixtures, new varieties of legumes stimulatory to the non-legume component have been developed, for example, late maturing varieties of soybean for corn+soybean mixtures (Lykhvar and Nazarova, 1970; Gulyaev *et al.,* 1970). Sokolova and Mikryukov (1970, 1971) studied white mustard + pea/barley mixtures. In white mustard + pea mixture, the root exudates of white mustard improved the phosphorus absorption and proportion of pea plants in the crop mixture, therefore, produced higher yield in mixtures than their sole crops. In pea + barley mixture, barley had no positive effect on pea but the ratio of barley in mixture increased in drought years compared with wet years. In white mustard + barley mixture, strong development of xylem occurred with a consequent decrease in the parenchyma of barley.

Table 26. Effect of mustard intercropping on yield attributes and yield of broccoli

Treatment	Head			
	No. of plants/ plot	Diameter (cm)	Weight (g)	Yield (t/acre)
Broccoli pure	73	13.55a	353.6a	441
2 mustard plants/m^2 + broccoli	89 ab	14.16a	332.8 a	5.02
4 mustard plants/m^2+ broccoli	101b	14.62a	358.35a	6.25
8 mustard plants/m^2+ broccoli	99 b	15.83 a	406.35b	6.74

Mean with the same letters are not significantly different, P = 0.05.
Source : Jimenez - Osorino and Gliessman (1987).

White mustard proved a better component of crop mixture with pea than with barley. In crop mixtures, legumes improved the phosphorus and potassium uptake of cereals by increasing the availability of soil phosphorus (Tomashevskaya and Lugovskaya, 1970). For example, in oat + lupine mixture, phosphorus uptake of oat was greater than in its sole crop (Rakhteenko and Egrova, 1971). Maize + horsebean mixtures produced high yields compared with sole crops due to greater exchange of root exudates between the component crops (Pronin *et al.,* 1972). In mustard + broccoli intercropping, broccoli gave very high yields, because mustard stimulated the growth and yield parameters (number of harvestable heads, diameter of the inflorescence and total biomass) of broccoli (Table 26) (Jimenez -Osornio and Gliessman, 1987). In buckwheat + mustard mixture the, root exudates of buckwheat inhibited the growth of mustard plants (Tsuzuki, 1980). In oat + alfalfa mixture, alfalfa cultivar `MN allelopathy' reduced grain yield of oat cultivars compared with other alfalfa cultivars (Nielsen *et al.,* 1981). In corn + water melon intercropping, corn pollens decreased the respiration and growth of watermelon seedlings (Cruz *et al.,* 1988). On the other hand, addition of 1-2 kg seed/ha of wild heliotrope (*Heliotropium europeum)* to the legume mixtures, reduced their weeds (30-70%) and pests (Grechkanev and Rodionov, 1971).

Root exudates play a vital role in productivity of crop mixtures, besides, volatiles from crop shoots may also influence nutrient uptake. Tomato + cucumber mixture excreted amino acids, aliphatic organic acids, phenolics and several unidentified compounds in nutrient culture (Gaidamak, 1971). Exudates of carrot roots and from leaves of beet, tomato and sweet potato contained volatile compounds (acetaldehyde, propionic aldehyde, acetone, methanol, ethanol and other unidentified compounds) but propionic aldehyde greatly reduced the growth of test species (Dadykin *et al.,* 1970). In wheat + rye mixtures, wheat growth was inhibited when its proportion was small, while the growth of rye was stimulated in all proportions. Since, the degree of stimulation or inhibition depends on the concentration of unidentified substances secreted from the root exudates of both species (Lastuvka, 1958). In oat + alfalfa mixture, the root exudates of oat reduced the growth of alfalfa but alfalfa exudates slightly increased the oat growth (Lambert, 1959). In general, root exudates may inhibit or stimulate the absorption, migration of ions and growth of plants. In maize + pea mixture, as the proportion of maize decreases its inhibition by pea increases and vice-versa (Lastuvka and Minarez, 1970). Kaurov (1970) reported greater dry matter production in bird'sfoot trefoil + yellow lupine mixture than their pure crops, because lupine improved the mineral nutrition of bird's foot trefoil plants. In buckwheat + lupine/mustard mixture, buckwheat yields were 12-35% higher than the pure crop, as the root exudates of mustard and lupine stimulated the growth and development of buckwheat. While in buckwheat + oat/lupine mixture, the root exudates of oat were inhibitory to buckwheat, conversely, buckwheat root exudates increased the productivity of oat by 10- 20% but decreased the growth and yield of lupine (Zabyalyendzick, 1973).

In maize + pea mixture, root exudates may inhibit or stimulate the growth, absorption and ion accumulation of plants, however, total N, P, K uptake was greater, than in sole crops and the nutrient uptake of maize was higher, than pea (Lastuvka, 1970; Lastuvka and Minarez, 1970). Similarly in bird'sfoot trefoil + yellow lupine mixture, lupine improved the mineral nutrition of bird'sfoot trefoil and bird'sfoot trefoil absorbed 32P more actively than lupine. The rate of 32p migration from lupine to bird'sfoot trefoil was greater than that from bird'sfoot trefoil to bird'sfoot trefoil plants (Kaurov, 1970).

According to Rakhteenko *et al.,* (1973 a,b,), the rate of exchange of root exudates in pea + oat and wheat + white mustard mixtures was 1.5 to 2 times and 3 to 7 times greater, respectively, than their pure crops. Substances in root exudates of pea and vetch stimulated the photosynthesis, absorption of 32_P, uptake of N, K and Ca by barley and oat plants. On the contrary, substances in root exudates of oat and barley inhibited these processes in pea and vetch plants. In pea + white mustard mixture, mustard root exudates improved the phosphorus absorption of pea plants than in mixture of pea with other crops (Sokolova, 1973). Volatile substances from the shoots of soybean, chickpea and bean reduced the 32_P uptake of maize plants, but combined volatile substances secreted from their roots and shoots stimulated its uptake and improved the maize yields (Petrova, 1977).

3.1.2 Crop-weed interactions

Crops exert allelopathic effects on other crops (Chapter 3, Section 3.1.1) and weeds. They may inhibit or stimulate the germination and growth of weeds in agroecosystems. The commonality of weed species within a crop ecosystem throughout a wide geographical area, indicate that crops biochemically promote certain weeds and inhibit others. Studies on the germplasm of cucumber, sunflower, oats and soybean etc. have shown greater allelopathic potential of their wild types compared with commercial cultivars. Because, the later ones have been selected for exceptional seedling vigour and other desirable growth and yield characteristics and the individuals with desirable allelopathic attributes may have been eliminated. The crop predecessors or wild types have greater allelopathic potential, since they were evolved in the presence of allelopathic and competitive influences from other species. Research on traditional crop mixtures have shown that crop mixtures whose canopy covered the soil throughout the growing season smothered the weeds, for example corn + squash, corn + rice + cassava and been + cassava mixtures (Piedrahita *et al.,* 1975: Harwood, 1976; Chacon and Gliessman, 1982; Gliessman and Altieri, 1982; Anaya *et al.,* 1987). Allelopathic effects of crops have been extensively reviewed (Putnam and Duke, 1974; Altieri and Doll, 1978; Rice, 1974, 1984; Hedin, 1985; Putnam and Tang, 1986; Einhellig and Leather, 1988) and could be used beneficially for weed control in field crops (Putnam *et al.,* 1983; Smeda and Putnam, 1984: Shilling, 1984; Shilling *et al.,* 1985, 1986a, b; James *et al.,* 1988). The crops inhibit or promote the germination and growth of weeds through (a) seed leachates, (b) root exudates and (c) crop residues or crop growth. However, the crop residues have major allelopathic effects on the weeds and could be exploited for weed control.

Table 27. Allelopathic effects of crop seeds leachates on the seed germination and growth of weeds

Donor crop	Recipient species		Reference
	Affected weed species	Inhibitory effect	
Barley	*Amaranthus hybridus* L., *Chenopodium album* L.	Inhibited seed germination	Went *et. al.* (1952)
	Sinapis arvensis L.	Decreased seed germination	Prutenskaya (1972)
Corn (Pollens)	*Cassia jalapensis* L.	Reduced seedling growth	Jimenez-Osorino *et al.* (1983)
Corn, wheat, oat, buckwheat, vetch	*Sinapis arvensis* L.	*	Prutenskaya (1972)
Sorghum	*Amaranthus retroflexus* L., *Abutilon theophrasti* Medic., *Echinochloa crusgalli* (L.) Beauv., *Ipomoea pendurata*, *Rumex crispus* L., *R. dentatus* L., *Setaria viridis* (L.) Beauv., *Sinapis arvensis* L., *Sorghum halepesne* L.	Reduced seed germination, root and shoot growth of seedlings	Panasiuk *et al.* (1986)
Tobacco	*Amaranthus hybridus* L.	Inhibited seed germination	Went *et al.* (1952)
Cucumber	*Panicum miliaceum* L.	Suppressed seed germination and seedling growth	Lockerman and Putnam (1981a, b)
Sugarbeet	*Agrostemma githago* L.	Inhibited growth	Funke (1941)

* Stimulated seed germination.

3.1.2.1 Seed leachates

The seed leachates of some crops particularly during germination, liberate inhibitors which may decrease or increase the seed germination and growth of certain weed species. The seed leachates of barley (Went *et al.,* 1952: Prutenskaya, 1972), sorghum (Panasiuk *et al.,* 1986), cucumber (Lockerman and Putnam, 1981 a,b), sugarbeet (Funke, 1941), tobacco (Went *et al.,* 1952) and maize pollens (Prutenskaya, 1972) inhibited seed germination and seedling growth of weeds (Table 27). On the contrary, germinating seeds of maize, wheat, oat, buckwheat and oat stimulated the seed germination of wild mustard (Prutenskaya, 1972).

The seed leachates of germinating barley seeds inhibited the germination of both common lambsquarter and smooth pigweed, while, seed leachates of tobacco decreased the seed germination of later only (Went *et al.,* 1952). Co-germination with sorghum seeds, inhibited seed germination of curly dock, barnyard grass and green foxtail but stimulated that of red sorrel compared with control, however, weed seedlings were necrotic and chlorotic (Table 28). The germinating sorghum seeds also reduced the root and shoot growth of curly cress, curly dock, barnyard grass, johnson grass, bigroot morning glory, redroot pigweed, red sorrel, velvet leaf and wild mustard (Panasiuk *et al.*, 1986). Corn pollen extracts inhibited the seedling growth of *Cassia jalapensis* (Jimenez-Osorino *et al.*, 1983). Seed leachates of cucumber 'PI 169391' inhibited the seed germination (46%) and dry weight (40%) of prosomillet seedlings and was more inhibitory than 'pioneer' cucumber. The inhibitory effects of seed leachates persisted upto 9 days of imbibition. The inhibitors were found present in the testa of cucumber seed and were polar in nature (Lockerman and Putnam, 1981a,b,). Substances produced from sugarbeet seeds reduced the growth of corn cockle (Funke, 1941). Barley seed leachates decreased the seed germination of wild mustard but germinating seeds of wheat, oat, maize, buckwheat and vetch stimulated its seed germination (Prutenskaya, 1972).

Table 28. Influence of germinating sorghum seeds on the germination of weed seeds

Weed species	Germination (%)	
	Control	Co-germinated with sorghum
Echinochloa crusgalli (L.) Beauv.	73a	68b
Rumex crispus L.	78b	71a
Setaria virdis (L.) Beauv.	81a	74b
Rumex acetosella (L.)	72b	85b

Two germination means in the same line with a common letter are not significantly different (P = 0.05).

Source : Panasiuk *et al.* (1986).

3.1.2.2 Root Exudates

The root exudates play a significant role in living plants and may inhibit or stimulate the seed germination or seedling growth of associated weeds (Table 29). The root exudates of rye (Borner, 1960), corn (Dzyubenko and Petrenko, 1971; Dzyubenko and Krupa, 1974), oat (Fay and Duke, 1977), wheat and oat (Martin and Rademacher, 1960), sorghum (Forney *et al.*, 1983; AlSaadawi *et al.*, 1985; Panasiuk *et al.*, 1986), alfalfa (Abdul-Rahman and Habib, 1989), lupine (Dzyubenko and Petrenko, 1971), soybean (Massantini *et al.*, 1977; Rose *et al.*, 1984), sunflower (Wilson and Rice, 1968) and buckwheat (Tsuzuki, 1980) inhibited the seed germination and seedling growth of weeds. On the contrary, root exudates of sorghum stimulated the seed germination of red sorrel Panasiuk *et al.*, 1986) and witchweed (Nctzy *et al.*, 1988).

Table 29. Allelopathic effects of crop root exudates on the germination and growth of weeds

Donor species		Recipient species		Reference
Crop	Phytotoxin	Affected weed species	Inhibitory effect	
Rye	ND	*Matricaria maritima*	Inhibited growth	Bonner (1950), Martin and Rademacher (1960)
Wheat	ND	*Matricaria maritima*	Inhibited seedling growth	Martin and Rademacher (1960)
Corn	ND	*Chenopodium album, Amaranthus retrofloxus*	Inhibited growth	Dzyubenko and Petrenko (1971), Dzyubenko and Krupa (1974)
Oat	Scopoletin	*Sinapis arvensis, Papaver rhoeas*	Reduced seedling growth	Martin and Redemacher (1960)
	Scopoletin	*Brassica kaber*	Chlorosis, stunted, twisted and poor growth	Fay and Duke (1977)
Sorghum	ND	*Amaranthus retroflexus*	Inhibited seed germintion and seedling growth	AlSaadawi *et al.* (1985)
	p-Benzoq-uinone	*Striga asiatica*	*	Netzy *et al.* (1988)
Sorghum-sudan grass hybrid	ND	*Lolium multiflorum*	Inhibited seedling growth	Forney *et al.* (1983)
Alfalfa	Caffeic, Chlorogenic, isochlorogenic p-coumaric, p-oH-benzoic ferulic acids	*Imperata cuylindrica*	Reduced root and shoot growth	Abdul-Rahman and Habib (1989)
Lupine	ND	*Amaranthus retroflexus, Chenopodium album*	Decreased growth	Dzyubenko and Petrenko (1971)
Soybean	ND	*Abutilon theophrasti*	Reduced dry weight	Rose *et al.* (1984)
	ND	*Helminthia echioides*	Inhibited leaf, shoot and root growth	Massantini *et al.* (1977).
Sunflower	ND	*Amaranthus retroflexus, Digitaria sanguinalis, Erigeron canadensis, Redbekia hirta*	Decreased dry weight of seedlings	Wilson and Rice (1968)
Buck wheat	ND	*Cyperus rotundus*	Inhibited growth	Tsuzuki (1980)

ND = Not determined. * Stimulated seed germination.

The root exudates of rye inhibited the growth of scentless mayweed (Borner, 1960). Gamma irradiation of sorghum increased the allelopathic activity of its root exudates against seed germination and seedling growth of redroot pigweed (AlSaadawi *et al.*, 1985). While *sorghum bicolor* 'SORGV IS 8768' root exudates contained four p-benzoquinones, which stimulated the germination of witchweed (*Netzy et al.,* 1988) and thereby decreased its infestation in the field. Sorghum x sudan grass hybrid 'De Kalb' root exudates reduced the growth of Italian rye grass (Forney *et al.,* 1983). Sorghum root exudates decreased seed germination, coleoptile and radicle length of barnyard grass by 20, 30 and 20% and that of redroot pigweed by 40, 30 and 20%, respectively. However, root exudates increased the seed germination (17%) and coleoptile length (11%) of green foxtail. Root exudates of sorghum exhibited greater phytotoxicity than seed leachates (Panasiuk *et al.,* 1986). The root exudates of alfalfa seedlings reduced shoot and dry weight of blady grass seedlings, when both were grown together in nutrient culture. The root exudates and residues contained caffeic, chlorogenic, isochlo-rogenic, p-coumaric, p-OH-benzoic and ferulic acids (Abdul-Rahman and Habib, 1989). Corn and lupine root excretions inhibited the growth of redroot pigweed and lambsquarter (Dzyubenko and Petrenko, 1971; Dzyubenko and Krupa, 1974). Martin and Rademacher (1960) observed that scopoletin liberated from the roots of oats inhibited growth of wild mustard and *Papaver rhoeas,* but root exudates of rye and wheat reduced growth of scentless mayweed. Two soybean lines out of 141 lines, severely reduced the leaf and root growth of *Helminthia echioides* (Massantini *et al.,* 1977). Root exudates of soybean decreased the dry weight of velvet leaf by 15% over control (Rose *et al.,* 1984). The root exudates of buckwheat diminished the growth of purple nutsedge (Tsuzuki, 1980).

3.1.2.3. Crop growth or crop residues

The growing crops suppressed the growth of certain weed species, while residues of some crops also inhibited the seed germination and seedling growth of weeds through release of phytotoxins. Crop residues left on the soil surface, suppress germination and growth of weeds by leaching allelochemicals, producing microbial phytotoxins during their decomposition and physically obstructing growth of seedlings. The crop residues of alfalfa (Abdul-Rahman and Habib, 1989), crimson clover and hairy vetch (White *et al.,* 1989), sunflower (Wilson and Rice, 1968; Hall, 1980, Hall *et al.,* 1982, 1983; Leather, 1983a), and sweet potato (Harrison Jr. and Peterson, 1986) decreased the germination and growth of weeds. Growing crops of barley (Mann and Barnes, 1945, 1947; Prutenskaya, 1974; Putnam and De frank, 1983), sorghum

(Panasiuk *et al.,* 1986; Weston *et al.,* 1989), rapeseed (Leather, 1983a), cucumber (Lockerman and Putnam, 1979), squash (Chacon and Gliessman, 1982; Gliessman, 1983 and Anaya *et al.,* 1987) and soybean (James *et al.,* 1988) suppressed the weeds (Table 30). While both growing crops and residues of rye (Rademacher, 1940; Barnes and Putnam 1981, 1983 a,b, 1986, Worsham, 1984; Shilling *et al.,* 1985, 1986 a; Barnes *et al.,* 1987), wheat (Rademacher, 1940; Steinseik *et al.,* 1982; Lible and Worsham, 1983a, b; Worsham, 1984; Shilling *et al.,* 1985), corn (Dzyubenko and Petrenko, 1971; Dzyubenko and Krupa, 1974; Anaya *et al.,* 1987), soybean (Maun, 1977; Rose *et al.,* 1984; Parkinson *et al.,* 1987) and beans (Altieri and Doll, 1978; Anaya *et al.,* 1987) inhibited the seed germination and growth of weeds. On the contrary, crop residues of wheat and pea (Purvis *et al.,* 1985 a,b) and sunflower (Leather, 1983a,b) stimulated seed germination and growth of certain weed species.

I. Cereals : Allelopathic effects of rye, barley, wheat, corn and sorghum have been studied on seed germination and growth of weeds. Rademacher (1940) was the first scientist, to report inhibitory effects of associated rye plants on dog fennel due to chemical depressing effect and unknown antagonism. Rye residues reduced seedling emergence of lettuce (58%) and prosomillet (35%), while extracts from soil having its residues, inhibited the growth of these species besides that of barnyard grass. Rye shoot tissues proved more phytotoxic than root tissues. Phytotoxicity increased as the seeds or seedlings of weed species came closer to rye residues (Barnes and Putnam, 1986). Shilling *et al.* (1985) studied the effects of immature desiccated mulch of rye and tillage on the germination and growth of broad leaved weeds. Rye mulch reduced the weed biomass of lambsquarter by 60% both in tilled or no-tilled plots. It decreased biomass of lambsquarter, redroot pigweed and ragweed by 41, 51 and 73% in no-till tobacco and by 99, 96 and 92%, respectively, in no-till soybean and sunflowers (Table 31). Because rye residue extracts contained two phytotoxins viz., ß-phenyllactic acid (ß-PLA) and ß-hydroxybutyric acid (ß-HBA), which were inhibitory to these weed species. ß-PLA was more toxic to weed species than ß-HBA. The isomeric ß-PLA proved 17 times more active than ß-PLA to the hypocotyl growth of redroot pigweed (Shilling *et al.,* 1986a). The residues of rye 'Wheeler' contained two phytotoxins; 2,4-dihydroxy-1,4 (2H)-benzoxazin-3-one (DIBOA) and a break down product 2(3H) benzoxazalinone (BOA), which inhibited the radicle growth of barnyard grass at high concentrations (Barnes *et al.,* 1987). Barley grown at normal or higher seed rates, smothered the spurrey weed (Mann and Barnes,1945) and twitch grass (Mann and Barnes, 1947). Durum wheat, rye and barley crops strongly inhibited the growth and development of wild mustard, but stimulated that of prosomillet (Prutenskaya, 1974).

Table 30. Allelopathic effects of live crop or crop residues on the germination and growth of weeds

Donor species			Recipient species		Reference
Crop	Source of phytotoxin	Phytotoxin	Affected weed species	Inhibitory effects	
1	2	3	4	5	6
Rye	Live crop	ND	*Anthemis arvensis*	Suppressed seed germination	Rademacher (1940)
		ND	*Ambrosia* spp., *Amaranthus retroflexus, Portulaca oleracea, Setaria viridis*	Reduced seedling emergence	Putnam *et al.* (1983)
		ND	*Ambrosia artemisifolia, Amaranthus retroflexus, Echinochloa crusgalli, Digitaria sanguinalis*	Reduced biomass	Barnes and Putnan (1983b)
	Crop residue	b-phenyllactic acid,	Ambrosia spp., *Amaranthus retrojlexus,*	Reduced biomass	shilling *et al.* (1985, 1986)
		b-hydroxybutyric acid	*Chenopodium album*	Reduced biomass	
	Crop residue, root leachate	2(3H)-Benzoxya zoline, 2,4-dihydroxy-1, 4(2H)-benzoxazin-3-one	*Echinochloa crusgalli*	Inhibited seed germination, shoot and root growth of seedlings	Barnes and Putnam (1981, 1986), Barnes *et al.* (1987)
		b-phenyllactic acid, b-hydroxybutryric acid, succinic acid	*Ambrosia* spp., *Amaranthus retroflexus, Chenopodium album*	Reduced growth and biomass	Worsham (1984)

Contd......

1	2	3	4	5	6
Wheat	Live crop	ND	*Anthemis arvensis*	Suppressed seed germination	Rademacher (1940)
	Straw extract	ND	*Ipomea hedracea, I. lacunosa, Abutilon theophrasti, Sesbania exaltata, Cassia obtusifolia, Echinochloa crusgalli* var. *frumetaceae*	Reduced seed germination and seedling growth	Steinseik *et al.* (1982)
	Straw extract	Ferulic acid	*Ambrosia* spp., *Ipomoea lacunosa*	Reduced seed germination and seedling growth	Liebl and Worsham (1983a)
	Plant extract	Ferulic acid, 2-methoxy-4- ethnylphenol	*Sida spinosa*	Inhibited seed germination and seedling growth	Worsham (1984)
	Crop residues	Ferulic acid	*Ipomoea* spp.	Decreased seedling biomass	Shilling *et al.* (1985)
Wheat, pea	Crop residues	ND	*Avena fatua, Avena sterilis* sub sp. *ludoviciana*	**	Purvis *et al.* (1985a,b)
Corn	Live crop	ND	*Amaranthus retroflexus*	Inhibited growth and dry matter	Dzyubenko and Petrenko (1971), Dzyubenko and Krupa (1974)
	Plant leachates	ND	*Portulaca oleracea, Echinochloa crusgalli, Amaranthus leucocarpus, Brassica campestris, Plantago major*	Inhibited root growth of seedlings	Anaya *et al.* (1987)

Contd....

1	2	3	4	5	6
Corn+bottle -gourd	Live crop	ND	*Polygonum* spp.	Reduced weed density and biomass	Jimenez-Osorino and Schultz (1981)
Barley	Live crop	ND	*Spergula arvensis*	Smothered weed	Mann and Barnes (1945)
	Live crop	ND	*Hydrocotyle sibthorpioides*	Smothered weed	Mann and Barnes (1947)
Barley, rye, Durum wheat	Live crop	ND	*Sinapis arvensis*	Inhibited growth and development	Prutenskaya (1974)
Rye, oat, wheat, sorghum, Sudan grass	Crop residues	ND	*Portulaca oleracea, Amaranthus retroflexus, Chenopodium album, Digitaria ischaemum*	Inhibited seedling emergence and dry weight of seedlings	Putnam and DeFrank (1983)
Alfalfa	Crop residaues	Caffeic, chlorogenic, isochlorogenic, p-coumaric, p-oH-benzoic, ferulic acids	*Imperata cylindrica*	Reduced germination and growth	Abdul-Rahman and Habib (1989)
Soybean	Live crop	ND	*Echinocloa crusgalli*	Suppressed plant height and dry matter	Maun (1977)
	Live crop	ND	*Cassia obtusifolia*	Reduced root and shoot growth	James *et al.* (1988)
	Crop residues, plant extract	ND	*Abutilon theophrasti, Setaria italica*	Reduced germination and seedling growth	Rose *et al.* (1984)
Soybean, cotton, groundnut	Live crop	ND	*Striga asiatica*	Stimulated seed germination and decreased population	Parkinson *et al.* (1987)
Beans	Crop residues	ND	*Amaranthus dubius*	Reduced seed germination	Altieri and Doll (1978)

Contd....

1	2	3	4	5	6
	Live crop	ND	*Echinochloa crusgalli, Portulaca oleracea, Brassica campestris, Amaranthus leucocarphus, Plantago major*	Inhibited radicle growth	Anaya *et al.* (1987)
Sunflower	Root, stem, leaves, flower extract	Chlorogenic acid, Isochlorogenic acid, scopolin	*Erigeron canadensis, Rudbekia hirta, Bromus japonicus, Haplopappus ciliatus, Digitaria sanguinalis, Amaranthus retroflexus*	Inhibited seed germination and seedling growth	Wilson and Rice (1968)
	Crop residues	Chlorogenic acid	*Amaranthus retroflexus*	Inhibited seed germination and seedling growth	Hall (1980), Hall *et al.* (1982, 1983)
	Crop residues	ND	*Abutilon theophrasti, Datura stramonium, Ipomoea purpurea*	Inhibited seedling growth	Leather (1983 a,b)
	Plant extract	ND	*Amaranthus retroflexus, Rumex acetosella, Brassica kaber*	*	Leather (1983a, b)
Rapeseed	Live crop	ND	*Abutilon theophrasti, Ipomoea purpurea*	Reduced seed germination and seedling growth	Leather (1983a)
Tobacco	Crop residues	ND	*Amaranthus retroflexus*	Inhibited seed germination and seedling biomass	Hall (1980)
Cucumber	Live crop	ND	*Panicum miliaceum*	Suppressed plant population and fresh weight	Lockerman and Putnam (1979)

Contd....

1	2	3	4	5	6
Squash	Live crop	ND	*Echinochloa crusgalli, Portulaca oleracea, Brassica campestris, Amaranthus leucocarpus, Plantago major*	Reduced radicle growth	Anaya *et al.* (1987)
Sweet potato	Crop residues	Phenolic acid	*Cyperus esculentus*	Reduced root and shoot growth and dry weight	Harrison and peterson (1986)
Buckwheat	Plant extract	ND	*Sagittaria pygmea*	Inhibited seed germination	Tsuzuki (1980)
Radish	Plant extract	ND	*Alhagi* spp., *Alopecurus myosuroides, Cochia maritima, Capsella bursapastoris, Daucus carota, Convolvulus arvensis, Cuscuta* spp. *Hirschfeldia incana, Ochthodium aegypticum, Sisymbrium polyceratum, Sorghum halepense*	Completely inhibited seed germination	Uygur *et al.* (1990)

Not determined, * Stimulated germination, ** Stimulated germination and growth.

Table 31. Effect of tillage and rye residues (above and below ground) on weed density, biomass and Weed control

Treatment	Amaranthus refoflexus*			Chenopodium album*			Ambrosia artemisifolia**			*Digitaria sanguinalis*** and *Panicum dichotomiflorum* control (%)
	Density (Plants 2.2 m^2)	Biomass (g)	Control (%)	Density (Plants 2.2 m^2)	Biomass (g)	Control (%)	Density (Plants 2.2 m^2)	Biomass (g)	Control (%)	
Remover mulch no till	8abc	2.4bc	55ab	25b	65.6cd	49d	21c	34.0c	68a	60c
Remover mulch till	4bc	0.8bc	58ab	132a	117.8bc	60cd	236a	118.3b	33c	71bc
Cut mulch, no till, replace mulch	3bc	0.8bc	63ab	120a	44.8d	70bc	186ab	108.2b	43bc	82ab
Remove mulch, till replace mulch	3bc	0.8bc	63ab	120a	44.8d	70bc	186ab	108.2b	43bc	82ab
Glyphosate, no till	2c	0.4c	74a	5b	44.3b	65bc	9c	5.9c	80a	84ab
Paraquat no till	1c	0.2c	79a	5b	25.9b	77ab	16c	12.3c	73a	81ab
No rye, till	26a	22.1a	8c	140a	199.9a	10f	140b	178.3a	10d	28d
No rye, no-till	21ab	12.0ab	41b	41b	124.3b	32e	15c	34.7c	55b	33d

Means within a column followed by the same letter are not significantly different as determined by Waller Duncan t=test (K ratio=100)

* Weed density, biomass and weed control 49 days aftter planting

** Weed Control at 31 days after planting

Source : Shilling *et al.* (1985)

Wheat straw aqueous extracts or leachates, inhibited the seed germination and seedling growth of weeds in order ivy leaf morning glory < velvet leaf < morning glory < hemp < sesbania < sicklepod < japanese < barnyard millet. The greatest inhibition of germination and growth was observed at 35°C (Steinseik *et al.,* 1982). Aqueous extracts of wheat straw inhibited the seed germination and root elongation of morning glory and ragweed, the phytotoxicity increased upto 70% in sunlight. The most inhibitory compound in these extract was ferulic acid, which completely inhibited the seed germination of crab grass at 5 mM and reduced seed germination and root length of morning glory by 23 and 82% and that of prickly sida with carpels by 85 and 82%, respectively, but had no inhibitory effect on prickly sida without carpels (Libel and Worsham, 1983b). Because, a bacterium living on the carpels of prickly sida decarboxylated the ferulic acid (4-hydroxy 3-Methoxy-cinnamic acid) to a more phytotoxic styrene derivative, 4-hydroxy-3-methoxy styrene (Shilling *et al.,* 1985). Planting no-till corn in immature desiccated wheat mulch, reduced morning glory spp. biomass by 79% than a non-mulch, tilled treatment. However, there was no effect of mulch in double-crop soybeans. No-till no-mulch plots decreased the weed biomass similar to tilled and mulch plots, this showed that elimination of tillage alone controlled weeds effectively (Shilling *et al.,* 1985). The mulch of autumn killed oat, rye, wheat and barley significantly suppressed the weed growth compared with their spring killed mulch (DeFrank and Putnam, 1983). Putnam and DeFrank (1979) desiccated wheat, rye, barley, corn, oat,sorghum and sudan grass by the herbicides (glyphosate or parquat) or by freezing. Wheat, rye, sudan grass and sorghum mulch reduced the weed growth biomass by 88, 88, 85 and 90%, respectively. In field studies, sorghum residues decreased the density of common purslane and crabgrass by 70 and 98%, respectively. All these crop residues inhibited the emergence and growth of weeds by 75% in the early stages. Wheat, oat, rye, sorghum and sudan grass residues reduced the infestation and biomass of common purslane and smooth crabgrass for 60 days. Sorghum residues present on the soil surface of muck and loamy sand soil, greatly inhibited the weed growth than its soil incorporation, owing to greater concentration of phytotoxins in the germination zone in the former (Putnam and DeFrank, 1983). Fall planted cover crops of rye `Wheeler' and wheat 'Yorkstar' greatly suppressed weeds of straw berry during early season and had no harmful effect on straw berry yields (Table 32). The suppression of weeds increased with increasing seed rates of cover crops (Smeda and Putnam, 1988).

Table 32. Effect of cover crops on winter and early summer annual weeds biomass an strawberry fruit yields

Cover crop treatment	Mean weed dry weight (Kg/ha)		Berry yield (Kg/plant)	Cover crop dry weight (Kg/ha)
	6 June	12 July		
None	378.3a	630.6a	2.48	–
Rye 146 Kg/ha	37.9b	73.1b	1.90	5052.7a
Rye 76 Kg/ha	37.3b	181.4b	2.15	4181.9ab
Winter wheat 146 Kg/ha	11.8b	150.7b	2.26	5550.8a
Winter wheat76 Kg/ha	52.0b	366.7ab	1.80	2142.9b

Means separated by Duncan's multiple range test, P=5%.
Source : Smeda and Putnam (1988).

Growing of maize considerably reduced the growth and dry matter accumulation, in the above ground parts of red root pigweed. The degree of weed inhibition increased with increasing productivity of maize (Dzyubenko and Krupa, 1974). Shoot and root leachätes of corn, inhibited the radicle growth in barnyard grass and root leachates were also phytotoxic to *Amaranthus leucocarpus* and *Plantago major* (Anaya *et al.,* 1987). According to Lovett (1983a), sorghum residues reduced the weed density and biomass of both annual grass weed barnyard grass *Echinochloa colonum*) by 60 and 40% and that of annual broad-leaved weed *Hibiscus trionum* by 24 and 17%, respectively, compared with control. Growing of forage sorghum- sudan grass hybrid prior to sowing of late summer no-till alfalfa, suppressed the weeds of alfalfa crop and increased its yields *(Forney et al.,* 1985). Allelopathic potential of sudex (Sorghum x sudan grass hybrid) on barnyard grass decreased with increasing age. Greatest allelopathic activity of sudex shoot tissue was observed in 7 days old plants. Small seeded broadleaf species were more inhibited than grass species. The sudex shoot tissue contained two major phytoinhibitors (p-hydroxybenzoic acid and p-hydroxybenzal-dehyde), with increasing age the percentage of later increased, while, that of former decreased. Seven days old tissues possessed a higher percentage of these phytoinhibitors than did older tissue (Weston *et al.,* 1989). Panasiuk *et al.* (1986) studied allelopathic effects of sorghum on ten weed species viz., curly cress, curly dock, barnyard grass, johnson grass, bigroot morning glory, redroot pigweed, red sorrel, velvet leaf and wild mustard. Interplanting of sorghum reduced the dry weight of barnyard grass, redroot pigweed and red sorrel during the first 60 days of sorghum growth but had no inhibitory effect on the interplanted green foxtail, morning glory and wild mustard. Sorghum interplanting reduced the shoot and root dry weight of red sorrel by 33 and 60%, respectively. Grain sorghum proved more inhibitory to weeds during their early seedling growth than at germination stage. Preceding sorghum crop, reduced the weed biomass in subsequent season by 30-60% compared with preceding crops of corn and soybean (Fig. 8). Therefore, some farmers in south Dakota obtain reasonable weed control, by planting sorghum and sunflower in the previous season (Einhellig and Leather, 1988).

II. Sunflower : Aqueous extracts of sunflower leaves, stem, root, inflorescence and plants inhibited the germination and seedling growth of crabgrass, redroot pigweed, *Erigeron canadensis, Rudbekia hirta, Haplopappus ciliatus* and *Bromus japonicum* (Wilson and Rice, 1968). Aqueous extracts from leaf and stem of 13 sunflower varieties showed mixed responses to seed germination of weeds, however, leachates of dry leaves and stem inhibited the growth of weed seedlings (Leather and Forrence, 1979). Application of sunflower residues to soil, reduced the seed germination and seedling biomass of redroot pigweed. The phenolic content of sunflower tissues, increased under nutrient deficiency (Hall, 1980) and was greater in field grown plants (69.7 mg/g tissue) compared with green house plants (44.16 mg/g tissue) (Hall *et al.,* 1982, 1983). Sunflower varieties during growth and their aqueous extracts inhibited the seed germination and growth of velvet leaf, thorn apple, morning glory and wild mustard (Table 33). The leaf extract of sunflower 'Hybrid 201' and 'Hybrid 8941' inhibited the seed germination of wild mustard (Leather, 1983 a). Some crops are also effective in minimizing the growth of weeds in the following season. Sunflower- oat- sunflower rotation over 5 - years period, lowered the densities of the broad leaved weeds and grasses than control plots. Weed density increased in all plots during the study, but in rotation with sunflower the increase in weed density was less than in plots without sunflower (Leather, 1983 b). Stem and leaf leachates of sunflower, inhibited the seedling growth of wild mustard, redroot pigweed, thorn apple and red sorrel, but stem extracts stimulated the seed germination of these weed species (Leather, 1983 b).

Table 33. Growth response of weed species to root exudates and leachates of sunflower

Weed species	Dry weight (% of control)		
	Leaf leachate	Stem leachate	Root exudates
Abutilon theophrasti	63*	42*	42*
Datura stramonium	74*	63*	58*
Ipomoea spp.	78	50*	180*
Brassica Kaber	73*	55**	—

*Significantly different from control at 5% level (Duncan's multiple range test).

Source : Leather (1983 a)

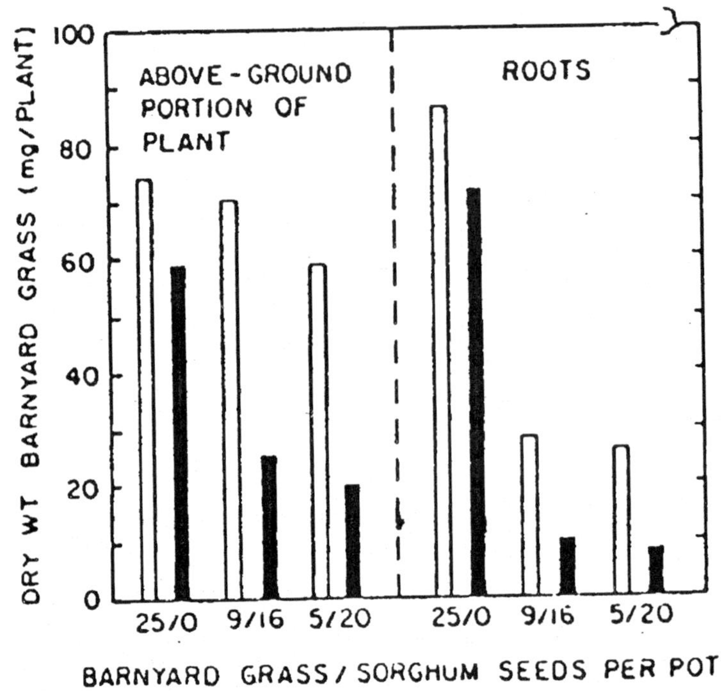

Fig. 8. Dry weight of barnyard grass five weeks after planting in pots by iself or interplanted in two different ratios with sorghum. White bars indicate average weights obtained from pots watered with tap water. Black bars represent average weights from pots watered with leachate obtained from watering pots planted with 25 sorghum seeds. (Source : Einhellig and leather (1988)).

III. Legumes : Soybean 'Vansoy' crop suppressed the plant height and dry matter of interplanted barnyard grass. The maximum suppression occurred, when it was planted with 6-10 weeks old soybean plants (Maun, 1977). Twenty soybean cultivars (selected from 280 cultivars) were studied for allelopathic effects on velvet leaf and foxtail millet. Soil incorporation of ground soybean residues @ 1% (w/w) inhibited the seed germination and dry weight of velvetleaf by 40% each and of foxtail millet by 82 and 65%, respectively. Extracts of these cultivars slowed the seed germination and dry weight accumulation in seedlings of these weeds (Rose *et al.,* 1984). James *et al.* (1988) evaluated 12 soybean cultivars in the field for interference on sicklepod weed. 'Essex' and 'Bragg' varieties reduced the sicklepod dry weight, but only 'Essex' remained unaffected by sicklepod. The genetic development of soybean cultivars, with improved tolerance of sicklepod infestations possibly could reduce weed control inputs. Continuous cropping of soybean for three years in plots heavily infested with witchweed, reduced its population in the subsequent corn crop (Table 34) (Parkinson *et al.*, 1987).

Decomposing roots of alfalfa and their associated soil considerably reduced germination (51-56%) and root and shoot length (88%) of blady grass. Decayed and undecayed mixtures of alfalfa roots and soil at 0.015 : 1 (w/w) inhibited reproduction of blady grass seedlings from rhizomes by 30 and 42%, respectively. The alfalfa root residues contained caffeic, chlorogenic, isochlo-rogenic, p-coumaric, p-OH-benzoic and ferulic acids and their highest amount (126 μg Phenolic acids/g soil) was found six months after decomposition in soil (Abdul-Rahman and Habib, 1989). Soil incorporation of crimspn clover and hairy vetch debris @ 0.8-6.7 mg/g soil considerably decreased emergence (60-80%) and dry weight of pitted morning glory. Germination and seedling growth of Italian rye grass, pitted morning glory and wild mustard decreased progressively with species dependent variation, when exposed to increasing concentrations (8.3-33.3g debris/1) of aqueous extracts of crimson clover and hairy vetch. The extracts completely inhibited the germination in mustard and rye grass. Hairy vetch proved more phytotoxic to test species than crimson clover (White *et al.*, 1989).

Table 34. Effect of continuous cropping of corn and soybean on *Striga* population

Corn cultivar	Continuous soybean		Continuous corn	
	Striga count (Plants/plot)	Yield (t/ha)	Striga count (plants/plot)	Yield (t/ha)
8322-2	10.2	4.8	98.3	4.3
8424-19	14.8	4.6	78.5	2.1
8338-1	25.6	4.6	207.5	2.6
8322-3	3.00	2.6	27.3	3.6
TZSR-W-1	19.5	3.0	83.5	3.0
Western yellow	11.0	4.4	23.6	0.8
LSD (0.05)	15.3	0.8	25.2	0.91

Source : Parkinson *et al.* (1987)

Purvis *et al.* (1985 a,b) studied the effect of mature residues of oilseed rape, sorghum, sunflower, wheat and field pea on the seed germination and growth of weeds viz., barnyard grass, perennial rye grass, prostate knapweed, wild oat, *Avena sterilis* sub sp. *ludoviciana*, *Laminum amplexicaule, Phalaris aquatica, Phalaris paradoxa, Lolium perenne, Valpis myuros, Hibiscus trionum, Bilder dykia* and *Convolvulus.* Wheat and pea residues stimulated the germination and growth of wild oat and *A. sterils* sub sp. *ludoviciana*. The residues of pea also stimulated the germination of *L. amplexicaule* but inhibited that of grassy weeds (*L. perenne, P. aquatica, P. paradoxa* and *V.*

myuros). Sorghum, sunflower and wheat residues promoted the germination of *L. amplexicaule* and *H. trionum* but inhibited that of grassy weeds. Wild oat plants produced dry matter in order wheat > pea > sorghum > sunflower > oilseed rape > no stubble but dry matter in other grassy weeds followed the reverse order (Fig. 9) (Purvis *et al.*, 1985b).

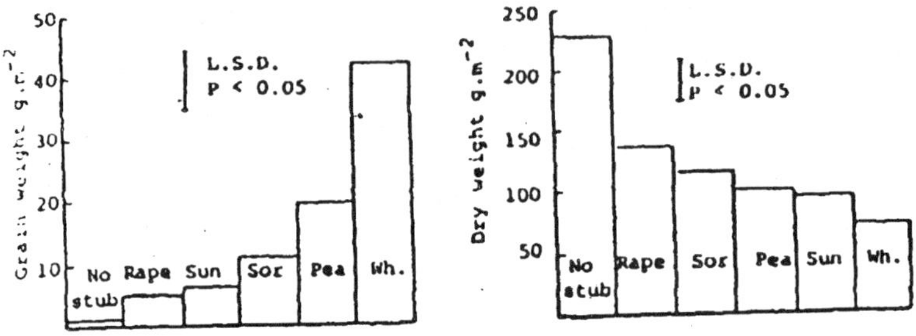

Fig. 9A & B. A. Wild Oat grain weight : Source (Purvis *et al.* 1985b).
B. Dry weight of grassy weeds excluding wild Oat.

IV. Other crops : Of the 526 accessions of cucumber, only 3% inhibited the seed germination of prosomillet and white mustard by 75% (Putnam and Duke, 1974). Under field conditions, cucumber 'PI 169391' suppressed fresh weight and density of prosomillet by 58 and 84%, respectively. This cultivar proved twice effective compared with 'PI 285605' and also reduced the weed density of associated weeds by 54% (Lockerman and Putnam, 1979). At Davis California, weed cover and biomass production were lower in unweeded corn-cowpea polyculture (193 g/m^2) than in monocultures of corn (418 g/m^2) or cowpea (1529 g/m^2). At Santa Cruz, the bean-collard mixture suppressed both the density and biomass of weeds as compared to slow growing collards planted alone. Growing crops with different chemical byproducts may also enhance the possibilities for biological weed suppression through the release of chemical inhibitors (Gilessman and Altieri, 1982). In unweeded plots, corn monocultures produced 20% lower yields than in its interplanting with bottle gourd, because interplanting of bottle gourd significantly reduced the biomass and density of *Polygonum* spp. (Jimenez-Osorino and Schultz, 1981). The interplanting of squash in corn-bean polycultures and in corn monoculture in south eastern Mexico helped in weed control (Fig. 10) (Chacon and Gliessman, 1982; Anaya *et al.*, 1987) and increased corn yields (Amador and Gliessman, 1982). Because, firstly beans planted with corn nodulated more and fixed nitrogen more actively, which directly became available to corn (Boucher and Espinose, 1982). This leads to net gains of nitrogen in the biomass of this agroecosystem, despite the removal of harvested yields (Gilessma, 1982). Secondly, squash forms a continuous canopy cover above the low lying weed species, thereby combining shade with phytotoxins and thirdly, it returned large biomass (8-10 t/ha) to the soil (Amador and Gliessman, 1982), which inhibited the weed growth. Therefore, total number of weeds in the corn-squash plots was smaller than in corn-bean and corn monoculture plots (Chacon and Gliessman, 1982). Aqueous extracts of dry leaves of squash inhibited the germination and growth of weeds (Gliessman, 1983). In fact, Mexican farmers grow squash mainly for weed control since many generations.

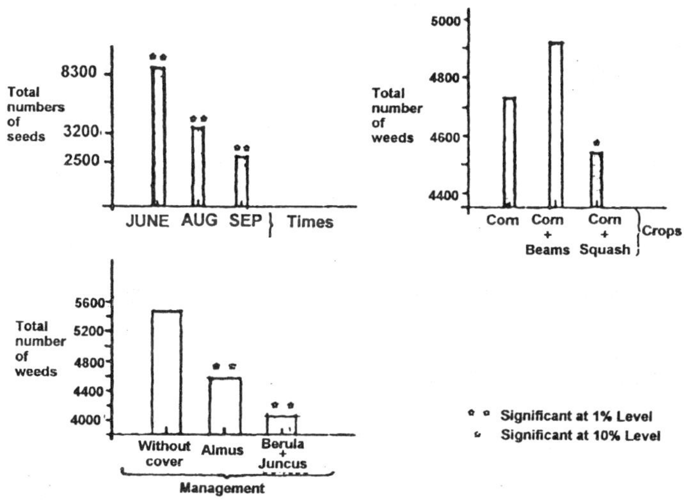

Fig. 10. Total number of weeds in times (months), crops and management.

Soils from sweet potato field, decomposing vines, roots and periderm extracts of sweet potato inhibited the root growth, shoot growth and dry matter of yellow nutsedge (Table 35). Sweet potato 'Regal' proved more phytotoxic than 'SC 1149-19'. The phytotoxicity of decomposing vines and roots persisted upto 12 weeks. The inhibitors were identified as phenolic compounds (Harrison and Peterson, 1986). Sweet potato skin contained natural allelochemicals, which inhibited the seed germination and root formation in many weeds. These allelochemicals in pure form were as potent as commercial herbicides (Duke *et al.*, 1989). The aqueous extracts of buckwheat inhibited the germination of purple nutsedge (Tsuzuki, 1980). Atlieri and Doll (1978) studied the phytotoxic effects of beans, corn and cassava residues on the seed germination of weeds. Bean residues reduced seed germination of common purslane *(Portulaca oleracea* L.) (39%), *Bidens pilosa* L. (47%), *Ipomoea hederifolia* L. (75%) and *Amaranthus dubius* Mart. Ex Thell (88%). While corn residues caused inhibition in *B. pilosa* L. (39%), *Solanum nigrum* L. (65%) and *I. hederifolia* (71%).

Thus bean residues were found more phytotoxic than corn residues. The juices of radish completely inhibited the seed germination of 11 weed species viz., *Alhagi* spp., *Alopecurus myosuroides*, Huds., *Cachia maritima*, *Capsella bursapastoris* (L.) Madic, *Cuscuta* spp. *Daucus carota* L., *Hirschfeldia incana* (L.) Largreze-Fossat, *Ochthodium aegyptiacum*, *Sisymbrium polyceratum*, hedge bindweed (*Convolvulus arvensis* L.), and Johnson grass (*Sorghum halepense* (L.) Pers.) The *in-vitro* sprouting of rhizome buds of Johnson grass decreased with increasing concentrations of the juices of radish in the soil (Uygur *et al.*, 1990).

Table 35. Effect of soil incorporated killed sweet potato plants on the growth of yellow nutsedge

Treatment	Tissue dry weight (mg)		
	Shoot	Root	Rhizome + tuber
Control	898a	169b	187a
Sweet potato `Sc 1149-19'	767ab	204a	159a
Sweet potato `Regal'	719b	99b	139a

Means within the column followed by same letter are not significantly different (LSD P = 0.05) according to Duncan's multiple range test.

Source : Harrison and Peterson (1986).

3.1.3. Crop allelopathy for weed control

Allelopathic potential of crops to suppress weeds growth has immediate utility for weed management strategies, if heritable factors for allelochemicals could be identified and incorporated into com-mercial cultivars. Putnam and Duke (1974) suggested (i) develo-pment of cultivars that would release allelochemicals as natural herbicides to provide satisfactory weed control, (ii) utilization of allelopathic plants which does not interfere with crop growth (Garb, 1961: Grechkanev and Rodionov, 1971) and (iii) that may be ploughed under to control weeds through root exudates or decomposing crop residues. Therefore, research on weed control has aimed to (a) identify cultivars and accessions with high allelopathic potential, (b) isolate allelochemicals and (c) to develop field techniques to capitalize on allelochemicals from crop plants (Einghllig, 1985b). The release of phytotoxins from decaying residues could be used to control weeds and plant pathogens (Baker and Cook, 1974). Putnam and his associates have found that crop residues of wheat, barley, oats, rye, sorghum and sudan grass reduced the density and biomass of annual weeds.

Spring planted rye reduced biomass of crabgrass (*Digitaria* spp.), ragweed (*Ambrosia* spp), and lambsquarter (*Chenopodium album* L.)

by 42, 90 and 98%, respectively, over plots without rye. While immature residues of fall planted and spring killed rye, reduced biomass of redroot pigweed (*Amaranthus retroflexus* L.) and bermuda grass *(Cynodon dactylon* (L.) Pers) by 55 and 74% respectively, compared with bare ground controls. However, the yields of pea sown in rye residues remained unaffected (Barnes and Putnam, 1983b). Surface residues of rye, inhibited emergence of ragweed, green foxtail (*Setaria viridis* (L.) Beauv.), redroot pigweed and common purslane (*Portulaca oleracea* L.) by 43, 80, 95 and 100%, respectively (Putnam *et al.,* 1983). Shilling (1984) reported that no-till planting of soybean and sunflowers in rye residues mulch inhibited the growth of ragweed, redroot pigweed and lambsqarter by 92, 96 and 99%, respectively. Tobacco planting reduced the growth of these weeds by 73, 51 and 41%, respectively, compared with tilled plots and no mulch. ß-PLA and ß-HBA present in the rye residues inhibited the root and shoot growth of lambsqarter and redroot pigweed. Rye mulch (grown from 198 kg seed/ha) with 6.72 kg diphenamid/ha provided complete weed control in tobacco and decreased broad leaved and grassy weeds by 22 and 29%, respectively. Wheat, oat, barley and rye mulches provided similar weed control in corn and soybean crops. In no-till mulched plots the weed control was 37 and 62% better than that in non-tilled non-mulched and tilled non-mulched plots, respectively. Eliminating tillage reduced the densities of morning glory *(Ipomoea* spp.), prickly sida *(Sida spinosa* L.) and sickle pod *(Cassia obtusifolia* L.) weeds (Shiling *et al.,* 1985). Winter rye sown in autumn and killed at soybean sowing controlled lambsquarter and redroot pigweed as effectively as 2,4-D (Warnes, 1985). Autumn sown rye killed by glyphosate or parquat nine days before tomato transplanting, effectively controlled weeds in tomato crop, compared with wheat and barley mulch. Thus rye mulch provided alternative to pre-plant herbicides for tomato crop (Smeda and Putnam, 1984).

Table 36. Effect of mulch, tillage and diphenamid on weed control and tobacco yield

Mulch type	Seed rate (kg/ha)	Diphenamid + glyphosatr (kh/ha)	*Weed control (%)*								Tobacco yield (kg/ha)
			Amar anthus sret roflexus		*Chenopod iumalbum*		*Ambrosia artem isifolia*		*Elusine indica* Digi taria spp.		
			ES	LS	ES	LS	ES	LS	ES	LS	
Rye	183	0.00+3.36	80abc	68bcd	72ef	67de	85ab	70cde	54bc	41cde	1302f
	188	6.72+3.36	99a	85ab	96ab	80bc	95a	86ab	80ab	63b	1989b
Wheat	202	0.00+3.36	85ab	75abc	77de	55f	96a	67de	52bc	35de	1405ef
	202	6.72+3.36	84ab	71abc	95ab	67de	97a	76bcd	71abc	51bcd	1865bcd
Barley	215	0.00+3.36	69abc	59cd	81cde	57ef	96a	57e	75abc	56bc	1435ef

	215	6.72+3.36	95a	72abc	90abc	72cd	82ab	87ab	72abc	59bc	1691cde
Oats	143	0.00+3.36	87ab	70ad	87bcd	60ef	75b	60e	67abc	59bc	1587def
	143	6.72+3.36	97a	82abc	99a	85b	89ab	85abc	95a	69ab	2001b
None	—	0.00+3.36	55bc	50de	65f	52f	84ab	62de	71abc	26e	1407ef
		6.72+3.36	81abc	70ad	94ab	77bcd	87ab	67de	94a	64b	1879bc
None	—	0.00+3.36*	20d	30e	Oh	Og	5d	20f	47c	39cde	827g
		6.72+3.36**	52c	90a	52g	97a	52c	100a	67abc	85a	2610a

Means within a column followed by the same letter are not signifiantly different as determined by Waller-Duncan t-test (K ratio = 100).
ES = Early season, LS = Late season, * Tilled and rebedded prior to transplanting, ** cultivated twice.
Source : Shilling *et al.* (1986 b).

Wheat straw mulch in corn reduced the density and dry matter of weeds. Wheat straw at 6 t/ha provided suppression of weeds similar to 1.0 kg atrazine. Combined application of straw mulch and herbicides proved most effective in controlling weeds than their separate applications (Prihar *et al.,* 1975). Shilling *et al.* (1986b) studied the effects of non-soil incorporated mulches of wheat, barley, oat and rye and application of diphenamid herbicide on weed control in tobacco crop planted in these mulches. None of the mulches had any harmful effect on the growth of tobacco. Rye 'Abruzzi' mulch plus diphenamed proved superior over wheat, barley and no-mulch and controlled 85% broad leaved weeds. Oat 'Brooks' and rye mulch killed 80% and 22% more broad leaved weeds than no-mulch (Table 36). Corn residues supplemented with low rates of herbicides (atrazine or alachlor) effectively controlled the weeds, but at higher rates of herbicides the effect of residues became less (Erbach and Lovely, 1975). The weed biomass was reduced equally in plots planted with sunflower, whether herbicide was applied or not. The weed control diminished in the second year because of its less emergence (13-45%) owing to autotoxicity from its (Sunflower) crop residues remaining after the first harvest (Leather, 1987). Of the 12 soybean cultivars studied, 'Essex' and 'Bragg' reduced the dry weight of sicklepod *(Cassida obtusifolia* L.) weed by 39% and weeds had no adverse effect on the dry weight of 'Essex' cultivar (Table 37) (James *et al.,* 1988). In a 6-year study, cover crops of wheat, barley, rye and sorghum when attained a height of 40-50 cm were killed by herbicides or freezing and the residues were maintained on the soil surface. This controlled 95% of the impo- rtant weeds for 30-60 days, through direct release of toxins and production of phytotoxic microbial products (Putnam *et al.,* 1983).

Table 37. Effect of soybean cultivars growth on the dry weight of sicklepod and vice-versa

Soybean cultivars	Dry weight per plant (g)		
	Sickle pod	Soybean	
		Alone	With sicklepod
S100	38	116*	55
CNS	38	101*	49
Palmetto	34	99*	55
Ogden	36	116*	65
Peking	33	64*	39
Jackson	30	78*	47
Lee	30	87*	51
Hill	42	64	52
Bragg	25	77*	47
Forest	34	98*	39
Essex	27	61	46
Wright	33	91*	48
Control	41	—	—
LSD (0.05)	12	16	16

* Significantly different from control at P = 0.05 using paired t-test.

Source : James *et al.* (1988).

3.2. Weeds

Since the beginning of agriculture, the weeds have been growing with crops. Probably because, weeds have evolved alongwith the crops or in some instances they are the ancestors of cultivated crops, therefore, many crops and weeds have same species. For example, wild races of wheat, rice, barley, maize, oat, sorghum, potato, radish, cabbage, lettuce and asparagus etc., are weeds. Besides, various practices of modern agriculture favour invasion by weeds, (a) row sown crops leave enough inter-row space for colonization by other species and (b) crop monocultures. A single plant specie generally fails to fully exploit the habitat, for example, it may not use the available sunshine, because its leaf area develops slowly or it may have too short growth cycle to consume all the available water or nutrients. Therefore, weeds invade such areas to utilize these growth resources. The growth and yield of plant species growing together in an area are influenced by the neighbouring plants, which may result in reduced size, chlorotic or nutrient starved conditions, wilting and death of less successful plant species. Weeds cause greater losses in crop yields than either insects or plant diseases. The weeds reduce the crop yields through (a) allelopathy, i.e., release of inhibitors from seeds, living plants and plant residues, (b) competition for growth resources (light, nutrients, water and space) with crops and (c) acting as an alternate host for insects and disease organisms. Allelopathic effects of weeds have been reported both on crops as well as on weeds, but only the former has been discussed in this review and the weeds included in the review have been listed in Table II, Chapter 7.

3.2.1. Weed-crop Interactions

Under field conditions, weed infestation is one of the major causes of yield reduction in crops. Historically, most investigators have attributed these losses to various forms of competition, between the weeds and crops and allelopathic interactions between them were not considered. However, findings after 1950's have shown that allelopathic interactions between the crops and weeds are also partly responsible for such losses in crop yields. DeCandolle (1832) for the first time reported the injurious effects of root exudates of canada thistle *(Cirsium arvense* (L.) Scop.) on the growth of neighbouring oat plants. Later on, allelopathic potential of numerous weeds on the crops have been reported (Schreiber and Williams, 1967; Buchholtz, 1971; Bell and Koeppe, 1972; Einhellig and Rasmussen, 1973; Rasmussen and Einhellig, 1975; Ashraf and Sen, 1978; Steenhagen and Zimdahl, 1979). Putnam and Weston (1986) listed about 90 weed species with allelopathic potential, while this book lists 129 such species (Table II, Chapter 7). Most of the weed species have inhibitory effects on crops, yet, some weed species also exhibited stimulatory effects on the seed germination, growth and yield of crops. The weeds influence the crop plants by releasing phytotoxins from their seeds, decomposing residues, leachates, exudates and volatiles. Complete weed control in field crops has been attained only in developed countries and in remaining ones weed infestation in crops is very high, thus under such conditions, weeds add substantial quantity of residues to the soil after crop harvest. Therefore, the weed residues are the major sources of phytotoxins in the soil.

3.2.1.1. Weed residues

The weed residues may exert allelopathic effect on crop plants similar to that of crop residues but detailed studies are lacking. Allelochemicals released from the weed residues may affect the crop plants (Appendix) in following manner : (I) inhibition of biological nitrogen fixation, (II) inhibition of nutrient uptake and (III) inhibition of seed germination, growth and yield.

(I). Inhibition of biological nitrogen fixation : Rice (1964) reported that aqueous extracts of lambsquarter (*Chenopodium album* L.) and crabgrass (*Digitaria* spp.) slightly inhibited the growth of nitrogen fixing and nitrifying bacteria. The inhibitors present in prostrate knotweed (*Polygonum aviculare* L.) inhibited the growth of *Rhizobium* and *Azotobacter* strains (Al Saadawi and Rice, 1982 b).

Aqueous extracts of *Avena ludoviciana,* reduced the seedling growth and nodulation in greengram (Bhandari *et al.,* 1982). Kanchan and Jayachandra (1979 a,b) found that growth and nodulation of kidney-bean `Burpees stringless' growing in association with *Parthenium hysterophorus* L. plants was retarded, the inhibition decreased with increasing distance from the weed. The weed leachates also reduced the nodulation in this crop. The plant extracts of *Aristida adscensionis* inhibited the nodulation in *Indigofera cordifolia* Heyne ex Roth. (Sarma, 1983). Soybean, navybean and snapbean grown in mowed quackgrass (*Agropyron repens* (L.) Beauv.) sod in the greenhouse and field, exhibited decrease in nodule number, nodule fresh weight and nitrogen fixation when compared to control soil (Table 38) (Weston and Putnam, 1985). Quackgrass inhibited the legume-*Rhizobium* symbiosis in snapbean `Bush Blue Lake'; indirectly by inhibiting root hair formation (Weston and Putnam, 1986). Shoot extract and residues of green foxtail (*Setaria viridis* (L.) Beauv.) increased the nodulation in soybean at lower concentration (1%) but decreased at higher one (2%) (Fig. 11) (Mallik and Tesfai, 1988). The ether extract of couch grass shoots severely reduced the root growth and root hairs and caused root browning in kidneybean, but had no inhibitory effect on the growth of *Rhizobium.* The inhibitors in extract indirectly influenced the *Rhizobium* symbiosis by reducing root growth and root hair formation in kidneybean. The inhibitors were two flavonoids, one was identified as tricin (*Weston et al.,* 1987).

Table 38. Influence of *Agropyron repens* (L.) Beauv. on nodule number, nodule fresh weight and N_2-fixing ability of legumes grown in the greenhouse

Treatment	Snapbean	Navybean	Soybean	Pea
	Nodule number/plant			
Live plants	24c	27c	51b	18
Dead plants*	39c	14c	62b	34
Control	195a	107a	83a	41
	Nodule weight (g)/plant			
Live plants	0.09c	0.08c	1.30	0.02
Dead plants*	0.16c	0.04c	1.60	0.05
Control	0.98	0.51	1.50	0.08
	Fixed N_2 (n moles/plant/ha)			
Live plants	66c	55c	1919	21
Dead plants*	183bc	42b	1749	21
Control	612a	413a	2237	43

**Agropyron repens* plants killed with Glyphosate. Mean within columns followed by different letters are different at LSD 0.05

Source : Weston and Putnam (1985).

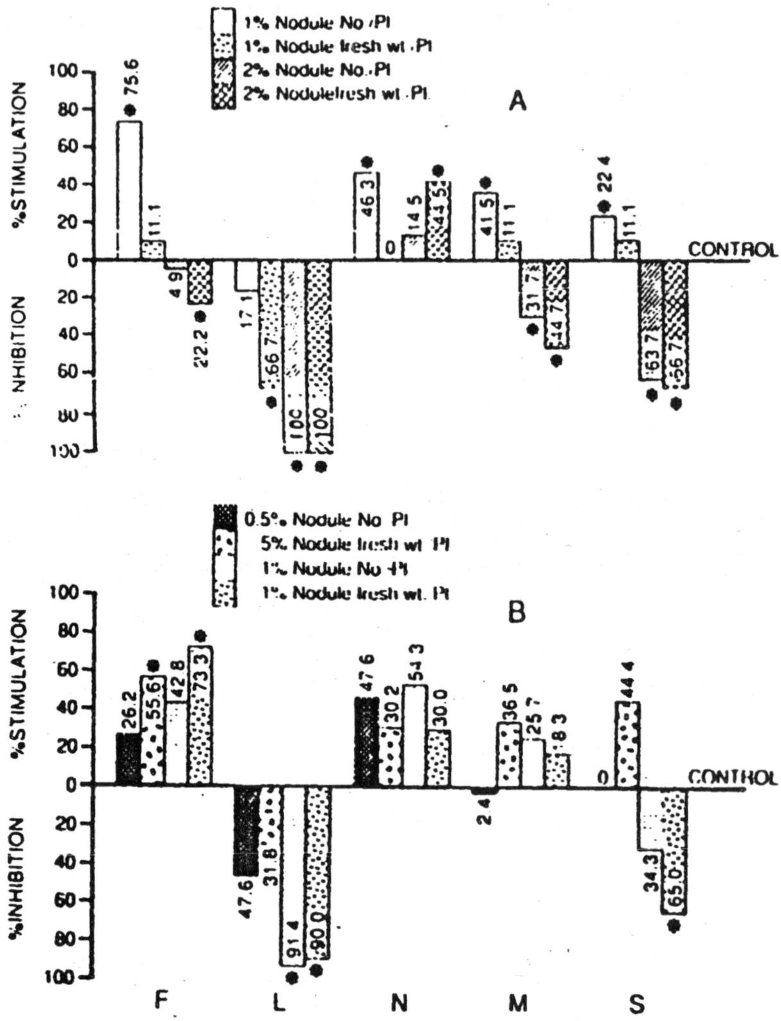

Fig. 11. Nodulation in soybean grown in nutrient solution (A) and in soil (B) amended with water extracts of plant residues and residan respectively of five plants at two levels. All values with asterisk are significantly different from control by test at 5% F = foxtn L = lambsquarters, N = nutsedge, M = smartweed, S = sunflower

(II). Nutrient uptake : The weed species have both inhibitory as well as stimulatory effects on the nutrient uptake of crop plants (Appendix, Table 39,40,56). Buchholtz (1971) observed in field studies that quackgrass reduced the nitrogen and potash uptake of maize plants. Inflorescence and seed extracts of 5-10% concentrations from large crabrass (*Digitaria sanguinalis* (L.) Scop.), fall panicum

(Panicum dichotomiflorum Michx.), *Ageratum conyzoides* L., *Cyperus iria* L., and *Paspalum dilatatum* Poir. inhibited the uptake of 32° in kidneybean (Table 39) (Bansal and Kalia, 1984). While the residues of yellow foxtail (*Setaria glauca* (L.) Beauv.) and common ragweed (*Ambrosia artemisifolia* L.) increased the potash uptake (21-48%) in maize and not in soybean (Table 40) (Bhowmik and Doll, 1984).

Table 39. Effect of aqueous extracts of seeds and infloresences of weeds on 32^{P} uptake in Urdbean seedlings

Weed species	Control	Cold water extract		Boiled autoclaved extract	
		5%	10%	5%	10%
	32P uptake (μg/g plant sample)				
Control	4293	—	—	—	—
Digitaria sangunialis (L.)	—	1764	568	2387	1385
Panicum dichotomiflorum	—	2187	634	1992	2075
Cyperus iria L.	—	1640	1678	2953	1399
Ageratum conyzoides L.	—	966	256	1246	322
Paspalum dilatatum Poir.	—	1487	285	1966	713

Source : Bansal and Kalia (1984).

Table 40. Influence of weed residues on the nutrient uptake (mg/plant) in corn and soybean 28 days after planting

Weed species	Corn			Soybean		
	N	P	K	N	P	K
Chenopodium album L.	27.4	11.4	37.1	30.6	17.7	32.1
Amaranthus retroflexus L.	23.3	9.1	31.0	28.7	10.8	31.4
Ambrosia artemisifolia L.	30.2	22.9	48.3	30.1	15.1	31.4
Abutilon theophrasti Medic.	31.3	12.9	35.4	30.7	11.5	28.8
Setaria glauca (L.) Beauv.	23.2	12.6	39.7	25.2	12.5	29.7
Control (No residue)	39.8	16.8	32.7	36.3	11.6	28.4
LSD (0.05)	6.3	3.1	6.5	3.3	2.0	4.1

Source : Bhowmik and Doll (1984).

(III). Seed germination, growth and yields : Most of the weed species has inhibitory effect on the seed germination, growth and yield of crops (Appendix), but some weeds have stimulatory effects too (Table 59).

3.2.1.1.1 Major weeds

I. Quack grass (*Agropyron repens* (L.) Beauv.) : It is an important perennial weed of field crops in many countries of the world (Holm *et al.,* 1979), whose allelopathic effects have been studied most (Appendix, Tables 41, 58). It causes serious reductions in the yields of maize, oat, soybean and potato (Isleib, 1960; Bandeen and Buchholtz, 1967; Kommedahl *et al.,* 1970), because its infested soil and its plant residues decreased the seed germination and growth of crops. Its allelopathy is attributed mainly to live plants and their residues. It has been found allelopathic to cereals (wheat, barley, oat, maize), legumes (alfalfa, peas, soybean, kidneybean), rapeseed, flax, sugarbeet, tomato and cucumber; however, maximum research has been done on alfalfa.

Aqueous extracts of leaves, stem, root, rhizomes and infested soil reduced the seed germination and seedling elongation of alfalfa `Ranger', wheat and barley compared with control. The toxicity of extracts was removed, when these were passed though columns of activated carbon (Table 41) (Kommedahl *et al.,* 1959; Ohman and Kommedahl, 1959, 1960). The living roots or rhizomes did not produce any toxin, which affected the germination or growth of test crops (Ohman, 1961; Ohman and Kommedahl, 1964). Harvey and Linscott (1978) found that quackgrass rhizomes generated ethylene, which may have allelopathic effects on germinating seeds. However, Toai and Linscott (1979) observed that water soluble toxins in the weed residues reduced the seed germination and seedling growth of barley `Proctor' and toxicity persisted for seven weeks after residue incorporation in the soil (Lynch and Penn, 1980). The toxicity was greater at higher temperature (35°C) owing to accelerated toxin formation from residues compared with lower temperatures (5-10°C), which prevented the formation of toxins (Toai and Linscott, 1979). The shoot and rhizome extracts proved less inhibitory to the germination than to seedling growth of alfalfa. The shoot extracts were more toxic compared with rhizome extracts and the toxicity was greater during the vegetative phase and decreased afterwards. The alfalfa varieties differed in their tolerance to these extracts (Touchette *et al.,* 1988).

Table 41. Effect of previously infested *Agropyron repens* (L.) Beauv. soil on the stand and dry weight of 90 days old crops

Crop	Dry weight (g)/plant		Stand on infested soil (% of control)
	Infested soil	Non-infested soil	
Alfalfa	0.2	0.5	56
Flax	0.2	2.0	81
Barley	1.1	4.2	52
Oat	1.8	4.1	76
Wheat	1.6	3.9	66

Source : Kommedahl *et al.* (1959).

Osvald (1948) reported that aqueous root extracts of quackgrass decayed the germinating seeds of oat, rape and peas and thus drastically reduced their seed germination and injured their seedlings. Its rhizome extract inhibited the root growth in seedlings of wheat, soybean alfalfa and pea (LeTourneau and Heggeness, 1957; Weston and Putnam, 1986). Kommedahl *et al.* (1959) reported that wheat, oat, barley, alfalfa and linseed crops stand and dry matter production remained poor in quackgrass infested soil. In heavily infested soil, the plant height and yield of wheat, maize, oat and soybean remained poor and inhibition persisted throughout the second growing season compared with control (Kommedahl *et al.*, 1970). Sugarbeet growth was also inhibited in its infested soil (Plhak, 1967). Welbank (1960) reported that its decaying residues inhibited the germination and growth of wheat, barley, rapeseed, kale and potato and the toxicity persisted upto 7 weeks after residue incorporation in the soil. The decaying rhizomes under waterlogged conditions produced large quantities of acetic and butyric acids (Lynch and Penn, 1980). Lynch *et al.* (1981b) found that fungus *Fusarium culmorum* present in its decaying rhizomes, stunted the crop growth and acetic and butyric acids produced during anaerobic rhizome decay caused rotting of barley seedlings and predisposed the cereal host plant to infection. Sowing of soybean, snapbean and navybean in mowed quackgrass sod decreased their seedling growth (Weston and Putnam, 1985). Gabor and Veatch (1981) isolated and identified glycosides in quackgrass residues, which inhibited root growth in maize, oat, alfalfa and cucumber seedlings. Hagin (1989) identified the phytotoxins in quackgrass as 5-hydroxy indole-3- acetic acid and 5-hydroxy tryptophan, which inhibited the seedling growth of maize and soybean.

Table 42. Effect of leaf material of *Parthenium hysterophorus* L. mixed with soil on tillering/branching and yield of crops

Treatment	Pearlmillet		Finger-millet		Tomato		Cowpea	Bean
	Tillers/ plant	Yield/ plant (g)	Tillers	Yield/ plant (g)	Branches/ plant	Yield/ plant (g)	Yield/ plant (g)	Yield/ plant (g)
Control	11 (3.3)	40 (0.2)	15 (0.8)	65 (5.1)	6 (0.66)	452 (199)	233 (33)	269 (76.3)
Leaf-mixed soil	12 (1.0)	49 (0.5)	10 (1.1)*	38 (11.7)	4 (0.5)*	277 (61)*	193 (10.6)*	186 (17)*

Figures in parenthesis indicate standard error.

* Significantly different from control at LSD 0.01.

Source : Kanchan and Jayachandra (1979 a).

II. Parthenium (*Parthenium hysterophorus* L.) : It is a serious weed in many countries (Holm *et al.,* 1979). It spread quickly throughout India and decreased the yields of several crops (Kanchan and Jayachandra, 1979a b; Khosla and Sobti, 1979). The mixture of its fruits and receptacles or their extracts inhibited the seed germination, coleoptile and radicle growth of wheat seedlings (Rajan, 1973). Its root, stem, leaf and inflorescence extracts reduced the seed germination and seedling growth of wheat, barley, sorghum, maize, pearlmillet, groundnut, pigeonpea, greengram, blackgram, pea, fenugreek, rapeseed, sesame, radish, cabbage and cauliflower (Appendix, Table 42). The leaf and inflorescence extracts proved more inhibitory compared with stem and root extracts (Rao *et al.*, 1977; Dube *et al.,* 1979; Mall and Dagar, 1979; Ambika and Jayachandra, 1980; Nath 1981; Dharmraj and Ali. 1985; Kohli *et al.,* 1985; Srivastava *et al.,* 1985). Its leaf and inflorescence extracts strongly inhibited the seed germination, seedling growth, cell survival and chlorophyll content of rapeseed. The aqueous extracts were more inhibitory than organic and inorganic extracts and completely inhibited the seed germination of rapeseed (Kohli *et al.,* 1985). The shoot and other plant parts extracts of parthenium considerably decreased the root growth of corn than wheat (Mersie and Singh, 1987a). However, the foliar spray of its root, shoot and inflorescence extracts had no inhibitory effect on wheat yields (Lad, 1983). The inhibitory effects of its extracts, were owing to the presence of inhibitors such as parthenin, coronopilin, caffeic acid, p-coumaric acid, alkaloids and sesquiterpene lactones in these extracts (Kanchan 1975; Ambika and Jayachandra, 1980; Jarvis *et al.,* 1985).

Kanchan and Jayachandra (1979 a,b) found stunted growth of crop plants (Wheat, `UP 301', fingermillet `Poorna', cowpea, beans `Burpees Stringless', tomato `Pusa Ruby') associated with this weed in the fields (Table 42). It released the growth toxins to the soil through leaching, decomposition of weed residues, root exudates and pollens. The inhibitors washed down from the aerial vegetative parts of the weed during rain, were adsorbed on the soil particles and made the soil inhibitory. Its leachates, markedly reduced the seedling growth of snapbean 'Burpees stringless' and wheat `UP 301'. Its dry leaves and roots mixed in the soil, inhibited the nodulation, growth and yield of snapbean and cowpea, branching in tomato 'Pusa Ruby', plant height and tillering in finger millet `Poorna' but stimulated the growth of pearlmillet `HB 1'. Muniyappa and Krishnamurthy (1981) reported that the addition of parthenium residues to the soil before sowing of test crops (wheat, barley, oat, rice, sorghum, fingermillet, groundnut, blackgram, cowpea, soybean and frenchbean) reduced the seed germination of all crops, except rice, oat and barley, however, inhibition was greater in pulses than in cereals. Besides, the dry weight, root length, number of leaves, number of tillers or branches and plant height of all test crops were considerably decreased. Its root exudates had highest concentration of inhibitors at rosette and flowering stage. The inhibitors released in root exudates or from decomposing weed residues remain active for 30 days (Kanchan and Jayachandra, 1979b). The trichomes covering the vegetative parts of parthenium contained inhibitors and were detached easily from dry parts. When these trichomes settled on leaf surfaces of wheat and soybean in large quantity, they reduced the chlorophyll and dry matter in test crops (Kanchan and Jayachandra, 1979b). Parthenium produced 3375 million pollen grains/m^2 area and were deposited on the aerial parts of nearby plants in large quantities (Kanchan, 1977). The artificial dusting of pollen grains or spraying their leachates on the stigma of snapbean, brinjal, tomato and chillies withered their flowers and inhibited the germination and elongation of their pollen tubes (Kanchan and Jayachandra, 1980a). The deposition of pollens on the leaves of snapbeans reduced their chlorophyll content and seedling dry weight.

III. Johnson grass *(Sorghum halepense* (L.) Pers.) : It is one of the ten worst weeds in the world, found in 37 countries (Holm *et al.,* 1979) and causes severe reductions in the yield of maize, sugarcane, soybean and cotton (Kleifeld, 1970; Millhollon, 1970; McWhorter and Hartwig, 1972; Roeth, 1973). Leaf, rhizome extracts, their incroporation in soil and root exudates inhibited the seed germination and growth of several crops (Appendix, Tables 43, 46). The aqueous extracts of its leaves and rhizomes, inhibited the seed germination and seedling growth of sunflower and tomato (Abdul-Wahab and Rice, 1967) and seedling growth of maize (Mikulas, 1984). Its extrat delayed the seed germination and inhibited root, shoot and leaf growth in wheat and chickpea (Table 43) (Angiras *et al.,* 1988; Singh *et al.,* 1983). Soil incorporation of its rhizomes decreased the seedling development of wheat, barley, soybean and mustard and toxicity persisted upto 4 months (Friedman and Horowitz, 1970; Horowitz and Friedman, 1971; Lolas and Coble, 1982). Its rhizome exudates decreased the seed germination and seedling growth of sunflower and tomato and root growth of radish seedlings (Abdul-Wahab and Rice, 1967; Nicollier *et al.,* 1985). The seed germination and seedling growth of test crops was inhibited owing to the presence of inhibitors such as chlorogenic, p-coumeric, fernlic p-hydroxybenzoic, p-hydroxyphe-nylacetic acids; dhurrin, vanillin, taxiphyllin and p-hydroxyben-zaldehyde in the leaf or rhizome extracts and rhizome decomposition products (Abdul-Wahab and Rice, 1967; Mainx and Tauscher, 1983; Patterson, 1984; Nicollier *et al.*, 1985). p-coumaric acid is present in its entire plant in the early stages of growth while cholorogenic acid and dhurrin are present in leaf and rhizome, respectively (Abdul-Klahab and Rice, 1967).

IV. Cyperus species : The weeds of this species (*C. esculentus* L., *C. iria* L., *C. rotundus* L., *C. serotinus* L.) are most serious weeds and are found in several countries of the world. These weeds caused various allelopathic effects in crop species (Appendix, Tables 39, 43, 44, 45, 49).

Table 43. Effect of aqueous extracts of weeds on germination and seedling growth of chickpea

Weeds	Germination (%)	Days to complete germination	Root length (cm)	Shoot length (cm)	No. of leaves/ shoot
Lantana camara L.	90	5.0	2.5	8.2	2.0
Cyperus esculentus L.	90	6.0	1.6	3.5	1.8
Ageratum conyzoides L.	80	5.3	3.6	4.6	2.3
Eupatorium adenophorum	80	5.7	1.4	2.7	1.7
Sorghum halepense Pers.	80	5.3	7.3	10.2	2.6
Echinochloa crusgalli L.	50	6.3	0.2	0.4	1.0
Water	90	4.0	16.8	14.8	3.0
CD at 1 %	–	1.0	1.0	1.5	0.1

Source : Angiras *et al.* (1988).

Table 44. Effect of soil infested with *Cyperus rotundus* L. on the germination of crops

Period of infestation (Weeks)	Germination as per cent of germination in non-infested soil		
	Mustard	Barley	Cotton
4	74 s a	120	100
6	58 ss	104	72 s
9	54 ss	78 s	58 ss
12	68 ss	73 ss	69 ss

[a] S ignificant differences from relevant control at S = LSD/0.05; SS= LSD/0.01

Source : Friedman and Horowitz (1971).

(i). Purple nutsedge (*Cyperus rotundus* L.) : It is major weed of *Cyperus species* and is one of the ten worst weeds in the world. It is found in 74 countries (Holm *et al.*, 1979) and considerably reduced the yields of rice, tomato, cabbage, cucumber, carrot, okra and garlic (Williams and Warren, 1975; Okafor and DeDatta, 1976; Zimdahl, 1980). Aqueous extracts of its tubers were tested on ten test crops (rice, sorghum, maize, pearlmillet, cowpea, blackgram, sannhemp, sesame, cotton) for effect on their seed germination and seedling growth. Except maize and cotton, the extracts inhibited the seed germination of remaining crops and seedling growth of all test crops (Singh, 1968). Similarly, aqueous extracts of its tubers also reduced the seed germination of sorghum, oat, soybean and sannhemp and survival of radish, onion and tomato seedlings (Lucena, 1974; Meissner *et al.,* 1979). *C. rotundus* infested soil, infested soil extract and soil incorporated tubers reduced the seed germination of barley, mustard and cotton and seedling growth of barley, sorghum, soybean cotton, cucumber and tomato (Table 44) (Friedman and Horowitz, 1971; Horowitz and Friedman, 1971; Lucena and Doll, 1976; Meissner *et al.,* 1979; Chivinge, 1985). While soil incorporated shoot residues stimulated the growth of sorghum and soybean (Lucena and Doll, 1975). Zimdahl (1980) reported that season long interference from this weed substantially decreased the garlic yield (89%) but not that of maize which received adequate fertilizer nitrogen. Thus, mode of interference from this weed was competition for nutrients with the crops. These allelopathic effects may be attributed to the presence of inhibitors (polyphenols and sesquiterpenes) in its tubers (Komai and Ueki, 1975, 1980; Komai *et al.*, 1977).

Table 45. Influence of *Cyperus esculentus* L. foliage and tuber residues on the dry weight of corn and soybean shoots and roots

Concentration of residue (%w/w)	Foliage residues				Tuber residues			
	Soybean		Corn		Soybean		Corn	
	Shoots	Roots	Shoots	Roots	Shoots	Roots	Shoots	Roots
% of Untreated control								
0	100 a	100 a	100 a	100 a	100 a	100 a	100 a	100 a
0.125	95 ab	55 b	93 ab	91 a	77 a	61 b	58 b	76 b
0.250	78 bc	43 b	93 ab	111 a	60 b	46 b	65 b	66 bc
0.375	72bc	39 b	83 ab	126 a	65 ab	58 b	59 b	70 bc
0.500	65 c	44 b	81 b	91 a	58 b	45 b	56 b	68 bc
0.675	79 bc	58 b	83 b	89 a	60 b	50 b	54 b	55 c

Means within column followed by the same letter are not significantly different at LSD 0.05.

Source : Drost and Doll (1980).

(ii). Yellow nutsedge (*Cyperus esculentus* L.) : It is a serious weed in 10 countries (Holm *et al.,* 1979) and reduces yields of several crops. Tames *et al.*(1973) found that its tubers contained inhibitory compounds (p-hydroxybenzoic, vanillic, syringic, ferulic, p-coumaric acids plus other unidentified ones) which inhibited the growth of oat coleoptiles and germination of sugarbeet, peas, lettuce, tomato and white clover. Drost and Doll (1980) reported that its foliage residues were very inhibitory to the root and shoot growth of maize and soybean (Table 45). The aqueous extracts of its tubers, reduced the growth of root and shoot in soybean. In general, yellow nutsedge proved more harmful to soybean compared to maize. Its aqueous extracts and soil incorporated residues, inhibited the germination and seedling growth in soybean (Table 49) (Mallik and Tesfai, 1988). While the extracts had no effect on germination but significantly decreased the root and shoot elongation and number of leaves in chickpea (Table 43) (Angiras *et al.*, 1988). Its extracts delayed the seed germination and seriously inhibited the root and shoot development in wheat (Singh *et al.,* 1988).

(iii). *Cyperus serotinus* L. : The methanol extracts of its tubers inhibited the seedling growth of rice and the inhibitors were identified as sesquiterpenoids (Komai *et al.,* 1981).

(iv). *Cyperus iria* L. : The extracts from the inflorescences and seeds inhibited the ^{32}P uptake in blackgram seedlings (Table 39) (Bansal and Kalia, 1984).

The rhizomes and roots of *Cyperus brevifolius* and *Cyperus kyllingia,* the common weeds of Hawaii contained allelopathic essential oils, which help in their spread to grasslands and lawns. *C. kyllingia* contained more essential oils than *C. brevifolius* but the former is rich in terpenes (α-cyperone, ß-selinene, α-humulene) and the later one in C_{17} to C_{25} n-paraffins, thus the former was more prevalent species (Komai and Tang, 1989).

(v). Amaranthus species : Its five species (*A. palmeri* (L) Wats., *A. retroflexus* L., *A. spinosus* L., *A. tricolour* L., *A. viridis* L.) have shown allelopathic effects towards field crops, however, *A. palmeri* has been studied most (Appendix, Tables 40, 46, 47).

Table 46. Effect of plant residues of *Amaranthus palmeri* L. and *Sorghum halepense* L. on the fresh weight of crops

Weed	Sorghum	Cotton	Cabbage	Carrot	Onion
	Fresh weight (kg/m²)				
Amaranthus palmeri L.	4.8	7.2	9.5	7.5 *	2.0*
Sorghum halepense L.	–	–	9.5	14.6	2.1
Control	5.2	9.7	10.4	14.7	6.3
LSD (0.05)	–	–	1.45	3.2	1.5

*Significantly different from control at LSD 0.05

Source : Menges (1987).

(i). Palmer amaranth *(Amaranthus palmeri* (L.) Wats) : Its residues release the allelochemicals through decomposition and through volatiles and the effect of volatile inhibitors on crops have been discussed in Section 3.2.1.5. Its residues inhibited the growth of onion, carrort and its own plants. These residues contained allelochemicals such as phytol, chondrillasterol, vanillin, 3-methoxy-4-hydroxynitrobenzene and 2,6-dimethoxy- benzoqui-none (Bradow, 1985; Fischer and Quijano, 1985). The wax ester consisted of a series of C_{36} to c_{56} homologues with C_{40}, C_{42}, C_{44} C_{46} and C_{48} homologues predominating. The major wax ester fatty acids were C_{16}, C_{18}, C_{20}, C_{22} and C_{24}. Similar trend in carbon number distribution were found between free and bound fatty alcohols with C_{22}, C_{24}, C_{26}, C_{28}, C_{30} and C_{32} homologues predomina-ting (Dailey Jr. *et al.,* 1989). Its residues in the soil reduced fresh weight accumulation in onion and carrot and markedly decreased the seedling establishment of carrot.

The inhibitory effects of weed residues against onion germination persisted upto 62 days after soil incorporation (Bradow and Connick Jr., 1987). Likewise, soil incorporated residues inhibited the growth of carrot (49%) and onion (68%) and phytotoxicity persisted for 11 to 16 weeks after soil incorporation (Bradow and Connick Jr., 1987; Menges, 1987). In further studies, soil incorporated residues, inhibited the root and shoot growth of grain sorghum, cabbage, carrot and onion seedlings; however, the toxicity was less towards carrot and onion compared with cabbage and sorghum (Table 46). Cabbage cultivars `Grand slam' and `Sanibel' differed in their sensitivity to decomposing weed residues. The residues concentration of 8000 and 16000 ppm in soil severely inhibited the growth of sorghum roots. Seedling growth of sorghum was most drastically inhibited by thyrus and leaf tissues than that from stem and root tissues (Menges, 1988). Three days exposure to three pentanone and light low molecular weight aliphatic alcohols and aldehydes identified in the volatiles of palmer amaranth, inhibited the seed germination of test species onion, carrot, tomato and the inhibition was dependent on expo-sure time and concentration (Bradow and Connick Jr., 1988a, b).

(ii). Redroot pigweed (*Amaranthus retroflexus* L.) : Its residues markedly decreased the plant height, shoot and root weight, leaf area duration, leaf weight ratio and total dry matter production and nitrogen and phosphorus uptake in maize and soybean and the allelopathic effects depended on atmospheric temperature (Table 47) (Bhowmik and Doll, 1982, 1983a). In field studies, its decaying shoot residues and soil infestation in the previous season decreased the growth of tobacco. After tobacco transplanting, its interference of more than 3-4 weeks decreased the plant height, fresh weight, leaf number and yield of tobacco compared with weed free crop (Lolas, 1986).

Table 47. Response of corn and soybean seedlings grown in soil incorporated with weed residues

Weed species	Corn			Soybean		
	Height	Fresh weight		Height	Fresh weight	
		Shoot	Root		Shoot	Root
			% reduction			
Dicotyledons						
Chenopodium album L.	36 bc	32 a	37 bc	40 bc	60 ad	69 ef
Amaranthus retroflexus L.	*38 c*	*27 a*	*42 c*	*61 c*	*80 d*	*96 f*
Ambrosia artemisifolia L.	*9 ab*	*15 a*	*22 abc*	*9 ab*	*33 ad*	*0 a*
Abutilon theophrasti L.	*22 abc*	*18 a*	*12 abc*	*30 abc*	*64 cd*	*31 cf*
Monocotyledons						
Cynodon dactylon (L) Pers.	11 abc	21 a	8 abc	11 ab	15 abc	9 ae
Panicum dichotomiflorum	*0 a*	*0 a*	*12 abc*	*0 a*	*10 abc*	*0 a*
Setaria faberi Herm.	9 ab	11 a	0 a	9 ab	0 a	0 a
Setaria viridis (L.) Beauv.	3 a	0 a	22 abc	18 ab	34 ad	11 bc
Setaria glauca (L.) Beauv.	0 a	0 a	1 a	10 ab	28 ad	0 a
No residue (control)	0 a	0 a	0 a	0 a	0 a	0 a

Means within a column followed by the same letter are not significantly different at LSD 0.05. Source : Bhowmik and Doll (1982).

(iii). *Amaranthus spinosus* L. : Its leaf and inflorescence extracts seriously decreased the vegetative and reproductive phases of mustard and wheat crops (Datta and Bandyopadhayay, 1981).

(iv). *Amaranthus tricolour* L. : Aqueous extracts from its dried leaves delayed the seed germination and severely inhibited the root and shoot growth of sorghum seedlings. The inhibition in root and shoot growth was directly related to the extract concentration (Mohnot and Soni, 1979).

(v). *Amaranthus viridis* L. : It grows widely in the tropical, subtropical and temperate regions of the world and is a notorious weed in India. Aqueous extracts of leaves, inhibited the seed germination of pearlmillet and shoot and root growth of pearlmillet and sesame seedlings. The order of growth inhibition in sesame seedlings was fruit wall > stem > leaf > root > seed and in pearlmillet seedlings was leaf > root > stem > seed > fruit wall. The burning or soil burial of its residues decreased the growth and yield of pearlmillet. However, burning or burial of weed biomass decreased the chlorophyll and carotenoid content of both test crops than that of control (Singhal and Sen, 1981). Rf values of the toxic chemicals were 0.15 in root, 0.75 - 0.85 in stem, 0.15 - 0.45 in leaf, 0.65- 0.75 in fruit wall and 0.65 in seed (Singhal and Sen, 1982).

VI. Polygonum species : Its four species (*P. aviculare* L., *P. lapathifolium* L., *P. orientale* L. and *P. pennsylvanicum* L.) have shown allelopathic effects on sorghum, rice, wheat, mustard, linseed, soybean, blackgram, pea, alfalfa, cabbage, carrot, and lettuce (Appendix, Tables 48, 49).

(i). Prostrate knotweed (*Polygonum aviculare* L.) : Water soluble compounds present in its leaves and stem interfered with the seed germination of *Medicago truncula**ta* and lettuce, cell division and early growth of meristem in Medicago seedlings (Kloot and Boyce, 1982). Its leaf leachates, slightly reduced the germination of both test crops (cotton and sorghum), but infested soil severely inhibited the seed germination of cotton and had no effect on sorghum. Its decaying roots (in the same concentration as found in the field soil) greatly reduced the seed germination of cotton and seedling growth of both test crops, while its decaying shoots had no harmful effect on the germination of test crops and reduced the seedling growth of only cotton (Table 48) (Al Saadawi and Rice, 1982a). Eight phenolic inhibitors were isolated from its living plants, residues and underneath soil. Some of the inhibitors also inhibited the growth of *Rhizobium* and *Azotobacter* strains (AlSaadawi and Rice, 1982b).

Table 48. Effect of infested soil, leaf leachates, root exudates and decaying plants of *Polygonum aviculare* L. on the shoot growth of cotton and sorghum

Treatment	Cotton			Sorghum		
	Dry weight/plant (mg)		Germination (% of control)	Dry weight/plant (mg)		Germination (% of control)
	Control	Test		Control	Test	
Infested soil (July 30)	1145 ±52	1096±69	91	107 ±71	1137 ±56	96
Infested soil (March 20)	358 ±19	284 ±28 a	54	544 ±22	366 ±50 a	96
Leaf leachates	1200 ±67	1270 ±67	81	1260±101	1254 ±40	96
Decaying shoots	890 ±23	563 ±53 a	25	804 ±47	696 ±37 a	61
Decaying roots	723 ±20	611 ±26 a	96	404 ±50	430 ±42	94
After 20 days of decomposition	525 ±16	411 ±11 b	—	705 ±42	414 ±56 a	—
After 40 days of decomposition	1721 ±52	1236 ±122 b	—	2144 ±151	1940 ±127	—
Root exudates	1545 ±381	1839 ±132	—	4048 ±404	3648 ±497	—

a Dry weights significantly different from control a – 0.01 level, b = 0.001 level. Source : Alsaadawi and Rice (1982 a).

(ii). *Polygonum orientale* L. : It is most troublesome weed of rice in India. Its leaf, stem, flower and root extracts reduced the seed germination of all test crops (rice, mustard, pea, lettuce), while, leaf and stem extracts also proved inhibitory to blackgram. Leaf extracts were more inhibitory to the seedling growth of test crops and mature leaves caused greater inhibition than younger ones, owing to the presence of inhibitory luteolin and apigenin glucosides. The foliar spray of leaf extract was more phytotoxic than its soil application on blackgram but foliar spray and soil application killed the mustard seedlings. This weed completely suppressed the seed germination and growth of all test species during the summer rainy season (Datta and Chatterjee, 1978, 1980).

(iii). Pennsylvania smartweed (*Polygonum pennsylvanicum* L.): Its aqueous plant extracts inhibited the seed germination, root and shoot growth of wheat seedlings (LeTourneau *et al.*, 1956), while its seed extracts decreased the seed germination of alfalfa but had no effect on radish, turnip, tomato, cabbage, carrot and pepper. Its plant extracts and soil incorporated residues, decreased the seed germination and nodulation in soybean at higher concentrations but caused stimulation at lower ones (Table 49) (Mallik and Tesfai, 1988).

Table 49. Response of soybean seed germination and shoot growth to aqueous extract (1%), nutrient solution (1%) and residues of weeds incorporated in soil @ 1% (Control = 100)

Weeds	Germination		Shoot growth			
	Aqueous extract	Soil incorporated	Nutrient solution		Soil incorporated	
			Shoot	Root	Shoot	Root
Setaria virdis L.	- 5.8	-13.2	+ 55.3	+ 80.9	+ 70.1	+ 65.2
Chenopodium album L.	- 29.0	- 96.2	+ 104.2	+ 133.3	- 63.0	- 41.3
Cyperus esculentus L.	- 14.5	- 10.6	+ 68.1	+ 76.2	- 3.9*	- 15.6*
Polygonum pennsylvanicum L.	- 8.7	- 42.5	+ 55.3	+ 76.2	+ 90.6	+100.0

+ and - indicate per cent stimulation or inhibition, respectively. All values without asterisk are significantly different form control LSD 0.05. Source : Mallik and Tesfai (1988).

(iv). *Polygonum lapathifolium* L. : Its interference in the growing flax crop reduced the seed yields by 28% compared with control (Grummer and Beyer, 1960).

VII. *Setaria* spp. : Its four species viz., *S. faberii* Herm., *S. glauca* (L.) Beauv., *S. italica* (L.) Beauv. and *S. viridis* (L.) Beauv have shown allelopathic effects on the seed germination and growth of maize, sorghum, pearlmillet, rapeseed, soybean, cabbage, radish and turnip (Appendix, Tables 40, 47, 49) (Gressel and Holm, 1964; Schreiber and Williams, 1967; Bell and Koeppe, 1972; Bhowmik and Doll, 1979; Hussain *et al.,* 1981; Mallik and Tesfai, 1988).

(i). Giant foxtail *(Setaria faberii* Herm.) : It is a common weed in Japan and U.S.A. (Holm *et al.,* 1979) and caused yield reduction in corn (12 to 13%) and soybean (12 to 38%) (Knake and Slife, 1962, 1965; Staniforth, 1965). Its decaying roots greatly inhibited the root growth (Schreiber and Williams, 1967), while its mature plants and leachates from dead residues decreased the shoot growth of maize (Bell and Koeppe, 1972). The aqueous extracts of its residues reduced the root and shoot growth of maize, residues incorporated in the soil inhibited the plant height and fresh weight of both maize and soybean seedlings (Bhowmik and Doll, 1982).

(ii). Yellow foxtail (*Setaria glauca)* (L.) Beauv.) : It is a major weed in 22 countries (Holm *et al.*, 1979), which reduced yields of corn (7-57%), sorghum (30%) and soybean (7%) (Staniforth, 1957, 1958, 1961; Feltner *et al.,* 1969). Its decaying roots reduced the root growth of maize (Schreiber and Williams, 1967). Its residues extracts, inhibited the radicle and coleoptile growth of maize and hypocotyl growth of soybean, while the extracts from decaying residues reduced their plant height (Table 47) (Bhowmik and Doll, 1982). Its soil incorporated residues, decreased the nitrogen and phosphorus uptake in maize and soybean seedlings (Table 40) (Bhowmik and Doll, 1984).

(iii). Foxtail millet (*Setaria italica* (L.) Beauv.) : Aqueous extracts from inflorescence, leaves, stem, root and soil rhizosphere, inhibited the germination and radicle growth of sorghum, pearlmillet and *Brassica campestris*. Its mulch reduced the fresh and dry weight and water content of these species (Hussain *et al.,* 1981).

(iv). Green foxtail (*Setaria viridis* (L.) Beauv.) : Its residues slightly decreased the seedling growth (plant height, fresh weight of root and shoot) of corn but caused greater decrement in sorghum (Table 45) (Bnowmik and Doll, 1982). Its soil incorporated residues and their extract slightly inhibited the germination of soybean but stimulated the seedling growth (Table 49). Its shoot extract and residues increased the nodulation in soybean at 1% concentration but decreased at 2% concentration (Mallik and Tesfai, 1988).

VIII. *Centaurea spp.* : Its four species viz., *C. diffusa* Lam., *C. maculosa* Lam, *C. nigra* L. and *C. repens* L. exhibited the allelopathic effects on wheat, barley, alfalfa, tomato and lettuce (Appendix) (Fletcher and Renny, 1963: Vezina and Doyon, 1983; Stevens, 1986;

Kelsey and Locken, 1987). Fletcher and Renny (1963) studied its three species and found that growth of tomato and barley was decreased in the infested soil or in soil added with powdered weed residue. Aqueous extracts from various plant parts indicated that leaves were most inhibitory to seed germination of test species (tomato, barley) and the order of inhibition was *C. repens* > *C. diffusa* >*C. maculosa* (Fig. 12). The inhibitor in the extract or plant residues was identified as Indole derivative, soluble in water and ether and was more inhibitory to root development than shoot.

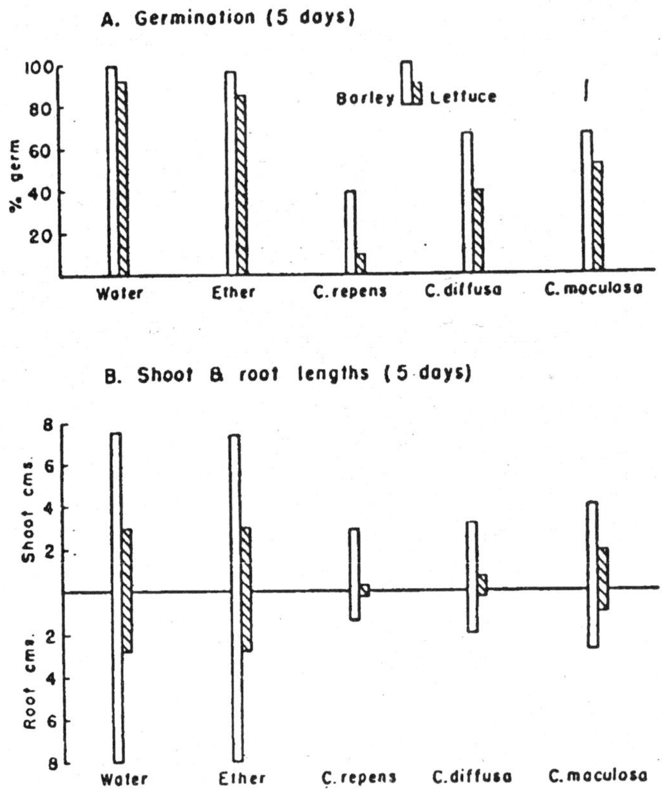

Fig. 12. Response of barley and lettuce to extracts of tree *Centaurea* spp.

(i). Spotted knapweed (*Centaurea maculosa* Lam.) : The aqueous and solvent extracts from its aerial tissues inhibited the root growth of lettuce. Its chloroform extract contained cnicin, a sesquiterpene lactone which inhibited the germination and root growth of lettuce and its own seedlings (Kelsey and Locken, 1987).

(ii). Black knapweed (*Centaurea nigra* L.) : Vezina and Doyon (1983) studied the effects of its residues from leaves, stem and roots on the germination and seedling growth of wheat and tomato. Soil added residues from stem and roots had no harmful effect on tomato growth, but those from leaves proved inhibitory. Cauline leaves had more toxic effect than from basal leaves, ether extracts of cauline leaves partially inhibited the seed germination and early seedling development in wheat and tomato.

(iii). Russian knapweed (*Centaurea repens* L.) : Its root contained polyacetylenes VIII-XIV; of these only IX was found active against root elongation in alfalfa, lettuce and red millet. Soil surrounding its root had sufficient concentration of polyacetylenes IX which greatly affect the neighbouring plants (Stevens, 1986).

IX. Digera species : Its three species viz., *D. alternifolia* (L.) Ascher, *D. arvensis* Forsk. and *D. muricata* (L.) Mart. have shown allelopathic effects on the germination and growth of pearlmillet, sorghum, sesame, pigeonpea, clusterbean, cotton, tomato, brinjal, onion and chillies (Appendix) (Sen *et al.,* 1969; Dubey, 1973; Sarma,1974; Ashraf and Sen, 1980).

Table 50. Effect of aqueous extracts from leaf, stem and root of *Digera alternifolia* (L.) Ascher on the seedling growth (mm) of pearlmillet and sesame.

Plant part	Concentration of extract (%)	Pearlmillet		Sesame	
		Radicle	Hypocotyl	Radicle	Hypocotyl
Leaf	1	15.2±3.6	12.1±2.8	4.1 ±2.2	8.2 ±3.6
	2	8.0 ±2.9	8.6 ±2.0	1.8 ±1.0	4.4 ±2.7
	3	3.6 ±2.5	6.5 ±3.1	*	*
	4	*	3.2 ±1.8	**	**
	5	*	*	**	**
Stem	1	18.6 ±4.2	13.0 ±3.6	6.7 ±2.9	8.5±3.0
	2	10.5 ±2.8	10.3 ±2.0	3.4 ±2.6	5.7 ±2.5
	3	6.9 ±2.5	9.0 ±2.5	*	3.9 ±2.2
	4	4.5 ±1.8	6.1 ±2.3	*	**
	5	3.6 ±1.6	2.5 ±1.5	**	**
Root	1	22.2 ±4.6	15.4 ±4.0	14.3 ±3.8	10.0 ±3.6
	2	13.7 ±3.8	14.3 ±3.2	4.2 ±2.8	8.7 ±3.1
	3	8.7 ±2.6	11.6 ±2.8	2.9 ±1.8	6.5 ±2.5
	4	5.8 ±2.0	8.1 ±2.8	*	*
	5	4.3 ±2.4	5.3 ±1.6	**	**
Control	—	42.7 ±4.3	21.2 ±2.6	45.6 ±4.3	20.0 ±2.1

* = Very little protrusion;

** = No protrusion, Data significant at 1 % level (P = 0.01).

Source : Ashraf and Sen (1980).

(i). *Digera alternifolia*(L.) Ascher : It appears in Indian crop fields with the first showers of summer rains. Its aqueous leaf extracts (1:0 to 1:250) inhibited the seed germination of cotton (36%), pigeonpea (44%), tomato (46%), pearlmillet (54%), brinjal (58%), chillies (60%), onion (62%), sorghum (62%) and clusterbean (62%) compared with control (Dubey, 1973). The aqueous extracts of its plant parts inhibited the germination and growth of pearlmillet and sesame; the order of inhibition was leaf > stem > root (Table 50). This weed proved more harmful to sesame than to pearlmillet. Its soil incorporated residues decreased the pigments content and yield of sesame compared with pearlmillet, while, burning of residues decreased the toxicity to test crops. No weeding, burning of weed and soil burial of weed reduced the grain yield per plant of soybean by 9.5, 72.9 and 86.6%, respectively, compared with weed free crop. While in pearlmillet, only soil incorporated weed residues caused 33% reduction in the yield (Ashraf and Sen, 1980).

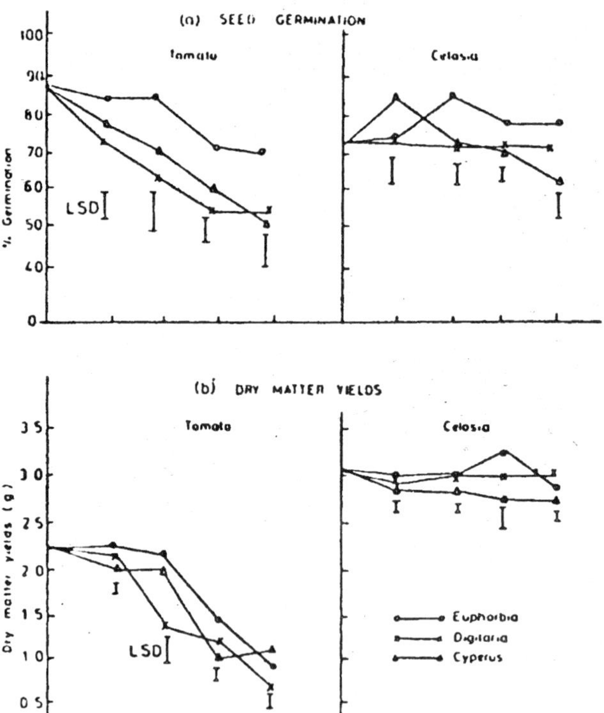

Fig. 13. Influence of the composition of tropical weeds in soils on crop seed germination and seedling dry matter yields. Vertical bars represent the least significant difference at the 5% level of probability. Source : Mohmmed-Saleem and Fawusi (1983)

(ii). *Digera arvensis* Forsk : It is a common weed of crops in Rajkot, India and produce toxins inhibitory to seed germination, radicle length and dry weight of pearlmillet (Sarma, 1974).

(iii). *Digera muricata* (L.) Mart : Aqueous extracts of its roots, stem and leaves did not influence the seed germination but inhibited the seedling growth of pearlmillet. The extracts were more toxic to root elongation than shoots. Leaf and root extracts proved more inhibitory than stem extract, the former ones did not inhibit the shoot elongation of pearlmillet (Sen *et al.,* 1969).

X. *Euphorbia* spp : Its four species viz., *E. esula* L., *E. geniculata* Ort., *E. granulata* and *E. heterophylla* L. have shown allelopathic effects on wheat, pearlmillet, sorghum, pea, tomato and pepper (Appendix) (LeTourneau *et al.,* 1956; Steenhagen and Zimdahl, 1979; Sugha, 1979, 1980; Hussain, 1980; Mohmed- Saleem and Fawusi, 1983). Aqueous extracts from fresh or dried leaves of *E. esula* inhibited the seed germination and coleoptile growth of wheat and radicle growth of pea, however, foliage ash was not inhibitory (LeTourneau and Heggeness, 1957). The weed infested soil or soil incorporated with its leaves, roots and litter severely reduced the growth of tomato, but surface applied litter had no such effect (Steenhagen and Zimdahl, 1979). The aqueous extracts from various plant parts of *E. geniculata* inhibited the seed germination of wheat in order of seed > root > leaf > stem (Sugha, 1979). Aqueous extracts of *E. granulata* decreased the seed germination of pearlmillet and lettuce (Hussain, 1980). Soil incorporated decomposing residues of *E. heterophylla* reduced the germination and seedling growth of sorghum, tomato and pepper. The leachates of this soil also severely retarded the growth of sorghum seedlings (Fig. 13) (Mohmed Saleem and Fawusi, 1983).

XI. *Chenopodium* spp. : Its two species *C. album* L. (Lambsquarter) and *C. murale* L. have proved allelopathic to germination and growth in wheat, rye, maize, chickpea, soybean, sugarbeet, mustard and nutrient uptake in maize and soybean (Appendix, Tables 40,47,49) (LeTourneau *et al.,* 1956; Bukolova, 1971; Caussanel, 1979; Bhowmik and Doll, 1980, 1982, 1984; Bhatia *et al.,* 1982; Goel, 1987; Mallik and Tesfai, 1988; Qasem and Hill, 1989). *C. album* is a major weed in 49 countries of the world and causes severe yield reductions in many crops including sugar beet (94%) and oilseed rape (99.8%) (Williams, 1964; Dawson, 1965). Its aqueous extracts inhibited the seed germination of wheat (LeTourneau *et al.,* 1956), reduced mitotic activity in meristematic cells, damaged the epidermis, cortex and vascular system of the roots and inhibited the root growth in wheat and rye (Bukolova, 1971). Aqueous extracts of its leaves and roots inhibited the root and shoot growth of maize (Caussanel, 1979; Bhowmik, 1982) and wheat (Bhatia *et al.,* 1982;

Porwal and Gupta, 1987). In pot and field studies, *C. album* reduced the plant height, fresh shoot and root weight, total dry matter and yields of maize and soybean (Table 47) (Bhowmik, 1982; Bhowmik and Doll, 1982). Its extracts did not affect the seed germination but inhibited the plumule growth of chickpea seedlings. The shoot extracts proved more toxic to seedlings than root extracts and shoot leachates (Goel, 1987). Mustard seeds presoaked in aqueous extracts from leaf and inflorescence of *C. murale* decreased the seed germination in petri plates but not in soil. The phytotoxin present in the residues was oxalic acid (Datta and Ghosh, 1982). Its soil incorporated residues decreased the germination and seedling growth in soybean (Table 49) (Mallik and Tesfai, 1988). *C. album* leachates reduced the dry weight and accumulation of N, P, K, Ca and Mg in shoots of tomato 'Alisa craig'. Soil incorporation of dried shoots 10 or 20 g/Kg soil decreased the dry weight of tomato plants (Qasem and Hill, 1989).

Table 51. Allelopathic influence of different parts of *Avena ludoviciana* and *Phalaris minor* Retz. on early seedling growth of wheat

Weed plant part	*Avena ludoviciana*			*Phalaris minor Retz.*		
	Root length (mm)	Epicoty length (mm)	Dry weight (mg)	Root length (mm)	Epicotyl length (mm)	Dry weight (mg)
Control	47.7	21.9	6.4	51.9	23.3	4.7
Leaf	54.8	30.4	6.1	64.5	29.9	5.5
Root	47.1	24.3	6.1	47.5	24.4	5.1
Straw	16.4	9.5	5.0	44.0	20.7	4.3
Glumes	21.4	11.0	5.7	44.9	20.9	4.7
CD (P = 0.05)	10.1	4.3	0.7	15.2	2.3	NS

Source : Bhatia *et al.* (1982).

XII. *Avena* spp. : Its two species *A. fatua* L. and *A. sterilis* sub sp. *Ludoviciana* are serious weeds in many countries (Holm *et al* ., 1979) and considerably reduce the yields of wheat, barley and flax (Friesen, 1961; Bell and Nalewaja 1968 a,b,c). These species have been reported allelopathic to wheat, barley, green gram, flax and lettuce (Appendix) (Bell and Nalewaja, 1968 a,b,c; Bhandari *et al.*, 1982; Bhatia, 1982; Chem *et al.*, 1982). Aqueous leachates of *A. fatua* straw inhibited the germination and seedling growth of *Centaurea melitensis* and *Silybum marianum* (L.) Gaertn owing to presence of ferulic, p-coumaric, p-hydroxybenzoic, syringic and vanillic acids (Tinnin and Muller, 1972). Aqueous extracts of its hulls inhibited the seed germination of lettuce and the inhibitors were vanillic acid, other phenolic compounds and short chain fatty acids (Chem *et al.,* 1982). Extracts from fresh plants of *A. ludoviciana* promoted seedling growth of wheat but that from dry straw and glumes caused inhibition (Table 51) (Bhatia *et al.,* 1982). In pot studies, aqueous extracts did not influence seed germination but reduced the seedling growth and nodulation in greengram (Bhandari *et al.,* 1982).

XIII. *Ambrosia* spp. : Its three species: *A. artemisiifolia* L. (common ragweed), *A. psilostachya* DC. (western ragweed) and *A. trifida* L. (Giant ragweed) have shown allelopathic effects on wheat, maize, sorghum, oat, rye, sunflower, soybean, alfalfa, tomato and radish (Appendix, Tables 40, 47). (LeTourneau and Haggeness, 1957; Rasmussen and Einhellig, 1979; Dalrymple and Rogers, 1983; Bhowmik and Doll, 1984). *A. trifida* extracts inhibited the germination of wheat (LeTourneau *et al.,* 1956), germination and seedling growth of sorghum and radish and five phytotoxins were isolated and characterized but not identified (Rasmussen and Einhellig, 1979). Top and rhizome extracts of *A. psilostachya* reduced the seed germination and seedling growth of wheat, oat, rye,alfalfa, yellow sweet clover and tomato (Dalrymple and Rogers, 1983). While residues of *A. artemisifolia* decreased the seedling growth of maize and soybean, but increased the potassium and phosphorus uptake in maize plants (Bhowmik and Doll, 1982, 1984). Its roots contained four polyacetylenes, a mixture of sesquiterpene hydrocarbons, methyl acetate and a mixture of ß-sitosterol and stigmasterol (Fischer and Quizano, 1985).

XIV. *Asclepias* spp : Its two species *A. syriaca* L. (milk weed) and *A. vaginalis* (L.) DC have been found allelopathic to maize, sorghum, soybean, wheat and sesame (Appendix) (LeTourneau *et al.,* 1956; Rasmussen and Einhellig, 1975; Singh and Sen, 1982). *A. syriaca* is a major weed in Canada, USA and Iraq (Holm *et al.,* 1979) and caused reductions in maize, sorghum and soybean yields (Evetts and Burnside, 1972; Cramer, 1981). Its shoot extracts markedly inhibited the seed germination, coleoptile and root elongation of wheat seedlings (LeTourneau *et al.,* 1956). while leaf extracts reduced the growth of sorghum seedlings and had no effect on maize seedlings (Rasmussen and Einhellig, 1975). The extracts of *A. vaginalis* (L.) DC completely inhibited the seed germination of sesame (Singh and Sen, 1982).

XV. *Citrullus* spp. : Its two species *C. colocynthis* (L.) Kuntz, and *C. lanatus* (Thumb.) Mansf. have shown allelopathy to wheat rice, pearlmillet, alfalfa, turnip and lettuce (Appendix). (Bhandari and Sen, 1972; Bhandari, 1974; Shaukat *et al.,* 1985). *C. colocyn-this* is a major weed of pearlmillet and sorghum in western Rajasthan, India and in Sind, Pakistan. Aqueous extracts of its roots, stem, leaves and fruit pulp delayed the seed germination and decreased root and shoot growth of pearlmillet seedlings, the order of inhibition was leaf > stem > fruit (Bhandari and Sen, 1971). Aqueous extracts of its shoot and root inhibited the seed germination and seedling growth of sorghum, pearlmillet and rapeseed. The shoot extract proved inhibitory to lettuce and alfalfa but reverse was true for root extract. Its decaying shoot and root residues were highly phytotoxic to germination and seedling growth of sorghum owing to presence of caffeic and vanillic acids. Artificial rain drip from its leaves also arrested seedling growth of sorghum and lettuce (Shaukat *et al.*, 1985). The fruit pulp of *C. lanatus* delayed the seed germination and suppressed the seedling growth of pearlmillet at higher concentrations but caused stimulation at lower concentrations (Bhandari and Sen, 1972).

XVI. *Crotolaria* spp : Its three species viz., *C. medicaginea*, Lamk, *C. pallida* and *C. verucosa* L. have proved allelopathic to wheat, rice, pearlmillet, sesame, greengram, blackgram, rapeseed and lettuce (Appendix) (Rao *et al.*, 1979; Singh and Sen, 1981, 1982: Datta and Dasmahapatra, 1984). *C. medicaginea* inhibited the root and shoot growth of sesame seedlings (Singh and Sen, 1982). The aqueous extracts of root, stem, leaf and flowers of *C. pallida* decreased the germination of wheat, rice, rapeseed, blackgram and lettuce owing to presence of phytotoxins, friedelin and quercetin-3-glycoside. The leachates from the plant parts were found more inhibitory than their extracts. As the leaves became senescent, their toxicity was increased (Datta and Dasmahapatra, 1984). Aqueous extracts of root, shoot and seed of *C. verucosa* L. inhibited the germination and seedling growth of pearlmillet (Rao *et al.*, 1979).

XVII. *Alhagi* spp. : Its two species *A. pseudoalhgi* (Bieb) Desv. and *A. sparsifolia* have been reported allelopathic to pearlmillet, chickpea and greengram (Appendix) (Mavlanov, 1982; Singhvi and Sharma, 1985). Seed, stem and leaf extracts of *A. sparsifolia* inhibited the seed germination and seedling growth of chickpea and greengram (Mavlanov, 1982). While the root, stem and top extracts of *A. pseudoalhagi* inhibited the germination and seedling growth of pearlmillet at higher concentrations but caused stimulation at lower concentrations (Singhvi and Sharma, 1985).

XVIII. *Bidens species* : Its two species *B. bipinnata* L. and *B. pilosa* L. have exhibited allelopathy through its residues and root exudates to maize, sorghum, beans, sunflower, onion, lettuce, cucumber, radish, squash and tomato (Appendix, Table 60) (Ruth *et al.*, 1986; Kahl, 1987). Aqueous extracts of leaves from young plant of *B. pilosa* stimulated the growth of maize seedling but it was not true for leaf extract from older plants near flowering (Kahl, 1987). The extracts of soil infested with *B. bipinnata* reduced the germination and growth of maize, sunflower, onion, lettuce, cucumber, radish, squash and tomato seedlings (Ruth *et al.,* 1986).

XIX. *Eupatorium* spp. : Its two species *E. adenophorum* Sprengel and *E. odoratum* L. showed allelopathic effects in wheat, mustard, chickpea and white clover (Appendix, Table 43) (Datta and Bandopadhyay, 1981; Tripathi *et al.,* 1981; Angiras *et al.,* 1988). *E. adenophorum* is a dominant weed in Meghalaya, India and it reduces density of certain plant species from the fields. Its aqueous extracts suppressed the germination, root and shoot growth of wheat, chickpea and white clover and reduced leaf number and drymatter in white clover. Its plants also decreased the fungi, bacteria and actinomycetes population in the soil (Tripathi *et al.*, 1981; Angiras *et al.,* 1988; Singh *et al.,* 1988). Leaf extracts of *E. odoratum* drastically reduced the growth of wheat and mustard seedlings (Datta and Bandopadhayay, 1981).

XX. *Trichodesma* spp. : Its two species *T. amplexicaule* Roth. and *T. sedgwiickianum* Banerji were found allelopathic to seed germination and seedling growth of pearlmillet (Appendix, Sen *et al.,* 1969; Bansal and Sen, 1982). Aqueous extracts of *T. amplexicaule* root, stem and leaf had no inhibitory effect on the seed germination, stimulated the seedling growth of pearlmillet (Sen *et al.,* 1969) but slightly reduced the seedling growth in sesame (Singh and Sen, 1981, 1982). Soil incorporated residues of *T. sedgwiickianum* reduced the grain yield of pearlmillet (8%) and sesame (46%) (Bansal and Sen, 1982).

XXI. *Ipomoea* spp. : Its three species viz., *I. aquatica* (L.) Forsk, *I. pestigridis* L. and *I. sindica* (L.) Stapf have shown allelopathic effects on pearlmillet (Appendix) (Singhvi and Sharma, 1984) and sesame (Singh and Sen, 1981, 1982). Aqueous extracts of *I. pestigridis* and *I. sindicus* species had less inhibitory effect on the growth of sesame seedlings (Singh and Sen, 1981,1982). Aqueous extracts of root, stem and leaf of *I. aquatica* inhibited the growth and drymatter production in pearlmillet seedlings, phenols were found in its leaves and roots but not in the stem (Singhvi and Sharma, 1984).

XXII. *Lolium* spp. : Its two species *L. multiflorum* Lam. (Italian ryegrass) and *L. temulentum* L. exhibited the allelopathic effects on rice, oat, lettuce and *Trifolium* spp. (Appendix) (Naqvi and Muller, 1975; Bansal and Singh, 1986). Leachates of *L. temulentum* live shoot, infested soil and its decomposing residues proved toxic to the seedling growth of oat, lettuce and *Trifolium* spp. (Naqvi and Muller, 1975). Soil incorporated roots, stem, leaves and inflorescence of *L. multiflorum* did not affect seed germination but severely decreased the growth of rice seedlings. Its roots proved most potent compared with other plant parts (Bansal and Singh, 1986).

XXIII. *Salvia* spp. : Its four species viz., *S. austriaca, S. leucocephalla, S. reflexa* and *S. syriaca* exert allelopathic effects through volatile terpenes and/or rain leachates (Appendix). Aqueous and ether washings of *S. reflexa* leaves decreased the length of seminal root in wheat seedlings and earlier washings were more potent than later ones (Lovett, 1980,1983a). Shoot and rhizome extracts of *S. syriaca* delayed the germination and growth of wheat seedlings, however, shoot extract proved more inhibitory than rhizome extract. Soil incorporation of its shoot residues considerably decreased the germination and growth of wheat seedlings (Qasem and Abu-Irmaileh, 1985).

XXIV. Canada thistle (*Cirsium arvense* (L.) Scop.) : It is a serious weed in 22 countries (Holm *et al.,* 1979) and causes severe reduction in wheat yields (Hodgson, 1958). Its live plants, extracts and residues inhibited the seed germination and seedling growth of wheat, barley, rye, maize, pearlmillet, sesame, soybean, alfalfa, turnip, linseed and cucumber (Appendix) (Hegelson and Konzak, 1950; Bukolova, 1971; Stachon and Zimdahl, 1980; Wilson Jr., 1981; Bhandari *et al.,* 1982; Bhatia *et al.,* 1982; Woodward and Glenn, 1983; Kovacks *et al.,* 1988). Its root and shoot extracts inhibited the germination and growth of wheat and linseed (Hegelson and Konzak, 1950). Its roots and foliage extracts completely inhibited the germination of turnip `Purple top', upto 65% in soybean 'Williams' and had no such effect on maize `Pioneer 3382' but decreased its root growth. These root and foliage extracts contained 185 ppm phenolics (Woodward and Glenn, 1983). Likewise, root and foliage aqueous extracts inhibited the germination in lucerne, coleopatile and root growth in wheat and barley, root growth in alfalfa and cucumber and dry matter production in wheat. However, their (root and foliage) ethanol extracts had no inhibitory effect on the germination of barley and cucumber but decreased the radicle growth (Kovacs *et al.*, 1988). The soil incorporated roots, foliage and litter reduced the growth of barley, cucumber, wheat, alfalfa, sugarbeet but had no effect on maize and kidneybean (Stachon and Zimdahl, 1980). The toxicity of residues in soil persisted upto 60 days and neither autoclaving nor fertilizer application decreased it. The leaf leachates of weed also reduced the growth of sugarbeet. (Wilson Jr., 1981).

XXV. Cogon grass (*Imperata cylindrica* L.) Beauv.) : It is one of the 10 worst weeds of the world and is most obnoxious in many tropical countries (Holm *et al.,* 1979). Its leaf extracts had no inhibitory effect on the germination of maize, sorghum, rice and tomato; slightly inhibited shoot growth of maize, sorghum and rice but markedly reduced the root growth of all test crops (Appendix). Its aqueous extracts stimulated the germination and seedling vigour of soybean and maize upto 5% concentration (w/v) but caused inhibition at 10% concentration (Singh *et al.,* 1989). While, soil incorporation of leaves reduced root and shoot growth of cucumber (Eussen and Soerjani, 1978). Aqueous extracts of its rhizomes and roots delayed seed germination and inhibited root growth of tomato. Leaf, rhizome and root extracts contained many phenolic compounds e.g. vanillic acid in root extract (Eussen, 1978). Aqueous extracts of leaves and culms contained scopolines, scopoletin, chlorogenic acid and isochlorogenic acid (Abdul-Wahab and Al-Naib, 1972). Eussen and Niemann (1981) identified *p*-and *o*-coumaric, gentisic, vanillic, benzoic, *p*-hydroxychloric acids, vanillin and *p*- hydroxybenzaldehyde growth inhibitors in the leaves. Aqueous extracts of its decaying rhizomes and infested soil contained ferulic, vanillic, *p*-coumaric and syringic acids (Amritlingam, 1984 cited by Lovett, 1986)

XXVI. Cocklebur (*Xanthium* spp.) : It inhibited the seed germination and seedling growth of wheat, maize, pearlmillet, chickpea, rapeseed, tobacco and lettuce (Appendix) (Ambasht and Rai. 1981; Cutler, 1985; Bushra *et al.,*1987). Its aqueous extracts decreased the seed germination of wheat (51.23%) and chickpea (39.54%) (Ambasht and Rai, 1981). Aqueous extracts of its leaves, inflorescence, fruits, stem and roots inhibited the seed germination of lettuce and radicle growth of pearlmillet, maize, rapeseed and lettuce seedlings, Rain leachates of its plants decreased the germination of pearlmillet and radicle growth of all test crops. Its root and shoot litter reduced the germination of rapeseed and radicle growth of all test crops. The shoot litter proved more toxic compared with root litter. The inhibitors present in its shoot were caffeic, p-coumaric, p-hydroxybenzoic, chlorogenic and ellagic acids (Bushra *et al.*, 1987). Its plants form carboxytractyloside, a hypoglycalmic agent which inhibited wheat colepotile growth, stunted maize seedlings and caused leaf deformity in tobacco and had fungistatic properties (Cutler, 1985).

XXVII. Thorn apple (*Datura stramonium* L.) : It is an important weed in USA, Australia, tropics and subtropics. It contains a number of alkaloids mainly scopolamine and hyoscyamine (Mothes, 1955), which are produced in the roots and translocated to the leaves and seeds (Dawson, 1948). In the early stages of growth, the alkaloids are evenly distributed throughout the plant but with age, became localized in fruits (Manske, 1950). Its leaf and seed washings reduced the germination and seedling growth of flax, owing to presence of scopolamine and hyoscyamine alkaloids, their concentration was higher in seed washings than in leaf washings (Lovett *et al.,* 1981). Under field conditions, leachings of seeds or seed capsule reduced the root elongation of sunflower, depending upon the type and amount of clay, organic matter content and soil texture. In podzolic soil the toxicity of alkaloids to seed germination persisted upto 15 weeks in the field and upto 35 weeks under controlled conditions (Tables 52, 64) (Lovett *et al.,* 1981; Harris, 1984; Levitt and Lovett, 1984; Levitt *et al.,* 1984), That is why Australian farmers complain failures of cotton, soybean and sorghum in the fields heavily infested with this weed, because total alkaloid concentration in the soil was 0.02% (Volume basis) even after seven months of its harvest (Levitt and Lovett, 1984) and this soil decreased the root growth of sunflower by 42-60% over control (Levitt *et al.,* 1984).

Table 52. Effect of *Datura stramonium* L. seeds and seed washings on radicle length of sunflower at 72 h in two soils

Treatment	Radicle length (mm)	
	Black earth	Lateritic podzolic
Sterile water	60.7	55.8
Datura stramonium L. seeds	51.0	32.0
Datura stramonium L. seed washings	42.8	22.5

(0.1% LSD = 6.80)

Source : Levitt and Lovett (1984).

XXVIII. Velvet leaf (*Abutilon theophrasti*) Medic : It is a major weed in United States which causes major yield reductions in soybean and cotton (Eaton *et al.,* 1976; Robinson, 1976; Chandler, 1977; Oliver, 1979; Hagwood *et al.,* 1980; Dekker and Meggitt, 1982). Residues or extracts from its leaves, stem and seeds inhibited the germination and seedling growth of radish, turnip, maize and soybean (Appendix, Tables 40,47,63) (Elmore, 1980; Bhowmik and Doll, 1982, 1984). Its leaf extracts depressed the seed germination of radish and inhibited seedling growth of soybean, owing to presence of phenolic compounds (Colton and Einhellig, 1980). Aqueous extracts of its residues, inhibited the radicle and coleoptile growth of maize and hypocotyl growth of soybean. The decaying residues proved more inhibitory to shoot growth and nitrogen and phosphorus uptake of test crops and the order of inhibition in three soils was sand > sand plus silt loam mixture (50:50) > silt loam (Tables 40, 47) (Bhowmik and Doll, 1979, 1982, 1984). Under field conditions weed plants decreased the flowering nodes, dry matter and seed yield in soybean (Dekker and Meggitt, 1982).

XXIX. *Celosea argentea* L. : It is most troublesome weed upto 1000 m altitude in India (Cooke, 1980) and severely reduces the yields of pearlmillet, sorghum, maize, blackgram and kidneybean (Pandya and Dash, 1982; Vyas, 1982; Pandya and Vyas, 1984; Pandya and Sidha, 1987). Aqueous extract of leaves, stem and roots inhibited the seed germination and radicle growth of pearlmillet (Pandya, 1975, 1976) and sesame, however, maximum inhibition occurred in sesame (Appendix). The order of inhibition was leaf > stem > root. Its burnt weed biomass, reduced the growth of pearlmillet but completely inhibited the seed germination of sesame (Ashraf *et al.,*1977; Ashraf and Sen, 1978). Its root, shoot and seed extracts inhibited the root and shoot growth of pearlmillet (Rao *et al.,* 1979) and wheat seedlings (Bhatia *et al.,* 1982). Its aqueous leaf extracts decreased germination, seedling growth and the amount of chlorophyll a,b and carotenoids in sorghum `CSH-1' (Mohnot and Soni, 1977; Soni and Mohnot, 1978) and proved inhibitory to rhizobium growth compared with root extracts (Bhaskar and Vyas, 1987).

XXX. Crab grass (*Digitaria* spp.) : It is a major weed in 40 countries (Holm *et al.,* 1979) and inhibited the germination and seedling growth of wheat, sorghum, maize, sunflower, kidneybean, tomato and pepper (Appendix, Table 39) (Schreiber and Williams, 1967; Bansal *et al.,* 1981; Mohmed-Saleem and Fawusi, 1983; Bansal and Kalia, 1984) and yields of cotton and groundnut (Hill and Santelmann, 1969; Robinson 1976). Its root and shoot extracts inhibited the growth of certain nitrogen fixing and nitrifying bacteria (Rice, 1964). Aqueous extracts of whole plant decreased the germination and caused tip burning of leaves in wheat (Bansal *et al.,* 1981). Aqueous extracts of seeds and inflorescence of *Digitaria sanguinalis* decreased the ^{32}P uptake in kidneybean and the uptake was inversely related to the extract concentration (Bansal and Kalia, 1984). Its decomposing roots reduced the root growth of maize plants (Schreiber and Williams, 1967). The decaying residues inhibited the germination and seedling growth of sorghum, tomato and pepper and the leachates of this soil retarded the growth of sorghum seedlings (Fig. 13) (Mohmed- Saleem and Fawusi, 1983).

XXXI. *Argemone mexicana* L. : Aqueous leachates of its leaves inhibited the germination of wheat, mustard and fenugreek and germination and seedling growth of sorghum, fingermillet, tomato and cucumber (Appendix). The leachates from flowers were more potent than from other plant parts. Soil incorporation of its dry leaves inhibited the growth of wheat seedlings for 10 days but promoted the seedling growth after 20 days. The concentration of inhibitors (cinnamic and benzoic acids) was 50- 60 mg/100 g plant drymatter and 0.2-0.3 mg/100 g dry soil in surface soil (upto 1 cm depth) under a natural stand (Leela, 1981). Its dried powdered plants added to water inhibited the seed germination of wheat, pearlmillet, rapeseed and radish; the inhibition was maximum (40-62%) in radish and least in wheat and rapeseed. The root and shoot growth was decreased in all test crops except in radish, whereas, stimulation was observed in the root growth (Sharma and Nathawat, 1987).

XXXII. *Lantana camara* L. : It is one of the 10 worst weeds of the world and is serious weed in 14 crops in 47 countries (Holm *et al.*, 1979). However, its allelopathic effects on crops have been studied least (Mersie and Singh, 1987 a,b). Its dried shoot residues mixed in soil inhibited the root and shoot elongation and dry matter accumulation in maize, soybean and wheat, however, inhibition was minimum in wheat (Appendix). Its leaf extracts contained 14 phenolic compounds viz; protocatechuic acid (1.5 mM), gentisic acid (2.9 mM), *p*-hydroxybenzoic acid (5.6 mM), vanillic acid (0.9 mM), caffeic acid (1.4 mM), syringic acid (11.3 mM), *p*-coumaric acid (26.4 mM). *m*-coumaric acid (3.3 mM), ferulic acid (7.0 mM), salicylic acid (2.0 mM), *o*-coumaric acid (1.7 mM), *t*-cinnamic acid (2.0 mM), vanillin (2.1mM), and methyl coumarin (2.8 mM). The phytotoxicity of its leaf extracts was due to complex interactions between these phenolic compounds (Jain *et al.,* 1989). Its plant extract delayed the seed germination and seriously inhibited the root, shoot and leaf development in wheat and chickpea (Table 43) (Angiras *et al.,* 1988; Singh *et al.,* 1988).

XXXIII. *Peganum harmala* L. : It is a perennial weed in arid areas of India and Pakistan. Aqueous extracts of leaves, stem, root and seeds inhibited the germination and seedling growth of pearlmillet at higher concentrations but lower concentrations stimulated the seedling growth (Appendix). The plants contained alkaloids and seedlings extracts were more potent than mature plants (Sharma *et al.,* 1982). Its aqueous extracts inhibited the germination and seedling growth of rapeseed. The root, stem, leaves and seeds contained the inhibitor `harmoline' which is insoluble in water but soluble in organic solvents (Yang *et al.,* 1987). Aqueous extracts from its shoots and infested soil reduced the germination and radicle growth of pearlmillet, rapeseed, lettuce and persian clover. The shoot extract contained caffeic, ferulic, p-coumaric and p-OH-benzoic acids (Hussain *et al.,* 1988).

XXXIV. *Kochia scoparia* (L.) Schrad : It is a major weed of sugarbeet in USA, Canada, Argentina (Holm *et al.,* 1979). Its plants contained caffeic, chlorogenic and ferulic acids, myricetin and quercetin which inhibited the radicle growth of radish seedlings (Appendix, Table 53) (Lodhi, 1979 a). Its 1.0 g fresh material in 60 and 240 ml nutrient solution reduced the growth of sorghum `Hybrid 701' and soybean `Corsoy', respectively (Iverson and Wali, 1982). In pot studies, 0.5 g debris per 80 g soil reduced the growth of sorghum seedlings, while 2.0g debris per 80g soil increased the leaf resistance and decreased the water potential of sorghum seedlings (Table 53) (Einhellig and Schon, 1982).

Table 53. Effects of dried *Kochia scoparia* (L.) Schreb. debris in soil on the growth of grain sorghum and soybean

Test plant	Trial	Rate of debris (g) per 80 g of soil			
		Control	0.5g	1.0g	2.0g
Dry weight (mg/plant)					
Sorghum	1 Shoot	96.6 ±2.3 a	86.9 ±2.7 ab	80.7±3.7b	78.2 ±4.6b
	Root	75.1 ±2.4 a	75.2 ±4.4a	49.2 ±3.3b	29.6± 2.7c
	Plant	171.7 ±3.5 a	162.1 ±5.8a	129.9 ±6.7b	107.9 ±7.1c
	2 Shoot	82.8 ±3.1 a	74.4 ±3.1b	76.6 ±2.9ab	66.0± 2.1c
	Root	58.6 ±2.2 a	49.9 ±2.0b	46.0 ±2.1ab	32.5 ±1.3c
	Plant	141.4 ±4.9a	124.3 ±4.8b	122.6 ±4.6b	98.5 ±3.1c
Soybean	1 Shoot	233.5 ±6.5b	270.3 ±8.3a	245.8 ±9.4ab	243.5 ±14.lab
	Root	59.7 ±2.5ab	66.4 ±3.0a	52.7 ±3.1 ab	54.2 ± 4.0b
	Plant	293.2 ±8.4b	336.7 ±10.8 a	298.4 ±12.2b	297.8 ±17.4 b
	2 Shoot	164.5 ±3.9b	182.5 ± 4.1 a	168.9 ±6.2 ab	143.6 ±5.6c
	Root	23.6 ±1.4 b	29.5 ±1.6a	25.4 ±2.3 ab	8.9 ±1.2c
	Plant	188.0 ±5.0 b	211.9 ±5.2 a	194.2 ±8.0 ab	152.5 ±6.5 c

All values in a row followed by the same letter are not significantly different (P=0.05), Duncan's Multiple Range Test.

Source : Einhelling and Schon (1982).

XXXV. *Ageratum conyzoides* L. : Aqueous extracts of seed, leaf, root and stem reduced the germination of wheat and the order of inhibition was leaf > root > stem (Appendix, Tables 39, 43), (Sugha, 1978, 1979, 1980). Its plant extracts inhibited the seed germination and root growth of wheat 'VL 421' and 'S-308' seedlings (Bansal *et al.*, 1981; Singh *et al.*, 1988). Its aqueous plant extracts delayed the germination and decreased the root and shoot elongation and number of leaves in chickpea (Table 43) (Angiras *et al.*, 1988). While its seeds and inflorescence extracts decreased the ^{32}P uptake in kidneybean (Table 39) (Bansal and Kalia, 1984). However, Singh *et al.*, (1989) observed that its plant extracts promoted the seedling vigour in root, shoot, length and fresh weight in soybean upto 5% concentration (w/v) and upto 10% concentrations in maize. Its extracts proved more stimulatory compared with extracts of cogon grass and *Commelina benghalensis* L.

XXXVI. Bermuda grass (*Cynodon dactylon* (L.) Pers) : Ethanol extracts of infested soil and infested soil free from its residues reduced seed germination, root and shoot growth of barley (Appendix, Table 47) (Horowitz and Friedman, 1971). Aqueous extracts of its rhizomes decreased the seed germination and root growth of pea and cucumber (Sapankevich *et al;* 1965). Its residues decreased the plant height and fresh weight of root and shoot in seedlings of corn and soybean (Bhowmik and Doll, 1982). Its root extracts inhibited the germination, root and shoot growth of tomato seedlings and germination and shoot growth of rice seedlings (Castro *et al.,* 1983, 1984).

XXXVII. *Phalaris minor* Retz. : It is a major weed of wheat crop in Punjab and Haryana states of India. Aqueous extracts of straw and glumes inhibited the growth of wheat seedlings but extract from its fresh plants stimulated the seedling growth (Appendix, Table 51) (Bhatia *et al.,* 1982). In pot study, application of its plants extract to soil had no inhibitory effect on seed germination in greengram but reduced the nodulation (Bhandari *et al.,* 1982). Likewise, soil incorporation of its root, stem, leaves and inflorecence, inhibited the seedling growth of rice 'Himdhan', however, seed germination was not affected (Bansal and Singh, 1986).

XXXVIII. *Croton bonplandinum* Baill : It is a widely distributed perennial weed in West Bengal, India and its leaves are repositories of germination and growth inhibitors (Datta and Sinha-Roy, 1973). Its leaf extract had no effect on germination of pea but caused severe inhibition in rice, mustard and lettuce (Appendix). Its leaf leachates decreased the root and shoot growth of rice, mustard and lettuce seedlings. The toxicity of leaf extracts is lessened near the plant maturity (Datta and Sinha-Roy, 1983)

XXXIX. Wild perennial buckwheat (*Fagopyrum cymosum* Meisen): Its aqueous extracts decreased the plant height, length of second leaf sheath and roots in rice seedlings (Appendix). Its plants reduced the growth of neighbouring rice and rapeseed plants compared with control (Tsuzuki *et al.,* 1977) The plants contained three phenolic acids viz., ferulic, caffeic and chlorogenic. Ferulic and caffeic acids inhibited the growth of rice seedlings at 10 ppm concentration and the concentration of caffeic acid like compounds in its roots exceeded 10 ppm. The root contained more ferulic and caffeic acids than shoots (Tsuzuki and Yamamoto, 1987).

XXXX. Peru (*Schinus molle* L.) : Aqueous extracts of its leaves, fruits and roots strongly inhibited the seed germination and seedling growth of wheat and cucumber (Appendix). The leaf and fruit extracts proved more toxic than root extract (Anaya and Gomez-Pompa, 1971). Its fruits extract decreased the germination, dry weight and seedling growth of pearlmillet at higher concentrations, but caused stimulation at lower ones (Chatterjee, 1975).

XXXXI. Spurrey weed (*Spergula arvensis* L.) : Its aqueous extracts were slightly inhibitory to growth of maize seedlings but proved most toxic to oat, alfalfa, lettuce and radish (Appendix, Tables

51,61) (Gliessman, 1983). Its extracts decreased the elongation of the second leaf sheath in rice plants, owing to presence of caffeic and ferulic acids. Its soil incorporated leaves and stem markedly inhibited the seedling emergence and growth of tomato, turnip and radish (Table 54) (Tsuzuki and Araki, 1984)

Table 54. Effect of *Spergula arvensis* L. residues buried in soil and their extracts on seedling emergence of crops

Treatment	Per cent emergence		
	Radish	Turnip	Tomato
Control	70	92	82
Residues buried	57	12	22
Residues extract	84	24	44

Source : Tsuzuki and Araki (1984).

XXXXII. *Tephrosia purpurea* (L.) Pers : It is a legume weed in arid zone of Rajasthan and Haryana, India. Its leaf extracts inhibited the germination and seedlings growth of greengram (Mohnot and Chowdhury, 1981), while plant extracts severely decreased the seedling growth of sesame (Singh and Sen, 1981). Aqueous extracts of its leaf, stem and root reduced the seed germination of sorghum (Chatterjee, 1975; Soni and Mohnot, 1981) and extracts of dry leaves strongly inhibited the seed germination and seedling growth of mothbean (Sethi and Mohnot, 1988).

XXXXIII. *Indigofera cordifolia* Heyne ex. Roth : Aqueous extracts of its root, stem and leaves had no harmful effect on germination and stimulated the seedling growth of pearlmillet (Sen *et al.,* 1969), but extracts of whole plants slightly inhibited the seedling growth of sesame (Appendix). (Singh and Sen 1981, 1982).

XXXXIV. *Fall Panicum (Panicum dichotomiflorum* Michx.) : Aqueous extracts of dried residues inhibited the hypocotyl elongation in soybean and radicle elongation in maize (Appendix, Tables 39, 47) (Bhowmik, 1982; Bhowmik and Doll, 1982). The aqueous extracts of its seeds and inflorescences decreased the ^{32}P uptake in kidneybean seedlings (Table 39) (Bansal and Kalia, 1984).

XXXXV. Russian thistle (*Salsola kali* L.) : It is common weed in nine countries (Holm *et al.,* 1979) and its leaves contained caffeic, ferulic, chlorogenic, isochlorogenic and neochlorogenic acids and quercetin, each of these compounds reduced the germination and root growth in radish (Lodhi, 1979b).

XXXXVI. *Galium mollungo* L. : Aqueous extracts of its shoots inhibited the germination of wheat and radish , seedling growth of sunflower, development of onion bulbs and caused necrosis in meristematic tissues of onion (Appendix) (Kohlmuenzer, 1965). In pots, its plants reduced the yield of flax by 18% (Grummer and Beyer, 1960).

Ageratum, Imperata, Commelina : Singh *et al.* (1989) studied allelopathic effects of aqueous extracts of cogon grass, *Ageratum conyzoides* L. and *Commelina benghalensis* L. on germination and vigour of soybean and maize. The extracts of cogon grass (*Imperata cylindrica* (L.) Beauv.) and *C. benghalensis* at 1% and 5% concentrations (w/v) stimulated the germination of soybean, while, extracts of all weeds caused inhibition in both test crops at 10% concentration. *A. conyzoides* extracts promoted the seedling vigour (length of roots, shoot and fresh weight) in soybean upto 5% concentration and at all concentrations in maize. Cogon grass extracts were stimulatory to the seedling vigour of both crops upto 5% concentration but the magnitude was less than *A. conyzoides*, on the contrary *C. benghalensis* extracts proved inhibitory to both crops at all concentrations.

3.2.1.1.2 Minor Weeds

These are not major weeds in large regions but may cause substantial losses to crops in certain areas. The extracts of *Antennaria microphylla* reduced the root growth of lettuce owing to presence of phenolic compounds such as hydroquinone, arbutin and caffeic acid (Appendix) (Manners and Galitz, 1986). The *Artemissia annua* plants contained artemisinin– a sesquiterpenoid lactone which at concentration of 33 mm inhibited the seed germination and seedling growth of lettuce but had no adverse effect on grain sorghum (Duke *et al.,* 1987). Leachates of *Calendula arvensis* L. reduced the germination and growth of wheat `Blue Silver' and pearlmillet `DP 2' and its capitula were found most inhibitory (Chagtai *et al.,* 1987). Aqueous extracts of *Commelina benghalensis* L. stimulated the germination of maize and soybean upto 5% concentration (w/v) and caused inhibition at 10% concentration. However, the extracts proved inhibitory to seedling growth at all concentrations (Singh *et al.,* 1989). Aqueous extracts of squash (*Cucurbita pepo* L.) plants slightly inhibited the germination of cowpea and cabbage and root growth of maize and cowpea, but severely reduced the root growth of cabbage (Table 55) (Gliessman, 1983). *Echinops echinatus* Roxb. is a principal weed in arid zones of India. Aqueous extracts of its leaf, stem and root inhibited the germination and seedling growth of pearlmillet and sesame. In field studies, its buried or burnt biomass reduced the growth and yield of sesame, but the burnt biomass increased the yield of pearlmillet (16.8%) (Jha and Sen, 1981). Aqueous extracts from

Erica australis L. and *E. arborea* L. had no inhibitory effect on the germination but inhibited the root and hypocotyl growth of red clover. Soil incorporated leaf and flower residues also inhibited the growth of red clover (Vietez and Vietez, 1979). Aqueous extracts from different plant parts of *Fumaria indica* Pugsly reduced the germination and seedling growth of wheat `Blue Silver' (Chagtai *et al.,* 1986). *Gomphrena decumbens* Jacq. is a common weed in sorghum and sesame crops in south India. Aqueous extracts of its leaves, stem and root inhibited the seedling growth of sorghum and sesame. Root extract was more potent than leaf and stem extract owing to higher Rf value of inhibitory compounds in root (0.8 - 1.0) than in stem (0.1 - 0.3) (Solomon and Bhandari, 1981). Aqueous extracts of lawn pennywort (*Hydrocotyle sibthorpioides* Lam.) had no adverse effect on the germination of rice but decreased that of rapeseed. The extract decreased the plant height, root length and leaf sheath length in rice `Toyotoma' seedlings. Soil incorporated residues inhibited the germination of rapeseed and tomato but had no effect on radish (Tsuzuki *et al.,* 1978). Rice grass (*Leersia hexandra* SW) and *Oryza perrenis* Moench are major weeds of rice, they inhibited the root growth of rice and lettuce seedlings. Rice grass showed greater phytotoxicity to test crops compared with *Oryza* because of higher concentration of phytotoxins. Application of their powdered plant material in soil decreased the seed germination of rice (Chou *et al.,* 1984). Aqueous extracts from leaves, stem and roots of *Lepidium sativa* inhibited the root and stem elongation and drymatter accumulation in wheat seedlings compared with control (Bhatia *et al.,* 1982). *Media glomerata* Hook is annual weed in USA and Canada and its leachates reduced the germination and growth of radish seedlings (Carnahan and Hull Jr., 1962).

Table 55. Allelopathic potential of water extracts of air dried leaves of *Cucurbita pepo* L. on germination and radicle growth of crops

Crops	Germination	Radicle length
	% of control	
Corn	100	85.4 a
Beans (*Vigna sinensis* L.)	96	79.0 a
Cabbage	92	43.0 b
Control	100	100.0

a = T test significant at LSD 0.05; b = Significant at LSD 0.01

Source : Gliessman (1983).

Aqueous extracts of dallis grass (*Paspalum dilatatum* Poir.) severely decreased the seed germination and seedling growth of wheat varieties `VL 421' and `S 308' (Bansal *et al.*, 1981) and 32P uptake in kidneybean seedlings (Table 39) (Bansal and Kalia, 1984). Aqueous extracts of *Plantago lanceolata* L. seed, leaf, root and stem inhibited the seed germination of wheat and leaf extract proved most potent (LeTourneau *et al.,* 1956; Sugha, 1979, 1980). Aqueous extracts from different plant parts of itch grass [*Rottboellia exaltata* (L.) L.f.] considerably delayed the seed germination of rice and greengram (Mercado and Manuel, 1978). Curly dock (*Rumex crispus* L.) is a weed of field margins and waste lands in USA. Aqueous extracts of its fresh leaves slowed the growth of sorghum and maize seedlings. Three phytotoxins present in the plants inhibited the seed germination of sorghum and radish. In field, its plants suppressed the growth and biomass of neighbouring plants (Einhellig and Rasmussen, 1973). Thatch grass (*Saccharum spontaneum* L.) is a perennial rhizomatous weed of crops in India. Aqueous leachates of roots and rhizomes had no effect on the seed germination in wheat but inhibited the elongation and drymatter production in seedlings of wheat `Kalyansona', HD 1954', `Narmada 4'. Rhizome leachates were more potent than root leachates (Amritphale and Mall, 1978). Leaf extracts of *Scoparia dulcis* suppressed the seed germination and seedling growth of wheat, rice, maize, pigeonpea, greengram, blackgram and fababean. At lower concentrations (>1%) seed germination was decreased only in maize and fababean and of all crops at higher concentrations (Trivedi and Trivedi, 1987). Aqueous extracts from different parts of *Silybum marianum* (L.) Gaertn. inhibited the seed germination and radicle growth of wheat `Blue silver' and the inhibitory order of extracts was stem < leaves < inflorescence < roots. The extracts of fruits and fruit parts were not toxic. In field, decomposing litter reduced the plant height, drymatter, earhead length and number of grains per earhead in wheat `Blue Silver' (Chagtai *et al.*, 1988). Aqueous extracts and litterbed of shoots and roots of *Stachaya parviflora* arrested the germination and seedling growth of *Sorghum sudanense* L. `Dale', chickpea, lentil, kidneybean, greengram, rapeseed, Chinese cabbage and Egyptian clover. The roots were found more potent than shoot (Farrukh *et al.*, 1986). Water soluble toxins of *Sonchus arvensis* L. inhibited the root growth, reduced mitotic activity of meristematic cells and damaged the epidermis cortex and vascular system in roots of wheat and rye (Bukolova, 1971). The aqueous extracts from its leaves, stem and root inhibited the root and epicotyl elongation and dry matter in wheat seedlings than control (Bhatia *et al.*, 1982). Aqueous extracts of *Tagetes minuta* L. infested soil delayed the seed germination of maize, sunflower, onion, carrot, cucumber, lettuce, squash and tomato. The extract from recently infested soil caused greater reduction in the plant height and dry matter of test species compared with infestation in previous years owing to presence of ∝-tertheine in its roots. (Ruth *et al.*, 1986).

3.2.1.1.3 Stimulatory weeds

Corn cockle (*Agrostemma githago* L.): When sown in mixture with wheat, it increased the grain yields of crop compared with wheat monoculture (Table 56) (Gajic, 1966). Aqueous extracts of *Lepidium virginicum* L. stimulated the seedling growth of maize (Kahl, 1987). Aqueous extracts of *Ludwigia adscendens* (L.) Hara. root, stem and leaf stimulated the seedling growth of pearlmillet. Phenols were found present in stem and leaves and not in roots (Singhvi and Sharma, 1984). Aqueous extracts from leaves and roots of *Rumex dentatus* L. and leaves of *Silene canoidea* L. increased the seedling growth of wheat over the control, while the stem and root extracts of *S. canoidea* proved inhibitory (Bhatia *et al.,* 1982). Aqueous extracts or leachates from *Anagallis arevensis* L. shoot and root had no harmful effect on the germination, stimulated the shoot growth but decreased the root growth of chickpea seedlings (Goel, 1987). In field, *Stizolobium deeringianum* Bort. plants increased dry matter of maize crop. Its aqueous extract had no inhibitory effect on germination of maize and cowpea but severely inhibited that of cabbage (Table 57) (Gliessman, 1983).

3.2.1.2 Root exudates

In crop fields, weeds suppress the growth of adjacent crop plants through excretion of inhibitory compounds in their root exudates. These compounds reduce the seed germination, root and shoot growth, nutrient uptake and nodulation (legumes). However, root exudates of bermuda grass *Cynodon dactylon* (L.) Pers.) and corn cockle (*Agrostemma githago* L.) stimulated the growth and yield of crops. In some weeds, the toxicity of exudates is high in the younger stage and became less with maturity, while reverse was true for others. The root exudates of johnson grass, quack grass, redroot pigweed, wild oat, *Cyperus* spp. *Chenopodium* spp., *Bidens pilosa, Celosia argentea* and *Polygonum* spp. caused severe reduction in the seed germination and growth of several crops (Table 59).

Table 56. Stimulatory effects of weed residues/extracts on the seed germination, growth and yield of crops

Donor weed			Recipient Crop		Reference
Weed species	Weed residue extract	Phytotoxin	Crops affected	Stimulatory effect	
1	2	3	4	5	6
Soil extract					
Bidens pilosa L.	Soil extract	ND	Maize	Stimulated growth	Kahl (1987)
Weed residues					
Cyperus rotundus L.	Aerial parts	ND	Maize, sorghum, soybean	Slightly stimulated growth	Lucena and Doll (1975), Chivinge (1985)
Parthenium hysterophorus L.	Weed residues	ND	Pearlmillet	Stimulated growth	Kanchan & Jayachandra (1979 a,b)
Setaria faberii Herm.	Weed residues	ND	Soybean	Increased soybean yields	Bhowmik & Doll (1982)
S. faberii Herm. *Ambrosia artemisifolia* L.	Weed residues	ND	Maize	Increased potassium uptake	Bhowmik & Doll (1984)
Plant extracts					
Agropyron repens (L.) Beauv.	Plant extract	S-hydroxyindole-3-acetic acid, 5-hydroxytryptophan	Wheat, rye, corn, kidneybean	Stimulated growth at low concentration	Lastuvka (1955), Hagin (1989)
Ageratum conyzoides L.	Plant extract	ND	Corn, soybean	Stimulated germination and growth	Singh *et al.* (1989)
Alhagi psuedoalhagi (Bieb) Desv.	Leaf, stem, seed extract	ND	Pearlmillet	Stimulated seed germination and growth at low concentrations	Singhvi and Sharma (1985)

Contd.....

1	2	3	4	5	6
Argemone mexicana L.	Plant extract	ND	Radish	Stimulated root growth at low concentrations	Sharma and Nathawat (1987)
Camelina sativa (L.) Crantz.	Leaf washings	ND	Wheat, linseed	Stimulated radicle growth	Lovett and Jackson (1980)
Euphorbia genticulata Ort.	seed, root, extract	ND	Wheat	Stimulated seedling growth	Sugha (1980)
Imperata cylindrica (L.) Beauv.	Leaf, root, rhizome extract	ND	Tomato	Stimulated shoot growth at low concentrations	Eussen (1978)
Indigofera cordifolia Heyne ex. Roth, Trichodesma amplexicaule Roth.	Leaf, stem, root extract	ND	Pearlmillet	Stimulated seedling growth	Sen *et al.* (1969)
Ludwigia adscendens (L.) Hara	Leaf, stem	ND	Pearlmillet	Stimulated seedling growth	Singhvi and Sharma (1984)
Solanum surratense Burm.	Fruit pulp extract	ND	Pearlmillet	Stimulated seedling growth at low concentration	Sharma and Sen (1971)
Live Plants					
Agrostemma githago L.	Live plants	Agrostemin, allantoin, gibberalin	Wheat	Stimulated growth and yield	Gajic (1966), Gajic *et al.* (1976)
Indigofera cordifolia Heyne ex. Roth.	Live plants	ND	Pearlmillet, sesame	Stimulated growth and yield	Bhandari and Sen (1979)

ND = Not determined.

Table 57. Effect of water extract of leaves of *Canavalia ensiformis* and *Stizolobium deerengianum* Bort. on the germination and radicle elongation of crops

Water extract	Corn		Beans		Cabbage	
	Germin-ation (%)	Radicle length+	Germi-nation (%)	Radicle length+	Germi-nation (%)	Radicle length+
Canavalia ensiformis	98.9	77.6**	100.0	66.3**	49.3**	35.6**
Stizolobium deerengianum Bort.	100.0	87.9*	98.7	92.9	46.7**	27.2**
Control	100.0	100.0	100.0	100.0	100.0	100.0

+ % of control, T test significant at 5 %, ** T test significant at 1%
Source : Gliessman (1983).

Table 58. Effect of leachates from quartz sand previously occupied by growing *Agropyron repens* (L.) Beauv. and oats on the stand of wheat `Selkirk' 21 days aftetr sowing

Leachates	pH	% Wheat stand
Control	6.5	92
Oats `Bonda'	6.5	90
Agropyron repens (L.) Beauv.	6.9	26

Source : Kommedahl *et al.* (1959).

I. *Quack grass (Agropyron repens* (L.) Beauv.): In mixture with wheat and rye, it reduced their growth and grain yield. In early planting of test crops, it killed the roots of germinating seedlings (Table 59) (Lastuvka, 1955). In the beginning, the root tips of wheat were damaged, thus limiting the formation of root hairs and root growth. The roots close to the quack grass were damaged most, owing to secretion of root exudates, which killed the secondary roots. Besides, root exudates also stimulated the growth of fungus *Stemphylium ilicis,* which attacked secondary roots and stem of wheat (Lastuvka, 1960). Its root exudates reduced the germination and seedling growth of wheat, compared with root exudates of oat and leachates of control pots (Table 58) (Kommedahl *et al.*, 1959). The root exudates also inhibited the root growth and drymatter of barley seedlings (Lovett and Jokinen, 1984). In mixture of quack grass and kidney bean, the quack grass reduced the nodule number, nodule fresh weight and nitrogen fixation compared with monoculture.

Aqueous extracts of quack grass reduced the root growth of kidneybean and thus reduced the nodulation and nitrogen fixation (Weston, 1986).

II. *Cyperus* spp : The root exudates of its two species *C. esculentus* L. and *C. rotundus* L. proved allelopathic to maize, sorghum, pearlmillet and soybean (Table 59) (McLaughlin, 1977; Drost *et al.*, 1980; Lall and Savongdy, 1981). When maize and *C. esculentus* were grown in connected pots and nutrient solution was circulated from pot to pot, its root exudates inhibited the maize growth compared with maize grown in separate pots (McLaughlin, 1977). However, Drost *et al.* (1980) did not observe such inhibitory effects of its root exudates on the growth of maize, sorghum and soybean. The root exudates of *C. rotundus* had no adverse effect on the seed germination but reduced the plant height and drymatter of pearlmillet (Lall and Savongdy, 1981).

III. *Chenopodium* spp : Root exudates of *C. album* L. and *C. murale* L. showed allelopathic effects in maize and wheat, respectively (Table 59) (Caussanel and Kunesch, 1979; Porwal and Gupta, 1987). The exudates of *C. album* during flowering stage inhibited the root elongation of maize seedlings. Exudates collected from field or pot grown plants were found more potent than that grown in nutrient culture, because the former contained two inhibitory fractions and later had only one. In both soil or nutrient solution exudates, the inhibitor was oxalic acid in concentrations of 2.5 to 5 x 10^{-3}M (Caussanel and Kunesch, 1979). The root exudates of *C. murale* reduced the shoot and earhead length and dry matter of wheat (Porwal and Gupta, 1987).

IV. Johnson grass (*Sorghum halepense* (L.) Pers) : Its root exudates, reduced the plant drymatter and 32P uptake of maize seedlings in nutrient culture (Table 59) (Beltrano and Montaldi, 1982). The exudates did not affect the root growth of radish and tomato but inhibited their germination and bacterial growth (Nicollier *et al.,* 1985). The exudates had no inhibitory effect on the germination, seedling growth and nodulation of soybean (Pope *et al.,* 1985).

V. Redroot pigweed (*Amaranthus retroflexus* L.) : The plants reduced, the plant height, leaf number, fresh weight and yields of tobacco, perhaps owing to its root exudates (Table 59) (Lolas, 1981). The root exudates severely reduced the seed germination but had no harmful effect on the drymatter production and nodulation in soybean (Pope *et al.,* 1985).

Table 59. Inhibitory/stimulatory effects of root exudates of weeds on the seed germination, growth and yield of crops

Donor Weed		Recipient crop			Reference
Weed species	Phytotoxin	Test crop	Inhibitory effect	Stimulatory effect	
1	2	3	4	5	6
Agropyron repens (L.) Beauv.	ND	Wheat, rye	Damaged and killed roots	Stimulate seedling growth at lower concentrations	Lastuvka (1955, 1960)
	ND	Wheat	Inhibited seed germination and seedling growth	—	Kommedahl *et al.* (1959)
	ND	Barley	Inhibited root growth of seedlings	—	Lovett and Jokinen (1984)
Amaranthus retroflexus L.	ND	Tobacco	Decreased growth	—	Lolas (1986)
Avena fatua L.	Scopoletin, vanillic acid	Wheat	Decreased growth	—	Schumacher *et al.* (1983)
Bidens pilosa L.	ND	Sorghum, maize, kidneybean, lettuce	Inhibited seedling growth	—	Stevens Jr. and Tang (1985, 1987)
Bothriochloa Pertusa (L.) Camus	*ND*	Pearlmillet, maize, giant foxtail, lett- uce, tomato, pepper	Inhibited seedling growth	—	Hussain *et al.* (1987)
Celosia argentea L.	Phenolics	Pearlmillet, blackgram	Root exudates of young weeds inhibited growth, nodulation, legheamoglobin and N fixation	Root exudates of old plants promoted root growth of pearlmillet	Pandya (1977), Vyas *et al.* (1983)

Contd.....

1	2	3	4	5	6
Chenopodium album L.	Oxalic acid	Maize	Inhibited root elongation	—	Caussanel and Kunesch (1979)
Cynodon dactylon (L.) Pers.	ND	Greengram	—	Increased growth and seed yield	Peswani (1981)
Cyperus rotundus L.	ND	Pearlmillet	Reduced growth	—	Lall and Savongdy (1981)
C. esculentus L.	ND	Sorghum, maize	Reduced maize growth	—	Laughlin (1977)
Polygonum aviculare L.	ND	Sorghum	Reduced root growth	—	Alsaadawi and Rice (1982a)
Parthenium hysterophorus L.	ND	Wheat, kidneybean	Decreased growth and nodulation in kidneybean	—	Kanchan and Jayachandra (1979 a,b)
Sorghum halepense (L.) Pers.	ND	Maize	Decreased growth and phosphorus uptake	—	Beltrano and Montaldi (1982)
	ND	Wheat	Inhibited growth	—	llison (1959)
	ND	Sugarbeet, okra, carrot, radish, tomato	Delayed germination of okra, tomato, radish	—	Nicollier *et al.* (1983)
Spergula arvensis L.	ND	Barley, lupine, radish	Inhibited growth	—	Tsuzuki *et al.* (1984)
Avena fatua L., *Phalaris minor Retz. Chenopodium album*, *C. murale* L.	ND	Wheat	Inhibited growth	—	Porwal and Gupta (1987)
Cassia obtusifolia L., *Cynodon dactylon Sesbania exaltata* (Ref.) Ryds., *Xanthium pennsylvanicum* Well.	ND	Soybean	*Cassia* inhibited germination, *Xanthium* and *Sesbania* reduced growth and nodulation	*Cynodon* stimulated germination	Pope *et al.* (1985)

ND = Not determined

VI. Wild oat (*Avena* spp.) : Its root exudates at 2-4 leaf stage reduced the growth of shoot, earlength, leaf and roots in wheat owing to presence of scopoletin and vanillic acid (Table 59) (Schumacher *et al.*, 1982, 1983; Porwal and Gupta, 1987).

Table 60. Effects of hydrophobic root exudates of *Bidens pilosa* L. on seedling growth of crops

Species	Treatment	Leaf area or plant height at days after seeding			Dry weight (mg) at 15 days after seeding
		5	9	14	
Leaf area (mm²)					
Lettuce	Control	76.3**	± 270.2**	1909.8**	57.5
	Eluate treatment	35.4	80.1	244.9	8.8
Bean (*Phaseolus vulgaris* L.)	Control	23.8**	112.0**	166.5**	1168.8
	Eluate treatment	16.0	59.3	84.1	737.5
Plant hight (cm)					
Maize	Control	6.2NS	18.6*	31.2**	536.3
	Eluate treatment	5.9	15.7	25.1	215.0
Sorghum	Control	6.4 NS	19.0*	33.2*	338.8
	Eluate treatment	5.1	9.7	15.0	100.0

± Differences between the treatment means (plant height or leaf area) compared statistically for each species at each date by student's 't' test. NS = Not significant. *, ** = Significant at the 0.05 and 0.01 levels, respectively.
Source : Stevens Jr. and Tang (1987).

VII. *Bidens pilosa* L. : Its root exudates reduced the leaf area, plant height and drymatter of all test crops (maize, sorghum, kidneybean and lettuce) and the dicot species were more sensitive to exudates than monocots (Table 59). Root exudates were most inhibitory to lettuce and least to maize seedlings. Its larger and older plants caused greater inhibition than smaller and younger ones (Table 60) (Stevens Jr. and Tang, 1985, 1987).

VIII. *Celosia argentea* L. : Its root exudates inhibited the root growth of pearlmillet seedlings and the toxicity of exudates decreased with age of the weed and became less in 50 days old plants. The root exudates also reduced the growth of *Rhizobium meliloti* and nodulation in some legumes (Table 59) (Pandya, 1977; Pandya and Pota, 1978; Vyas 1982; Pandya *et al.*, 1984). In mixture of *C. argentea* with pigeonpea 'T-15-1', kidneybean '15-7' and greengram 'Gujarat 2' it decreased the nodulation in pigeonpea and kidneybean, while, stimulation was observed in greengram (Pandya *et al.*, 1984).

IX. *Polygonum spp* : Root exudates of *P. pennsylvanicum* L. had no inhibitory effect on soybean in green house studies using a recirculating nutrient solution (Table 59) (Gressel and Holm, 1964). While the root exudates of *P. aviculare* L. retarded the root growth of sorghum and had no effect on cotton plants (Al Saadawi and Rice, 1982 a).

X. *Digitaria ischaemum* (Schreb.) Muhl. : Its root exudates had no inhibitory effect on the germination, drymatter and nodulation in soybean (Table 59) (Pope *et al.,* 1985).

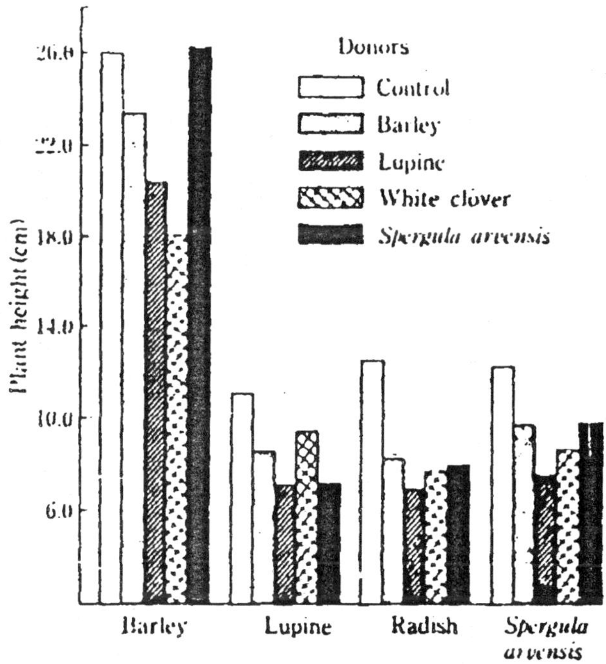

Fig. 14. Effect of exudates from root of donor plant height.
Source : Tsuzuki *et al.* (1984)

Minor Weeds : The root exudates of *Bothriochloa pertusa* (L.) Camus a weed in Pakistan reduced the germination, root growth and drymatter of pearlmillet, rapeseed, tomato, lettuce and pepper (Table 59) (Hussain *et al.,* 1987). The root exudates of sicklepod (*Cassia obtusifolia* L.) and Ivy morning glory (*Ipomoea hederacea* (L.) Jacq.) decreased the seed germination but had little effect on the drymatter and nodulation in soybean. While root exudates of *Sesbania exaltata* (Raf.) Rydb. and common cocklebur (*Xanthium pennsylvanicum* Wall.) slightly inhibited the germination, drymatter and nodulation in soybean (Pope *et al.,* 1985). *Parthenium hysterophorus* L. root exudates decreased the growth and nodulation in neighbouring plants of kidneybean. The maximum exudation of inhibitors occurred at rosette and flowering stages and the inhibitors remained active in soil for about 30 days. Its root exudates decreased the growth of wheat seedlings (Kanchan and Jayachandra, 1979 b). The root exudates of *Phalaris minor* Retz. reduced the growth of wheat seedlings (Porwal and Gupta, 1987). Spurrey weed (*Spergula arvensis* L.) root exudates depressed the seedling growth of barley, white clover and radish (Fig. 14, Table, 61) (Tsuzuki *et al.,* 1984). Thorn apple (*Datura stramonium* L.) seedlings excrete inhibitory alkaloids through their roots as early as on 6th day of germination, which inhibited the germination and growth of crops in their vicinity (James, 1946; Mothes, 1955).

3.2.1.2.1. Stimulatory weeds : Root exudates of bermuda grass (*Cynodon dactyton* L. Pers.) stimulated the seed germination and drymatter and had no adverse effect on the nodulation in soybean (Pope *et al.,* 1985). Its root exudates also promoted the growth and yield of greengram (Peswani, 1981). In field, grain yields of wheat increased when sown in mixture with corn cockle (*Agrostemma githago* L.) (Gajic, 1966). Its root exudates stimulated the wheat growth even under conditions of heterotrophic nutrition and in a sterile environment, owing to presence of agrostemin, allantonin and gibberellin in its seeds (Gajic and Nikocevik, 1973).

Table 61. Effect of root exudates of donor plants on the root dry weight of receiver plants

Donor plant	Control	Receiver plants			
		Barley	Lupine	White clover	Spergula arvensis L.
Barley	1.11 (100)	0.59* (53)	0.32** (29)	0.16** (29)	0.50** (45)
Lupine	0.42 (100)	0.23 (55)	0.06** (14)	0.41 (98)	0.18* (43)
Radish	0.58 (100)	0.27 (47)	0.18 (31)	0.19 (33)	0.26 (45)
Spergula arvensis L.	0.05 (100)	0.02 (40)	0.02 (40)	0.02 (40)	0.03 (60)

*,** = Significant at 5%; 1 % level, respectively, ( ) per cent against the value of control.

Source : Tsukuzi *et al.* (1984).

3.2.1.3 Leaf Secretions

Leaf washings of *Camelina sativa* (L.) Crantz, stimulated the root growth of flax and wheat owing to secretion of benzylamine from leaves which is stimulatory <100 ppm concentration (Lovett, 1980; Lovett and Jackson, 1980).

3.2.1.4 Seed leachates/ extracts

The seeds or seed coats of certain weed species contain inhibitory compounds, which are released mainly during germination. These compounds inhibit the seed germination and root and shoot growth of crops sown in their vicinity (Table 62).

I. Velvet leaf (*Abutilon theophrasti* Medic): Aqueous extracts of its ground seeds markedly inhibited the seed germination of alfalfa, radish and turnip due to release of free amino acids from its seeds (Table 63) (Gressel and Holm, 1964). Its seed extract as well as germinating seeds inhibited the seed germination and radicle growth in turnip 'Purple Top' partly owing to presence of phenolic acids (Elmore, 1980). The seed extracts contained six flavonoid compounds (delphindin, cyanidin, quercetin, myricetin, (+)-catechin and (-)-epicatechin), which play dual defensive role against competing seedlings and seed attacking fungi (Paszkowski and Kremer, 1988).

Table 62. Inhibitory/stimulatory effects of weed seed leachates on the germination and growth of crops

Donor weed		Recipient crop		Reference
Weed species	Phytotoxin	Crops affected	Inhibitory/stimulatory effect	
Abutilon theophrasti Medic.	Amino acids	Tomato, radish, lettuce	Inhibited germination	Gressel and Holm (1964)
	Amino acids, phenolic acids	Turnip	Inhibited germination and root growth	Elmore (1980)
Avena fatua L.	ND	Wheat	Inhibited germination and growth	Evetkovic (1980)
Chenopodium album L.	ND	Wheat, alfalfa, seed clover	Inhibited germination	Stefureac and Fratilescu -Sesan (1979)
Datura stramonium L.	Scopolamine, hyoscyamine	Linseed	Inhibited germination and root growth	Lovett *et al.* (1981)
	ND	Sunflower	Inhibited seedling growth	Levitt and Lovett (1984), Levitt *et al.* (1984)
Galium aparine L.	Asperulosidic acid, Deacetyl asperulosidic acid	Alfalfa	Inhibited germination and growth	Komai *et al.* (1986)
Heracleum laciniatum	ND	Lettuce	Inhibited germination and growth	Juntilla (1975)
Heliotropium eichwaldi	ND	Greengram	Inhibited seed germination	Srivastava (1969)
Sinapis arvensis L.	ND	Alfalfa	*	Stefureac and Fratilescu-Sesan (1979)

ND = Not determined   * Stimulated seed germination

II. Thorn apple (*Datura stramonium* L.) : Its seed leachates depressed the germination and radicle elongation in flax owing to presence of tropane alkaloids (scopolamine and hyoscyamine), however, radicle elongation was more sensitive to leachates than seed germination (Tables 62, 64) (Lovett *et al.*, 1981). Alkaloids washed from its seeds inhibited the early growth of sunflower in black and lateritic podzolic soils (Table 51). The phytotoxicity persisted upto 20 weeks in lateritic podzolic soils under controlled conditions and upto 56 weeks in a *Datura* infested black soil in the field (Levitt and Lovett, 1984). The alkaloids from seeds remained phytotoxic to sunflower upto 15 weeks in soil under field conditions and retarded the metabolism of food reserves, particularly starch in germinating seeds of sunflower (Table 65) (Levitt *et al.,* 1984).

Table 63. Bioassay of germination-inhibiting ability of *Abutilon theophrasti* Medic, seed or seed extracts with turnip seed

Test Conditions	Performance of turnip seed	
	Germination	Radicle length
	% of control	
On paper used to germinate *A. theophrasti* seed	89	96
With *A. theophrasti* seed	99	83[a]
With *A. theophrasti* seed incubated for 24 hours	98	83[a]
With *A. theophrasti* seed extract :		
Alcoholic	37b	3[b]
Water	4b	2[b]
With fraction of *A. theophrasti* seed extract:		
Neutral	107	108
Acid + neutral	36.5b	3.3[b]
Basic + neutral	100.8	31.1[b]
Basic	97.	82[b]

[a,b] Significantly different from control at 5% and 1%, respectivly

Source : Elmore (1980).

Minor weeds : The secretion from wild oat (*Avena* spp.) seeds inhibited the germination and early seedling growth of wheat, while, the seed secretions from wheat increased the germination of wild oat (Evetkovic, 1980). Aqueous extracts of old seeds of *Citrullus colocynthis* (L.) Kuntz. inhibited the root and shoot growth of pearlmillet seedlings, but extracts of fresh seeds caused stimulation (Bhandari, 1974). Aqueous extracts of large crabgrass (*Digitaria* *sanguinalis* (L.) Scop.) seeds, decreased the germination of cabbage `Badger Ballhead', turnip `Purple Top' and radish `Round Black spanish' (Gressel and Holm, 1964). The germinating seeds of *Galium aparine* L. inhibited the germination and growth of lettuce seeds placed together, owing to liberation of iridoidylucosides (Komai *et al.* 1986). Likewise, germinating seeds of *Heracleum laciniatum* inhibited the seed germination of lettuce and radish seeds kept in the same petri plate (Juntilla, 1975). An inhibitor extracted from the seeds of *Rottboellia exaltata* L. inhibited the germination of maize and soybean (Mercado and Manuel, 1978). Germinating seeds of lambsquarter, *Sinapis arvensis* L. and *Thalpi arvense* L. decreased seed germination of wheat `Dacia' seeds kept 1.5 cm apart. Lambsquarter (*Cheno-podium album* L.) seeds reduced the germination of lucerne while that of *T. arvense* caused stimulation in white clover and lucerne (Sterfureac and Fratilescu-Sesan, 1979).

Table 64. Effect of *Datura stramonium* L. seed leachates concentrations on germination and radicle elongation of linseed

Seed leachate concentration*	Germination **	Radicle length **
	% of control	
Control	100	100
31	106	101
62	84	87
125	87	89
250	82	89
500	73	68
1000	62	58
2000	53	41

* *Datura Stramonium* L. seeds leachate in 20 ml distilled water.

** Mean of four seed growing conditions.

Source : Lovett *et al.* (1981)

Table 65. Effect of time on the inhibition of sunflower radicle elongation by seeds of *Datura stramonium* L.

Treatment	Time (weeks)							
	0	5	10	15	20	25	30	35
Radicle length (mm)								
Control	80.2	61.9	68.3	52.9	51.2	53.3	50.8	52.0
Seeds	66.9	48.9	55.9	41.3	51.6	51.3	53.8	52.9
LSD 0.05	6.87	7.75	6.16	7.03	7.07	8.10	8.09	8.03

Source : Levitt *et al.* (1984)

Table 66. Inhibitory/stimulatory effects of volatile inhibitors from weeds on the germination and growth of crops

Donor Weed		Recipient crops		Reference
Weed species	Phytotoxin	Crops affected	Inhibitory/stimulatory effect	
Heliotropim europeum	ND	Buckwheat, radish	*	Grechkanev and Rodionov (1971)
Amaranthus palmeri (L.) Wats.	2-octanol, 2-novanone, 2-heptanone, 2-heptanol, 2-undec-anone, 2- hexone, 2-butanone, 2-pent-anone, 3-methyl-2-butanone, 3-hydroxy-2-butanone	Soybean, oat, alfalfa lettuce, tomato, onion, carrot	Inhibited seed germination	Bradow and Connick Jr. (1987 1988a) Connick Jr. *et al.* (1987)
Salvia reflexa Hornem.	α-pinene, β-pinene, cineole	Wheat	Retarded germination and growth	Lovett (1986)
Stevia eupatoria	ND	White clover	Inhibited root growth	Lovett (1982a)

ND = Not determined, * Stimulated germination and growth

3.2.1.5 Volatiles

The volatiles of palmer amaranth (*Amaranthus palmeri* (L.) Wats.), mintweed (*Salvia reflexa* Hornem) and *Stevia eupatoria* had inhibitory effect on crops, while, that of wild heliotrope (*Heliotropium europeum*) stimulated the crop growth (Table 66). The volatiles released from air dry residues of *Stevia eupatoria,* decreased the root elongation of white clover seedlings in a closed system (Lovett, 1982b). The volatiles emitted from the soil incorporated or soil surface residues severely inhibited the seed germination of carrot, tomato and onion (Bradow and Connick Jr., 1987). Mintweed is a principal weed in some countries including USA and Australia (Holm *et al.*, 1979). Volatiles released from its leaves retarded the germination and seedling growth of wheat in a closed circulating system and volatiles contained monterpenes including α-pinene, ß-pinene and cineole (Lovett, 1986). Volatiles released from the dry leaves and flowers of palmer amaranth reduced the seed germination of tomato, onion and carrot. The volatiles contained 2-octanone, 2-nonanone, 2-undecanone, 2-heptanone, 2-hexanone, 3-methyl-2-3 butanone, 2- pentanone, 3-hydroxy-2-butanone and 2-butanone phytotoxins (Bradow and Connick Jr., 1987, 1988a, b). While the volatiles from leaves, stem and seeds of wild heliotrope stimulated the germination and root growth in radish and fodder bean (Grechkanev and Rodionov, 1971).

CHAPTER - 4

EXPLOITATION OF ALLELOPATHY FOR CROP PRODUCTION

It has been established that allelopathy could be used to increase crop production, to reduce crop production expenses and to diminish the current reliance on synthetic agrochemicals that degrade the environmental quality. The strategies to capitalize on allelopathy may be (a) avoidance of negative impacts, (b) exploitation of stimulatory effects, (c) management and development of allelopathy crops to suppress weeds, (d) development of allelochemicals as pesticides (herbicides, insecticides, nematicide fungicides) and growth regulators and (e) combinations of these approaches (Einhellig, 1985b). Several scientists have speculated on such opportunities to utilize allelopathy (USDA, 1977; Putnam and Duke, 1978; Klein and Miller, 1980; Einhellig, 1981; Newman, 1982; Putnam, 1983a; Rice, 1983; Einhellig and Leather, 1988) and some of these goals are feasible to achieve. However, in this review use of allelopathic practices or allelochemicals for (a) weed control, (b) insect management (c) nematode management and (d) as bioregulators have been discussed.

4.1 WEED CONTROL

Indiscriminate use of herbicides for weed control has resulted in very serious ecological and environmental problems as under. (1) Increasing incidence of resistance in the weeds to important herbicide such as S-triazines (Forney *et al.,* 1985) and dinitroanilines (Mudge *et al.,* 1984). (2) Shifts in weed population (I) to species that are more closely related to the crops that they infest e.g. wild oat (*Avena fatua* L.) in oat, shattercane (*Abelmoschus esculentus* L. Monech) in cotton (*Gossypium hirsutum* L.) and red rice in rich both *Oryza sativa* L. and (II) minor weeds have become dominant e.g. in crop fields many major weeds suppress *Cyperus rotundus* and keep its population under control but after their chemical control, the competitors of *C.*

rotundus minimised, thereby, it becomes dominant weed. (3) Greater environmental pollution (FAO 1970) and health hazards (I) particularly from groundwater contamination e.g. in US, the ground water contains sizeable quantity of alachor and atrazine widely used herbicides (Ahrens, 1989) and (II) from inhalation during handling and application of herbicides. (4) Toxic residues of herbicides pollute the environment and later on may prove hazardous to even future generations. The serious ecological questions about the reliance on herbicides for weed control has been questioned.

As the traditional methods of discovering and developing new herbicides have become more difficult and expensive, interest in allelopathy for weed control and in natural products as sources of new herbicide chemistries have increased. Besides public awareness and demand for environmentally safer herbicides, with less persistence, more specific targets and less potential for contaminating groundwater, makes searches for new weed control strategies like allelopathy and natural products as herbicides more attractive.

Crops are being raised since ancient times without damage to the environment, but the use of herbicides during the short span of last fifty years have raised serious doubts about their continuous use. Allelopathic suppression of weeds is receiving greater attention in recent years as a possible alternative weed management strategy (Altieri, 1988, Einhellig and Leather, 1988). Present understanding of the plant biochemistry, physiology, morphology and inter-and intra-plant biochemistry, physiology, morphology and inter-and intra-plant interactions and chemistry of natural products have shown that crops and allelochemicals may be used for weed control over-coming the above problems associated with herbicides. This review, outlines the potentials of allelopathy and allelochemicals for weed control.

4.1.1 Allelopathy

Allelopathy may help in weed management (Zimdahl, 1987) through (a) suppression of weed seed germination and seedling emergence by potential allelopathic spp., (b) suppression of weeds growth through enhancing allelopathic potential of crops particularly in monoculture and (c) use of plant residues and allelopathic crops in rotations and in intercropping systems. Besides tree litter could also be used for weed suppression.

4.1.1.1 Crops

Crop allelopathic interactions may provide weed control in the crop, using various techniques such as (a) use of phytotoxic crop residues as mulches and cover crops, (b) use of allelopathic plants in crop rotation crop mixture/intercropping or smother crops, (c) use of trap crops to control parasitic weeds and (d) breeding crops for greater suppression effect on weeds.

4.1.1.1.1. Phytotoxic mulches and cover crops

The toxicity of crop residues poses problems for subsequent crops, particularly in multiple cropping systems. Based on such findings, Putnam and Co-workers initiated researches in 1973 at Michigan State University to utilize allelopathic effects of crop residues to control weeds. It was found that residues of wheat `Tecamseh' and rye `Balba' desiccated by herbicides or freezing reduced the weed growth upto 88 per cent. Generally three approaches for exploiting the residues for weed suppression have been evolved (Worsham, 1989) viz., (a) fall planting followed by winter kill, (b) fall or spring planting and subsequent mowing or herbicide desiccating and (c) double crop systems, where a summer crop is generally planted into harvested small grain or legume residues.

I. Sorghum : Sorghum and sudan grass mulch applied to apple orchard in early spring decreased weed biomass by 85 and 90 percent, respectively (Putnam and DeFrank, 1979). *Sorghum bicolor* residues proved highly toxic to annual grass weeds like *Echinochloa colonum* and barnyard grass compared with annual broad leaved spp. (Lovett, 1983 a). Putnam (1986) suggested that cover crops for cool season vegetables viz., Pea or cole crops may be sown in autumn and killed with contact herbicides in the spring i.e. at the sowing or transplanting of vegetables. Sorghum sown in late summer and allowed to freeze in winter controlled the weeds in these crops. Putnam and DeFrank (1983) evaluated the weed suppressing ability of desiccated residues of wheat, barley, oat, rye, sorghum and sudan grass. Sorghum residues reduced the population of common purslane and smooth crabgrass by 70 per cent and 80 per cent, respectively. Sorghum-sudan grass residues completely inhibited the germination of smooth crab grass for 60 days period. Residues of wheat, barley, oats and rye also consistently decreased the total weed biomass and inhibited early season weed growth upto 75 per cent. Crop residues on the soil surface decreased the emergence of lambsquarter, common purslane and redroot pigweed compared with its soil incorporation. The sorghum, rye and wheat residues, released toxic compounds from their residues (Barnes and Putnam, 1983a, Shilling *et al.,* 1985). In general, large seeded vegetables (maize, cucumber, pea, snapbean) grow normally or were sometime stimulated by crop residues, whereas, several spp. of small seeded vegetables (carrot, lettuce, onion etc.) were severely injured. Green house experiments with two soils, confirmed that sorghum residues have both weed suppressing and crop stimulating effects. Sorghum sudan grass hybrid sudex is generally grown as summer cover crop because of its rapid growth and ability for excellent suppression of weeds (Forney *et al.,* 1985). Since sudex show potential as a weed suppressive green manure crop, therefore, Weston *et al.* (1989) conducted studies to determine the stage of plant growth with maximum allelochemical production and to identify allelochemicals. Greatest allelopathic potential of sudex tissue was at 7 days of age and it caused greater inhibition in broad leaf spp. (tomato and curly cress) than in grass spp. (*Setaria italica* and *Echinochloa crusgalli*). The phytoinhibitors were identified as p-hydroxyberzoic acid and *p*-hydroxybenzaledhyde (enzymatic breakdown products of cyanogenic glycoside dhurrin) and their I_{50} values of *Lepidium sativum* seed bioassays were 140 and 113 μg/ml, respectively. Seven days old sudex tissue had maximum percentage of these phyto inhibitors on dry matter basis than older tissue. With increase in sudex age, the percentage of p-hydroxybenzaldehyde increased and that of p-hydroxybenzoic acid decreased.

As sudex age increased from 7 to 28 days, allelopathic potential of tissue decreased.

II. Rye: Rye is most studied cover crop, for allelopathic weed suppression from its crop residues. Its high biomass production of shoots and roots, winter hardiness and phytotoxicity of the residues, makes it very effective crop in no-tillage soil conservation systems. Robertson *et al.* (1976) found that rye residues suppressed weed growth as compared to sod or conventional tillage plots. It has been used to control the weeds in fruit orchards (Olmstead, 1981). Spring planted cool season grasses may be used to control weeds in warm season vegetables and autumn-planted cover crops for perennial cropping systems (orchards, strawberry and asparagus plantings) (Putnam, 1983b, Putnam *et al.,* 1983). Rye proved best, because it produced more biomass and toxins prior to desiccation. Shilling *et al.* (1985) studied the allelopathic effects of desiccated residues of rye and wheat on the germination and growth of maize, soybean, sunflower and tobacco and their weeds. In maize crop, wheat mulch reduced *Ipomoea* spp. biomass by 79 per cent, While in tobacco crop, rye mulch decreased biomass of redroot pigweed, lambsquarters and ragweed by 51, 41 and 73 per cent, respectively. In soybean and sunflower crops, rye mulch reduced the biomass of lambsquarters, ragweed and redroot pigweed by 99, 92 and 96%, respectively. Further studies indicated presence of β-phenyllactic acid (β-PLA) and β-hydroxybutyric acid (β-HBA) in the rye mulch and ferulic acid in wheat mulch. β-HBA and β-PLA at 8 mM inhibited the hypocotyl and root growth of these weeds, while, ferulic acid inhibited seed germination and root length of pitted morning glary and prickly side.

Chou and Patrick (1976) identified nine acids from ether extracts of decaying rye residues in soil. Phenyeacetic, 4-phenylbutyric, venillic, ferulic, *p*-coumaric, *p*-hydroxybenzoic, *o*-coumaric and salicylic acids all inhibited the growth of bioassay plants. The compounds isolated from water extracts of above ground rye mulch, inhibited the weed growth in laboratory bioassays. Shilling *et al.* (1985, 1986a, 1986b) found β-phenylactic acid (PLA) and β-hydroxybutyric acid (HBA) caused 20 to 60 per cent inhibition of common labsquarter and redroot pigweed. Barnes *et al.*,(1986) isolated two hydroxamic acids, 2,4-dihydroxy -1, 4 - (2H)- benzoxazin – 3 - one (DIBOA) and 2(3H) - benzoxazolinone (BOA) with phytotoxicity to a large numbers of test weed spp. These two compounds were more phytotoxic than PLA or HBA to large crabgrass, barnyard grass, prosomillet and redroot pigweed at concentrations of 67-250 ppm and DIBOA remained toxic in soil for extended period following soil application (Barnes *et al.*, 1987). Nair *et al.* (1990) and Chase *et al.* (1991) isolated a microbially produced 2,2`-oxo-1,1-azobenzone (AZOB) from soil enriched with BOA and showed that DIBOA, BOA and AZOB were inhibitory to barnyard grass and garden cress at concentrations of 67-250 ppm. In these tests, AZOB was more toxic to test spp. than DIBOA and 10-fold toxic than BOA. In further studies, AZOB at 100 and 200 ppm produced 38-49 per cent more inhibition in test spp. (garden cress, barnyard grass, cucumber and snapbean) than DIBOA (Chase *et al.*, 1991). This indicates that a variety of natural products contribute to the herbicidal activity of rye residues.

The collective allelochemical action of rye mulch on weeds suppression in the field is outstanding. Barnes *et al.* (1986) reported that weed biomass in a cover crop of living rye was reduced 90 per cent over unplanted controls. A mulch of 40- day old spring-planted rye gave 69 per cent reduction. Shilling *et al.* (1985) found that rye mulch and root residues gave over 90 per cent reduction in early season biomass of lambsquarter, redroot pigweed and common ragweed in no-till planted soybean, sunflower and tobacco compared to tillage and no rye. Rye, sown in autumn and treated with glyphosate immediately before maize sowing gave complete weed control in maize crop due to allelopathy. This system not only provided weed control but also reduced contamination of soil and ground water with persistent herbicides (Zink and Hurle, 1990). Killed mulch of rye and *Trifolium subterranean* controlled 80-90 per cent sicklepod, morning glory, prickly sida and pigweed in soybean, tobacco, maize sorghum and sunflower (Worsham, 1991). In field trials, *Amaranthus retroflexius, A. spinosus, A. hybridus* and

Chenopodium album were controlled 80-100 per cent upto four weeks after sowing into killed cover crops of rye and *Trifolium subterranean* without use of herbicides. Only rye mulch suppressed *Brachiara platyphylla*. *Amaranthus* spp. increased with increase in soil pH and total phenolic acid content under various mulches.

III. *Wheat* : Cold aqueous extracts of wheat straw affected the germination and growth of annual broad leaved weed spp. (velvet leaf, ivyleaf, morning glory, prickly sida, hemp, sesbania) but the annual grass weed barnyard grass remained unaffected (Steinsiek *et al.,* 1980). Lovett *et al.* (unpublished, quoted by Lovett, 1982b) reported that the residues of field pea, wheat, oilseed rape reduced weed population by 71, 53 and 33 per cent, respectively. Libel and Worsham (1983 a,b) reported significant reduction in morning glory and prickly sida in field studies involving wheat mulch and isolated ferulic acid as most phytotoxic compound from foliar wheat extracts. Kim *et al.* (1987) identified 12 simple free phenolic acids in straws of wheat, barley and rye. Ferulic acid content was maximum (20%) followed by p-coumaric, sinapic, protocatechuic and caffeic acids etc. and polyphenols such as scopoletin and rutin, fatty acids like linoleic and organic acids were present. These allelochemicals exert their inhibitory effect on the growth of lettuce seedlings. The inhibitory order of aqueous and alcohol extracts of crop straw were in order : barley >wheat >rye. In pot studies decomposed wheat straw, hemp and pea straw secreased the growth of *Elymus* (Muminovic, 1991). In field studies mature crop residues of rice `Sarjoo 52', wheat `HD 1553' pea `Rachna', chickpea `Type 3' mustard `Varuna', linseed `Neelam' was incorporated into the soil at a depth of 5-6cm @ 5th/ha followed by rice sowing next day. The data on weed density and biomass recorded at 60 days after crop sowing showed that all crop residues inhibited the germination and growth of weeds. Wheat and pea residues inhibited germination of broad leaved weeds *(Ammania baccifera, A. multiflora* and *Phyllanthous niruri*). Rice residues reduced the population of *Echinochloa colonum* and broad leaved weeds by 40 and 56 per cent and biomass by 39 and 64 per cent, respectively (Khan and Vaishya, 1992).

IV. Sunflower: Soil incorporation of sunflower residues significantly reduced the number of dicot weeds by 66 percent compared with control (Anaya, 1989). In pot experiments sunflower straw depressed the plant height of wild oat, *Agropyron repens,* branyard grass, *Ambrosia artemisifolia* and lambsquarter and depressed the biomass of last three weed App. (Muminovic, 1991).

V. Ipomoea : *Ipomoea tricolor* is a major summer cover crop for weed control during the fallow periods in sugarcane in Molelos state, Mexico. Its aqueous and organic extracts exert allelopathic suppression on seed germination and seedling growth of *Amaranthus* *leucocarpus* Watts and barnyard grass owing to pytotoxic mixture of resin glycosides (Anaya *et al.,* 1990). In further studies, major phytogrowth inhibitor was identified as tricolorin-A, which in bioassays inhibited the seedling growth of *A. leucocarpus* and barnyard grass at IC_{50} values ranging from 12 to 37 M (Miranda and Mata, 1993).

VI. Legumes: The residues of several annual legumes have also shown weed suppression. White *et al.* (1989) reported inhibition of several weeds from field residues and leachates of crimson clover and hairy vetch. Teasdale (1988) showed some weed suppression from hairy vetch residues. In fields studies at Beltsville and Ithaca, the residues of hairy vetch and rye reduced weed emergence with increase in biomass residue in no-tillage system. Decay models developed from these studies predicted emergence levels of 21 and 62 per cent of control level at a residue rate of 350 g/m^2 at Beltsville and Ithaca, respectively (Teasdale and Mohler, 1992). Their further studies, showed that light interception and allelopathy are responsible for the weed suppression from hairy vetch residues (Teasdale, 1993, Teasdale and Daughtary, 1993). Enache and Ilinicki (1988) concluded that subterranean clover has potential for controlling weeds in corn. Else and Ilnicki (1989) studied growth and species composition of weeds in four mulch and tillage systems, with and without maize. A living subterranean clover, provided near perfect weed control with or without corn. Evidence of allelopathic activity was found in extracts of clover leaves and in dead residues. Corn presence reduced weed biomass in the other three treatments viz., a bead rye mulch, bead common chickweed mulch and tillage. Worsham (1992) reported that killed heavy mulches of subterranean clover among the legumes and rye among grasses provided 80-90 per cent early control of sickle pod, morning glory, prickly sida and pigweed in crops of soybean, tobacco, maize, sorghum and sunflower. In field trials, *Amaranthus retroflexus, A. spinosus, A. hybridus* and *Chenopodium album* were controlled 80-100% upto four weeks after sowing into killed cover crops of rye and subterranean clover (Worsham and Blum, 1992). In pot studies, pea straw depressed the biomass of Johnson grass and barnyard grass weeds (Muminovic, 1991). Sahid *et al.* (1993) studied the allelopathic potential of five species of legume cover crops viz., *Calopogonium caeruleum, C. mucunoide, C. pubescens, P. javanica* and velvetbean *(Mucuna cochinchinensis)* in bioassays, pot culture and field studies on *Asystasia intrusa* and *Paspalum conjugatum* weeds. *C. caeruleum* and velvetbean exhibited greater phytotoxic effects on the germination and growth of these weeds in bioassays. In pot culture, soil incorporation of *C. caeruleum* and velvetbean residues decreased the dry weight of both these weeds. Pigeonpea leaf litter significantly controlled the weed population (m^2) to 40 as compared with 322 in control and weeds biomass (gm^{-2}) to 25 than 105 in control. The grassy weeds population was controlled to one-seventh of control for two months (Hepperly and Diaz, 1983).

The mulching of crop residues, provide weed control for 30 to 75 days and depends upon soil and weather conditions. It persists longers, when rains occur soon after the desiccation of crops followed by dry conditions. Experience with local conditions, crops tolerance and integration with appropriate tillage practices are pre-requisties for relying on crop residues for weed control. Crop injury from mulch related weed control may be overcome through adjustment of time between mulching and planting and use of tolerant crops.

4.1.1.2 Crop rotations, crop mixtures, intercropping

Barley, rye, maize, sorghum, sudan grass, buckwheat, proso-millet, sweet clover, sunflower, rapeseed, soybean, alfalfa, cowpea and hemp somthered various weed species (Robbins *et al.,* 1942) through crops interference. Farmers often plant these crops to suppress the growth of weeds. The smothering crop exert weed suppression through competition for growth resources and growth inhibition through allelopathic interactions (Newman, 1982).

4.1.1.2.1 Monocultures, Crop rotations

The strongly interfering smother crops had been included in rotations by the farmers for centuries. There can also be carry over weed suppression from crops in certain rotations. Altieri and Doll (1978) found that residues of *Tagetes patula* L. inhibited germination of several weeds with little or no harmful influence on corn. They also reported residual effect of beans, corn and cassava on several weed species. Residual effects on weeds have also been reported for sorghum, barley, oat, wheat and rye (Putnam, 1988b), sunflower, sweet potatoes. soybeans (Einhellig and Leather, 1988) and forage sorghum-sudangrass hybrid (Forney *et al.,* 1985). In some of these crops, weed suppression was also observed in the succeeding crops. Studies have shown that several cereal cover crops effectively suppress weeds in the succeeding or companion crop in no-tillage systems. 45-60 cm tall crops of sorghum, wheat, barley, oat, rye kelled by natural freezing or contact herbicide reduced the density and biomass of several weed species. The residues of sorghum and oat effectively suppressed the population and biomass of common purslane and smooth crabgrass (Putnam and De Frank, 1983). More research is needed on applicable crop sequences to gain full benefit from the allelopathic weed suppression.

I. Barley: Overland (1966) investigated the efficacy of this crop for weed suppression and reported its allelopathic effects in interference. However, in the absence of competition, barley still inhibited the germination of *Amaranthus hybridus* L. and *C. album* (Went *et al.,* 1952), suggesting the involvement of phytotoxins (Overland, 1966). The inhibitory activity of barley was selective among broad-leaved plants, *Stellaria media* L. being more severely inhibited than *Capsella bursapastoris* (Overland, 1986). The barley produces phenolic compounds (Borner, 1960) and two alkaloids viz., gramine and hordenine (Mudd, 1961, Mann *et al.,* 1963). Overland (1966) showed that the former inhibited the growth of chickweed, even when present in small concentrations. Liu and Lovett (1993a,b) confirmed the allelopathic potential of barley and reported that germination of white mustard *(Sinapis alba)* was delayed and radicle length was significantly inhibited at density of 0.5 barley seeds/cm^2. In another study, when the two species were sown together, radicle elongation of white mustard was increasingly inhibited with time. Barley released allelochemicals for at least 70 days after its germination, which inhibited the growth of *S. alba*. The hordenine was released from the roots as root exudates upto 60 days after barley germination. The amount reached maximum of 2 mg/plant/day at 36 days and then declined The hordenine and gramine decreased root length and vigour of root tips. Transmission EM examination of *S. alba* radical tips exposed to hordenine and gramine showed damage to cell walls, increase in both size and number of vacuoles, autophagy and disorganization of organelles. In further studies, Hoult and Lovett (1993) reported significant differences in the ability of barley cultivars `Proctor', Skiff', and `CH 14001' to produce hordenine and gramine.

II. Sorghum : Sorghum has been reported allelopathic to several weed species (De Frank and Putnam, 1977; Forney *et al.,* 1983; Alsaadawi *et al.,* 1985; Panasiuk *et al.,* 1986). The living plants and residues of sorghum and sudan grass reduced the population of common purslane by almost 100 per cent and that of smooth crabgrass by 50 per cent (De Frank and Putnam, 1977). Forney *et al.* (1985) reported that growing a forage sorghum-sudan grass hybrid prior to late-summer planting of no-till alfalfa suppressed weeds and enhanced alfalfa growth. Einhellig and Leather (1988) monitored the weed biomass in strip cropping of grain sorghum, maize and soybean (grown without herbicides) in the following year. Weed cover in early spring in former sorghum strips was less than 30 per cent of that in prior soybean strip. At mid-summer, plots where grain sorghum had grown the year before, had about 60 per cent of the weed mass of others. This decrease in biomass resulted from the suppression of broad leaved weeds, with no effect on grasses. It was found in laboratory studies that grain sorghum inhibits early seedling growth and development more than germination. Some farmers in South Dakota, strip plant grain sorghum and sunflower without herbicides and find reasonable weed control. The summer sorghum delayed and reduced the wild oat emergence in the following winter season. The presence of a small quantity of sorghum stubble inhibited wild oat emergence and reduced root and shoot dry matter and leaf number over no stubble treatment (Jones, 1992). In field studies, grain sorghum crop reduced weed density and growth in the following crop year compared with that after maize and soybean. Sorghum inhibited the broad leaved weeds and had no effect on grassy weeds. The inhibitory effect of sorghum on weeds was attributed to allelopathy (Einhellig and Rasmussen, 1989). Einhellig and Souza (1992) purified sorgoleone, a *p*-benzoquinone from the root exudates of 5-day old sorghum `Dekalb hybrid KK 28' seedlings. In laboratory studies, sorgholeone considerabley reduced the root and shoot growth of *Eragrostis* spp., *Lemna Minor* and other test weed spp. through inhibition of photosynthesis (Einhelling *et al.,* 1993).

III. Wheat : In incubation studies, wheat straw extract inhibited the seed germination (24 per cent) of *Amaranthus hybridus* and stimulated that of *Rottboelia cochinchinensis* and had slight inhibitory effect on *Nicandra physaloides, Acanthospermum* and stimulated that of *Rottboelia cochinchinensis* and had slight inhibitory effect on *Nicandra physaloides, Acanthospermum hispidum, Urochloa panicoides, Chenopodium album, Tagetes minuta* and *Portulaca oleracea* (Ramakudzibga, 1991). Schreiber (1992) studied the influence of continuous corn, corn-soybean and soybean-wheat-corn rotations for 12 years (1980-91) on the weed density of giant foxtail. Its stands were reduced following wheat in no-tillage due to allelopathic influence of wheat straw. The 2,4 Dihyroxy-7- methoxy-1,4-benzoxain-3-one (DIMBOA) has been identified as the main hydroxamic acid of wheat. DIMBOA and its decomposition product 6-methoxy-benzoxazolics-2-one (BOA) inhibited root growth of wild oat by 50 per cent at concentrations of 0.7 to 0.5 mM. BOA inhibited the seed germination of wild oat at all concentrations tested and stimulated root growth in wild oat at 1.5 mM and inhibited above this concentration (Perez, 1990). The wheat and rye release hydroxamic acids through root exudates. In bioassay studies with the root exudates of wheat and rye seedlings, the root exudates of rye inhibited root growth of wild oat but that of wheat had no effect (Perez and Juan, 1991).

IV. Rye: In field studies, rye `Forrajero-Baer' exuding hydroxamic acids through its roots, restulted in 83 per cent less weed biomass than wheat and 76 per cent less than forage oat *(Avena strigosa* L.). Forage oats reduced broadleaf weed biomass by 40 per cent compared to wheat but failed to suppress wild oats. Rye interference reduced wild oat biomass by 84 percent and 86 percent compared to wheat and forage oat, respectively. In bioassays with 2-1-β-glucopyranosyl-4-hydroxy-1,4 dihydroxy-1, 4-benzoxazin-3-one (DIBOA-glc), the main hydroxamic acid found in rye, its aglucone (DIBOA) and its breakdown product BOA inhibited root and coleoptile growth of wild oat seedlings at concentrations of 0.25 mM. It was concluded that hydroxamic acids from rye `Forrajero-Baer' suppressed the weeds (Perez and Juan, 1993).

V. Pearlmillet: Sarmah (1992) studied the smothering effect of pearlmillet, sorghum, maize, cowpea and clusterbean grown for fodder on the summer weeds under field conditions. At harvest (60 days after sowing) the sorghum considerably smothered the weed gorwth over the fallow (non-crop control), while pearlmillet completely controlled the weeds growth. The residual suppression effect of pearlmillet was also observed on the weed density and growth in the succeeding sorghum crop.

VI Sweet Potato: Sweet potato is considered an extremely successful competitor against some weeds under field conditions, including yellow nustsedge (Harrison and Peterson, 1986), but not competitive with others. It interferes with crop growth and its whole plant residues inhibited the growth of sweet potato vine cuttings and cowpea (Walker and Jenkins, 1986). Several constituent compounds viz., fatty alcohols, fatty acids, caffeic and chlorogenic acid in its tissues are well documented allelochemicals (Rice, 1984). Other compounds have been identified but no attempt has been made to test these as allelochemicals. Peterson and Harrison (1991a) found evidence of a high concentration of allelochemicals in its root periderm extracts and observed quantitative differences in the content of two cultivars. The hexane extracts were inhibitory to many weed spp. and I_{50} estimates were 0.5, 0.6, 2.8, 4.4, 5.1, 9.6, 15.7, 21.0 and 25.8 mg/ml for velvetleaf, prosomillet, black night shade *(Solanum nigrum* L.) goose grass, sweet potato, tall morning glory, *Eclipta alba, Cassia occidentalis* and pigweed, respectively. In field studies, sweet potato `Regal' at the end of growing season decreased the yellow nutsedge shoot dry weight/m^2 up to 90 percent than nutsedge grown alone. While nutsedge did not influence sweet potato growth. In green house experiment, designed to minimise competitive effects of sweet potato, yellow nutsedge growth was reduced more than 50 per cent by sweet potato at 8 and 12 weeks after planting. The polar fraction of sweet potato periderm tissue was highly inhibitory to yellow nutsedge root growth (Harrisson and Peterson, 1991). In further studies, sweet potato periderm tissue contains 1.2% active inhibitor, it is labile in methanol solution and various seed species exhibited extremely wide range of sensitivity. The I_{50} values were 0.16, 0.013 and 0.011 mg/ml for redroot pigweed, velvet leaf and prosomillet, respectively (Peterson and Harrison, 1991b) Reinhardt *et al.* (1993) reported that in glass house studies in the soils from control and plots of three sweet potato varieties `Bosbok' Brondal' and `Koedoe' indicator plants (Yellow nutsedge, purple nutsedge, common pigweed carrot, I `Kaapse Mark', tomato `Heinz 1370', cucumber `Special Rust Resistant', radish 'White Icicle, `onion `Pyramid, lettuce `great lakes ' and oats `SWK 001') were grown. The soils from sweet potato `Brondal' plots significantly decreased the dry matter in yellow nutsedge plants than that grown in `Bosbok`, `Koedoe' or fallow plots. This indicates presence of substances inhibitory to yellow nutsedge growth in the soil after `Brondal' sweet potato. Compared with fallow plots, the growth of carrot, cucumber, lettuce, oats, onion and tomato plants was stimulated. The related species *Ipomoea aquatica* and *I. tricolor* have been proved allelopathic.

VII Sunflower: Sunflower-oat rotation over five years period significantly lowered the density of grassy and broad leaved weeds than in control plots (Leather, 1983a). Although weed density increased in all plots over the five seasons but the rate of increase was less in sunflower plots. There was little difference among the sunflower cultivars. In further studies, weed biomass was equivalent in plots planted with sunflower, whether EPTC (S-ethyl dipropyl carbamothioate) herbicide was applied or not, clearly showing the efficacy of sunflower mediated weed control (Leather, 1987). In sunflower-wheat rotation field trials, sunflower decreased the density and dry weight of wild oat and *Cirsium arvense* in the following wheat crop (Cernusko and Boreky, 1992). The potential of sunflower as source of allelochemicals is well known (Varela, 1992) and bioassays of leaf aqueous extracts have shown strong inhibition and stimulation in germination and root length of test plant species. The leaf aqueous extracts of sunflower `SH-222' were found to contain five new guaianolides, the annuolides A-E have been tested as allelochemicals. All these guaianolides possess allelopathic activity over dicotyledon species and are likely to be involved in the allelopathic action of sunflower cultivars (Macias *et al.,* 1983).

VIII. Tomato: *Solanum nigrum* and barnyard grass are most troublesome weeds of tomato. The root plus shoot extracts of wild types of tomato inhibited root and shoot growth of both weeds. At 10-fold dilution, the extract stimulated shoot growth of weeds but still inhibited root growth of *S. nigrum.* Residues of six tomato accessions and one commercial cultivar inhibited weed seedling growth, wild types having stronger inhibitory effects than commercial cultivar (Perez-Pico and Ashton, 1984). In field studies, tomato experts suppression effect on some weed species, probably through root exudates. However, the aqueous root extract of tomato did not influence germination of barnyard grass, pigweed and wild mustard (Abdullah *et al.*, 1990).

IX. Cruciferous Crops : *Brassica campestris* crop reduced the weed density in the same field in following year, therefore, used as weed controlling crop by the tarahumara Indians in North Mexico (Chacon and Gliessman, 1982). It was owing to inhibitory effects of its residues and excretions on the germination and growth of other plants (Bye, 1979). There are reports on use of winter and spring cruciferous crops as intermediate crops to reduce population of weeds. Bell (1970) and Bell and Muller (1973) reported that post harvest cropping of cruciferous crops in maize fields reduced weed population by 90 to 96%. Likewise, intermediate crop of winter/spring rape reduced weed population by 40 to 47% (Uteush, 1979).

X. Legume crops

(i). Alfalfa : Alfalfa produces allelopathic saponins, which may be a major cause of the reduction in yields of subsequent crops. The detailed chemical structure of many alfalfa saponins are known with 30-40 compounds, with different properties of inhibition or stimulation. The alfalfa plant parts differ greatly in type and quantity of aglycones present, among which medicagenic acid glycosides are most active biologically and are dominant in alfalfa roots (Oleszek *et al.*, 1992). In wheat, maize, soybean, sorghum reduced yields are commonly associated when subsequent crop is sown after old alfalfa stands, even if soil moisture and fertility is adequate and pests are fully controlled. In bioassay studies, six saponins from alfalfa `Cimarron' inhibited the root and shoot growth of barnyard grass and cheat (*Bromus secalinus* L.) at all the concentrations (10 to 5000 ppm) tested, while, in coffee weed, pigweed and dandelion (*Taraxacum vulgare*) stimulated the root and shoot growth at 10 and 100 ppm and caused inhibition at 1000 and 5000 ppm concentrations. More research is needed before alfalfa saponins could be used for weed control (Waller *et al.*, 1993).

(ii). Pigeonpea : Hepperly and Diaz (1983) reported that pigeonpea litter gave excellent weed control for two months in the following crop. Its leaf litter significantly controlled the weed density m^2 to 40 compared with 322 in control and weed biomass g/m^2 to 25 compared to 105 g in control. The leaf litter decreased the population of grassy weeds to one-seventh of those in control.

(iii). Velvetbean : The velvetbean (*Mucuna pruriens* L. DC) `Utilis' increased the graminae crops yields in crop mixture and smothers notorious weeds like nutsedge and *Imperata cylindrica* (Lorenzi, 1984). It is an allelopathic plant in seedling growth tests and effectively smothered the weeds in crop fields (Fujii *et al.,* 1988). The weed emergence and population in fields after continues cropping of velvetbean, upland rice, egg plant and tomato was minimum (50%) in the velvetbean plot, compared to other crops plots where *Cerastium glomeratum* was the dominant weed. The emergence and hypocotyl length of lettuce seedlings were not affected but radicle growth was reduced significantly in plot with continuous cropping of velvetbean for 5 years, however, no such reduction occurred after one year cultivation of velvetbean. The crude methanol extracts of velvetbean leaves markedly reduced the growth of *C. glomeratum* and lettuce (Fujii *et al.,* 1991 a). The velvetbean leaves and roots were found to contain L-3,4-dihydroxy phenyealanine (L-DOPA) in high concentrations (0.5 to 1.5%) of the fresh weight. The younger leaves contain more L-DOPA than mature leaves. L-DOPA strongly inhibited the growth of *C. glomeratum, Spergula arvensis,* moderately inhibited the growth of compositae but had little effect on graminae and leguminosae plants. Since fresh leaves of velvetbean contain 1% of L-DOPA and it produces 20 to 30 tonnes of fresh leaves and stems per hectare, therefore, about 200 to 300 kg L-DOPA is produced and added to soil per hectare per year. Therefore, the L-DOPA is responsible for weed control in velvetbean (Fujii *et al.,* 1991 b).

XI. Miscellaneous Crops : The sole crops of clusterbean and greengram caused greater suppression of weed species (*Borreria articularis,* purple nutsedge, *Digera alternifolia, Euphorbia granulata, Gisekia pharnaciodes, Polycarpa corybosa, Pulicaria crispa*) compared with sesame and pearlmillet. However, intercropping of clusterbean and greengram with sesame or pearlmillet proved less effective compared with their sole cropping but were superior than sole crops of sesame and pearlmillet (Sen, 1987).

4.1.1.1.2.2. Living Crop interference

Some crops are effective in smothering growth of companion weeds. For example, presence of leachates of sugarbeet, lupine, corn, wheat, oat, pea, buckwheat, hairyvetch, (*Vicia vilosa* Roth), millet (*Panicum* spp.), barley, rye and cucumber inhibited the growth of a number of weeds (Rice, 1984) and allelopathy has been implicated in these studies. The suppression effect of crops on weeds may be the combination of competition and allelopathy (Newman, 1982). The crops of barley, rye, maize, sorghum, sudan grass, buckwheat, prosomillet, sweet clover, sunflower, rapeseed, soybean, alfalfa, cowpea and hemp smothered various weeds (Robbins *et al.,* 1942).

I. Barley : Barley proved most potent, because its roots exudates inhibited the germination and growth of *Stellaria media, Capsella bursapastoris* and tobacco but had no effect on wheat. Live barley plants and their root exudates were more inhibitory than aqueous leachates of dead roots, this supported the concept of an active metabolic secretion of the allelopathic substances. The inhibitor was identified as alkaloid gramine (Overland, 1966). In pot culture studies, in barley pots the growth of *Sinapis arvensis* and *Thlaspi arvense* weeds was stronger in the beginning but were smothered at later stages. Many weeds remained suppressed during early growth but after its maturity their growth increased markedly e.g. *Viola arvensis.* The higher seed rate (7.5 kernels/dm^2) increased suppression of all weed species than lower seed rate (3.0 kernels/dm^2). The weeds *Chenopodium album, V. arvensis, S. arvensis, T. arvense* and *Lamium amplexicaule* produced more dry matter in wheat pots than in barley pots (Dock Gustavsson, 1989). Barley suppress the growth of weeds through competition for environmental resources and inhibited germination of *Amarathus hybridus* and *C. album* through phytotoxins. The inhibitory activity was greater in broadleaved weeds like *Stellaria media* than in *Capsella bursapastoris* (Overland, 1966). The allelopathic potential of barley has been confirmed during germination and later stages of growth (Liu and Lovett, 1992, 1993 a,b,c) in Section 2.1.2.(I).

II. Sweet Potato : Sweet potato (*Ipomoea batatas*) is very efficient crop in controlling weeds through interference, where a very aggressive competition potential is complemented with a high allelopathic effect (Anaya *et al.,* 1990). In field studies, sweet potato `Regal' reduced the growth of yellow nutsedge >50%, 8 to 12 weeks after planting and upto 90% till maturity over nutsedge grown alone (Harrison and Peterson, 1991). In Mexico, sweet potato and *Ipomoea tricolor* and different species of beans viz., velvetbean, *Stizolobium pruriens* efficiently control weeds through interference (Gliessman, 1983, Caamal and del Amo, 1987). *I. tricolor* is used to add organic matter and to control weeds. Its ethyl extract contained an inhibitor, which at 100 ppm caused inhibition of 80% in radicle growth of *Amaranthus leucocarpus* and *Echinochloa* spp. (Anaya, 1989 a).

III. Maize : The aqueous extract of its pollens strongly inhibited the growth of *Bidens pilosa, Rumex crispus* and *Cassia jalapensis* (Jimenez *et al.,* 1983). This indicates great potential of maize pollens to control weeds (Anaya *et al.,* 1987). The pollens are produced in abundance at flowering and significantly inhibited the growth of weeds. Ethanol extract of pollens contained two chromatographic fractions 3A and 3B, which inhibited the radicle growth of *Amaranthus leucocarpus* to 85 and 40%, respectively compared to control (100%).

IV. Sunnhemp (*Crotolaria juncea*) : Leafy spurge (*Euphorbia esula*) has infested millions of hectares crop land, range land and conservation areas in central plains of USA and Canada. Attempts to control this weed with herbicides, grazing animals, insects and plant pathogens have met with limited success. In sand culture, two seeds of sunnhemp inhibited the growth of leafy spurge by 60 to 80% and the regrowth from root and crown buds was also inhibited (Leather *et al.,* 1989). The partially purified seed extract containing 5- hydroxy-norleucene inhibited growth of leafy spurge at a concentration of 25 ppm (Cole, 1991).

V. Velvetbean : Velvetbean (*Mucuna deeringianum* Borr. Merr) plants significantly declined the weed population with time. It suppressed broad leaved weeds and declined their population by 78% at 120 days after sowing, grasses declined by 58% but cyperaceous weeds increased by 34% (Anguilar-Erazo, 1984). In field studies, there was minimum weed density in plots of oats, *Vicia* spp. and *Lolium multiflorum* owing to allelopathic suppression effect of these crops on weeds (Medeiros *et al.,* 1990).

4.1.1.1.2.3. Crop mixtures, intercropping

The role of allelochemicals in interference between intercrop components and weeds has scarcely been explored, yet this type of weed control is useful in such cropping systems (Putnam and Duke, 1974; Fay and Duke, 1977; Lockerman and Putnam, 1979). The question of allelopathic control of weeds in intercropping systems has been complicated by the possibility of interference between the crop species. For allelopathy to be effective for weed control in intercrops, there should be selectivity in the toxins released from the crops and weed species must be more susceptible to such phytotoxins than crops. Gliessman (1982) reported several examples of companion plantings in crops to suppress weeds in Mexico.

I. Squash : The interplanting of squash in maize is done mainly for weed suppression. The interplanting of squash (*Cucurbita pepo*) by peasants in maize/bean polyculture or in maize or bean monoculture provides weed control (Amador, 1980). Squash is planted mainly owing to its ability to control weeds. It forms a continuous plant cover over the low growing weed species, eliminating them and restoring soil productivity through addition of large quantities of organic matter (8 to 10 t dry matter/ha). Laboratory bioassays partly implicate allelopathy in this inhibition. Gliessman (1983) studied the effect of squash leaf extract on radicles elongation of traditional crops of maize and cowpea and to newly introduced cabbage. The extract proved most inhibitory to cabbage than to maize and cowpea. Therefore, farmers in south eastern Mexico interplant squash into maize/cowpea intercrops for effective weed control (Le Tourneau *et al.*, 1956; Gliessman *et al.*, 1981) not only because of the shade from squash leaves but also owing to selective allelochemical inhibition.

Gliessman (1983) also found that extracts of certain `non-weed' species grown by farmers in maize/cowpea fields (Chacon and Gliessman, 1982) had only weak inhibitory effects against maize and cowpea but strong inhibitory effects against cabbage, a new introduction to this region. He proposed that non-crop species also offer possibilities for selective allelopathic weed control.

II. *Stizolobium* spp. : The legumes *Stizolobium deeriangianum* and *Canavalia ensiformis* interplanted in maize provides effective weed control. In the southeastern parts of Mexico, many peasants intercrop the legumes such as *Stizolobium pruriens* (Gliessman and Garcia, 1979) with maize. It incorporates nitrogen into the soil, increases corn yields, significantly reduces both number and biomass of weeds, its forage is fed to livestock and seeds are used in human consumption. In laboratory studies, its dry leaves and root leachates during flowering stage inhibited the growth of dicotyledonous and monocotyledonous weeds (Escarzaga, 1987).

III. *Heracleum laciniatum* Hornem: *H. laciniatum* suppresses the growth of plants under its canopy. In meadows, it inhibited the growth of various plant species without competition. Water percolating through its pots slightly inhibited the growth of *Poa pratensis* and *Phleum pratense* but had no effect on its own growth. The soil samples collected from meadow soil from mid June to late October inhibited the germination of seeds of *P. pratense* than in ploughed soil. Allelopathy was partly responsible for weed suppression (Myras and Juntilla, 1981). Its seeds contain biological inhibitors and the ethyl acetate fraction contained three main groups of inhibitors having following Rf values : (1). 0.60 to 0.80, (2). 0.20 to 0.40 and (3) 0.0 to 0.10 (Juntilla, 1976). The group (1) inhibited seed germination and root growth of lettuce and hypocotyl growth of *Avena*. The group (2) contained the furanocoumarins, pimpinellin, bergapten, isobergapten and angelicin, synthetic pimpinellin and sphondin inhibited seed germination of lettuce. Inhibitors present in groups (1) and (3) were not identified.

Grechkanov and Rodionov (1971) reported benefits from mixing of 1-2 Kg seeds/ha of wild heliotrope (*Heliotropeum europeum* L.) in several legumes. This plant not only reduced weeds from 30 to 70% but also controlled other pests.

Intercropping is a major practice in Latin America, tropics, Africa, Asia and particularly India. Intercrops are always superior to monocultures in weed suppression but their effectiveness varies greatly. Liebman (1988) cited a number of impressive examples.

Although allelopathy is involved in weed suppression in intercropping systems but little information is available. Putnam (1986) employed allelopathic cover crops both as preceding crops to tomato and sweet corn and as intercrops to suppress weeds between their rows. Both sweet corn and tomato performed well in these trials, where early intercrops of broccoli and pea or perennially intercropped white clover reduced weed population upto 90%. Sweet corn yields were reduced (18 to 33%) but the pea and broccoli intercropping system produced land equivalent ratios (LER) twice to that of sole sweet corn. Tomato yields were reduced upto 40% but several tomato-intercrop combinations produced higher LER's. In 3-years field studies, cover crops of barley, rye and *Vicia faba* were planted in monoculture and intercropping after the harvest of summer crop. These crops grow during the winter season and are ploughed in the soil in the month of March or early April, thereafter, summer vegetables are planted in the end of May. Barley + *V. faba* and rye + *V. faba* provided almost complete weed control and the later was most effective. The mean weed biomass (g/m^2) followed the order rye + *V. faba* (1.8) > barley + *V. faba* (7.2) > rye (8.5) > barley (13.4) > *V. faba* (51.2) and control (188). It was attributed to the release of allelochemicals from root exudates during crop growth and from decomposing crop residues (Gliessman, 1989).

To make this concept successful, more extensive research is needed on (a) applicable crop sequences, (b) the role of soil and climatic factors, (c) long-term impacts on the weeds seed population and (d) ways to integrate allelopathic crops with other weed control practices (Einhellig and Leather, 1988).

4.1.1.1.3. Trap crops to control parasitic weeds

Generally agricultural soils contain many thousands of dormant weed seeds per square meter. Some of these seeds germinate within a year after release from the weed plant, while others may remain dormant and survive for several years or even decades before germinating. A major factor in seed longevity is the role of allelochemicals in seeds which prevent seed decay by microbes. Only a fraction of weed seeds in a soil germinate at a time and conventional weed control practices do not kill ungerminated weed seeds. Therefore, the use of chemical germination stimulants has been proposed to deplete the soils of dormant weed seeds (Egley, 1986) and some success has been achieved with certain allelochemicals. This book includes control of parasitic weeds viz., witch weed (*Striga asiatica* L. Kuntze), *Orobanche aegyptica* Pers and dodder (*Cuscuta campestris*) through use of allelochemicals or trap crops. There is need to isolate and identify allelochemicals in trap crops which control these weeds.

I. Witch weed : It parasitizes the roots of corn, millet, sorghum, sugarcane and several warm season grasses. It is widespread in semiarid tropics of Asia, southern Africa, Australia and in limited areas of north and south Carolina, USA and causes 10-100% losses in crops. Its tiny seeds require ageing, a warm moist 'conditioning period' and exposure to an exogenous stimulant before germination occurs. Otherwise, the seeds may remain dormant in soil for 20 years or more. In the field, root exudates of host plants and some non-host plants contain one or more compounds that stimulate its germination. Its seedlings die, unless, its seeds germinate within a few mm of the host root so that attachment can occur. This requirement makes the parasite vulnerable to control, through stimulation of seed germination in the absence of host crop and thus leading to the death of seedlings.

Andrews (1945) and Hosmani (1978) cited cotton, sunnhemp, pigeonpea, greengram, blackgram, lucerne, sunflower and sesamum as trap crops, however, their efficacy has not been studied. A natural germination stimulant of '*Striga*' was identified in the roots of non-host plants of cotton and termed '*strigol*'. It was synthesized and found to stimulate witchweed seed germination at concentrations as low as 10^{-12} M. This compound is also effective as soil treatment but the high costs of its synthesis makes it impractical. Certain strigol analogues have shown potential as soil treatment in Africa. Some analogues of the strigol D-ring (butenolides) viz., allyloxybutenolide germinated 30-60% seeds of witchweed at a concentration of 10^{-6} to 10^{-11} M compared with strigol, while sec-butyloxybutenolide was less active (20 to 40% germination at 10^{-9} M). This indicated that butenolide analogues of strigol were less effective in stimulating seed germination than the multi-ring analogues (Pepperman and Bradow, 1988). Ethylene has been found very effective as germination stimulant and single injection reduced population of dormant weed seeds by 91% and is widely used in USA, however, its application is difficult with small farmers. Chang *et al.* (1986) isolated a germination stimulant from sorghum for *Striga* and was identified as the unstable dihydroquinone form of 2-hydroxy-5 methoxy-3 (8'Z, 11'Z)-8',11',14'-pentadecatriene)-p-benzoquinone. Since its active form is highly unstable in pure form, therefore, the possibility of its use as a soil treatment to reduce weed seeds is very unlikely. It is possible that modification in the molecular structure could be made to make it more stable without causing loss of biological activity. Fischer *et al.* (1989) found that dihydroparthenolide promoted witchweed seed germination at a concentration of 10^{-9} M. Therefore, to determine the specific structural requirements for this activity, a series of natural and synthetic germacranolides, eudesmanolides and pseudoguaianolides were tested as witchweed stimulants. All the 10 germacranolides tested induced 40 to 65% germination of witchweed seeds at a concentration of 10^{-5} to 10^{-9} M. The structure of these compounds is similar to active molecule (+) -strigol (Fischer *et al.*, 1990).

II. *Orobanche aegyptiaca* Pers. : It is root parasite which infests many countries and causes severe damage to many crops of various botanical families. Many expensive methods have been used for its control but they provide partial control. Many workers have used trap crops, catch crops and sometimes false host plants in crop rotation to decrease its infestation in the fields. The trap and catch crops stimulate germination of *Orobanche* seed and are not parasitized themselves. Flax is an efficient trap crop for *O. crenata, O. cernua* and *O. ramosa,* however, it failed to control *O. aegyptiaca.* Vetch (*Vicia sativa* L.) `Sadot' was not parasitized by *O. aegyptiaca* and efficiently controled it (Kleifeld *et al.*, 1994).

III. Dodder (*Cuscuta campestris*) : It is a major weed in alfalfa and some other crops and causes great losses in crop yields. Abdul-Rahman and Habib (1986) studied the influence of extracts of bermuda grass (*Cynodon dactylon*), Johnson grass, pigweed, (*Amaranthus albus*) and various doses of herbicides viz., DCPA, glyphosate, metribuzin against dodder on alfalfa. The extracts of weeds significantly controlled dodder similar to DCPA and glyphosate. The per cent control increased with increase in extract concentration. The highest concentration of bermuda grass was as effective as the higher doses of metribuzin and was less phytotoxic to alfalfa than herbicides. It was found that the weed extracts contained phenols and was highest in bermuda grass and lowest in pigweed.

4.1.2.1.4. Breeding of weed suppressing crop varieties

One potential technique of exploiting allelopathy in weed management is the transfer of allelopathic characteristics from wild types or unrelated plants into commercial crop cultivars. If the new allelopathic character does not have undesirable effects, this technique could increase the ability of the crop to compete naturally against weeds. It has been established that potential for suppression of weeds could be enhanced through selection of crops for allelopathic character (Arai *et al.,* 1976). Superior weed suppressing genotypes have been reported in cucumber (Putnam and Duke, 1974), oat (Fay and Duke 1977), rice (Dilday, 1992, Fujii, 1992), sunflower (Leather 1983a), soybean (Massantini *et al.,* 1977), pearlmillet (Narwal *et al.,* 1992) and *Brassica* (Sarmah *et al.,* 1992) (Table 67). Putnam and Duke (1974) reported that `wild-types' of existing crops possessed high allelopathic potential and this character has been reduced or lost in the breeding of crop varieties for other characteristics viz., high yield etc. Although insect and disease resistance has been incorporated into the crop varieties through crossing with `wild-types', no such effort has been made to develop crops with higher allelopathic potential to smother the weeds even though the character exists in the `wild-types'.

Table 67. Weed smothering crops

Crop	Genotype	Reference
Cucumber	PI 169391	Putnam and Duke (1974)
Oat	PI 266281	Fay and Duke (1977)
Pearlmillet	HHB 67	Narwal *et al.* (1992)
	88004 A x 833-2	
Brassica juncea	H 8689	Sarmah *et al.* (1992)
B. napus	HNS 8902	Sarmah *et al.* (1992)
B. carinata	BCCN 5	Sarmah *et al.* (1992)
Soybean	ND	Massantini *et al.* (1977)
Sunflower	ND	Leather (1983)
Sorghum Sudan grass	ND	De Frank and Putnam (1977); Forney *et al.* (1985); Panasiuk *et al.* (1986); Einhellig and Leather (1988); Sarmah (1992).
Sweet potato	ND	Harrison and Peterson (1986)

ND = Not determined

The first step in such a process would be to select plants with greatest potential for allelopathy. Putnam and Tang (1986) and Einhellig and Leather (1988) believed that present cultivars have low allelopathic capacity except in sunflower. Several varieties of sunflower inhibited the germination and growth of associated weeds greater than that found in several biotypes of native sunflower. Allelopathic characteristics are more likely to occur in crop predecessors or 'wild types' that has evolved in the presence of allelopathic and competitive influences from other species (Putnam and Tang, 1986). Till now there is limited research in this direction and screening of germplasm in oat, rice, sorghum, soybean, cucumber, sunflower, pearlmillet and *Brassica* has been done for allelopathic types. Theoretically, such a character could be incorporated into a cultivar through conventional plant breeding methods or other genetic recombination strategies.

I. Oats : Fay and Duke (1977) evaluated the amount of scopoletin exuded from 3000 accessions of *Avena* spp. and found that some of these accessions exuded three times more scopoletin compared to a standard cultivar `Garry'. They suggested that the wild types of crop varieties once possessed the allelopathic potential to smother weeds but over the time this character has been inadvertently selected against. Oat `PI- 266281' plants caused severe chlorosis, stunting and twisting in wild mustard compared with `Garry' oat.

Table 68. Rice germplasm accessions with potential allelopathic activity to broadleaf signalgrass, ducksalad and purple ammannia (1988-1991)

Weed Species	Germplasm Designation	Country of origin	Radius of Activity (cm)	Weed control(%)
Broadleaf	Daudzai	Pakistan	25.0	90
signalgrass	IR32919522	Phillippines	25.0	90
	UNMD MGVR	Indonesia	25.0	80
	RIKUU 132	Japan	12.5	60
Ducksalad	India AC 1423	India	17.8	85
	Tono Brea 439	Dominican Rep.	16.5	85
	Tsai Yuan Chon	Taiwan	15.2	90
	Donduni Kunluz	Afghanistan	15.2	85
Purple ammannia	IR 104456	Phillippines	17.5	77
	Cuba 65V58	United States	17.5	72
	Cuba 6558A	United States	17.5	70
	IR75693	Phillippines	14.0	80
	IR52 1673	Phillippines	14.0	80

Check cultivars (Rexmont and Palmyra) had zero allelopathic activity
Source : Dilday *et al.* (1992)

II. Rice : Dilday (1992) reported that 347 accessions out of 16134 from 99 countries showed allelopathic activity against five aquatic weeds viz., ducksalad, signalgrass, redstem flatsedge and barnyard grass. Some of the accessions repelled weeds and maintained weed free area with radius of 12-25 cm around their base. Since rice is planted in rows spacings of 10-25 cm, therefore, allelopathic activity would overlap the space between the rice plants or rows and should control these problem weeds in rice. The breeding for allelopathic traits showed promise and success has been achieved upto third generation and research is continuing. Dilday *et al.* (1992) further reported that out of the tested accessions, 347, 161 and 6 accessions demonstrated allelopathic activity to ducksalad, purple ammannia and broad leaf signal grass, respectively (Table 68). Some accessions from India and Bangladesh exhibited allelopathic activity to barnyard grass and *Cyperus iria.* Likewise, Fujii (1992) studied root exudates of 189 rice strains from world over for the suppression effect on weeds in Plant Box method using lettuce as acceptor test spp. Improved cultivars of Japonica rice showed little allelopathic activity, while native varieties of Javanica type and red rice strains showed strong inhibitory activity.

III. Sorghum: *Sorghum bicolor* has great weed controlling potential (Putnam *et al.*, 1983). AlSaadawi *et al.* (1985) screened its 100 accessions from Bulgaria in sand culture for their allelopathic potential to control *Amaranthus retroflexus* weed through root exudates. Twenty five accessions reduced seed germination of *A. retroflexus* by 82% and growth by 85% of the control. The neutral fraction of root exudate was more inhibitory than acidic and basic fractions. Two cultivars `219' and `260' inhibited germination, radicle and hypocotyl growth by 25-30% over the control. The cultivars whose root exudates were non-toxic, exhibited considerable toxicity in their aqueous extracts and decaying materials. However, the degree of toxicity was less than toxic root exudates. Therefore, these cultivars of sorghum could reduce the requirement of herbicides. In further studies, seeds of sorghum cultivars were exposed to 0.5, 1.0 and 1.5 K rad of gamma rays. The seedlings from these seeds produce more inhibitory root exudates, aqueous extracts and decaying residues against growth of *A. retroflexus*. These results have opened a new possible approach for increasing allelopathic potential of crops without causing genetic damage. The inhibitory compounds in sorghum are mostly phenols (Guenzi and McCalla, 1966) and they at concentrations of 10^{-4}, 5×10^{-4} and 10^{-3} M reduced seedling growth, chlorophyll a and b, total chlorophyll and uptake of N, P, K, Fe and Mo in cowpea.

IV. Cucumber : Putnam and Duke (1974) screened 526 accessions of cucumber (*Cucumis sativus*) from 41 countries for allelopathic activity against *Panicum miliaceum* and *Brassica hirta* weeds. Only 3% of these accessions inhibited more than 75 to 87% growth of these weeds. Subsequent research in the field on the allelopathic potential of the more effective accessons showed that higher rainfall decreases allelopathic effects and the results of weed inhibition were not consistent. The `PI 169391' reduced the population and fresh weight of *P. miliaceum* by 84 and 58% with a reduction of 54% in total weeds density. `PI 285605' was half effective in suppressing *P. miliaceum* growth. These both accessions have greater allelopathic activity against weeds than commercial variety `Pioneer' (Lockerman and Putnam, 1979). PI 169391' contained inhibitors in seeds which suppressed germination of *P. miliaceum* (Lockerman and Putnam, 1981a).

V. Soybean: A large number of soybean accessions were investigated for their allelopathic activity on two weeds viz., *Helminthia echiodes* and *Alopecurus myosuroides* (Massantini *et al.,* 1977). Likewise, Rose *et al.* (1984) screened 280 cultivars of soybean to suppress the weeds *Abutilon theophrasti* and *Setaria italica.* Twenty most competitive cultivars of soybean were grown in green house to determine their allelopathic activity in competing with weeds. Root exudates from soybean cultivars grown in sand culture reduced the dry weight of 4-week old velvetleaf by 15% but foxtail-millet was not inhibited. Incorporation of 1% ground soybean drymatter into clayloam soil inhibited germination and dry weight of velvetleaf by 46% each and that of foxtailmillet by 82 and 65%, respectively. The aqueous extracts of some soybean cultivars plants also inhibited germination and dry weight of velvetbean and foxtailmillet.

VI. Miscellaneous: In these field studies accessions of pearlmillet, *Brassica* spp. have been studied but information on allelopathic potential of promising varieties is not available due to lack of facilities for allelopathy research.

(i). Pearlmillet : Narwal *et al.* (1992) screened 13 genotypes of pearlmillet in the field for their smothering effect on summer weeds. All the genotypes reduced the density of major weeds (*Trianthema portulacastrum, Amaranthus* spp.) and minor weeds compared with control. Two genotypes `HHB-67' and `88004A x 833-2' caused maximum decrease in weed population (72.3 and 71.1%) and in weed biomass (85.5 and 87.7%) over the control. These genotypes had the greatest suppression effect on *T. portulacastrum,* the major weed of irrigated crops in Haryana, India.

(ii). Brassica spp.: Sarmah *et al.* (1992) determined the suppression effect of 11, 10 and 8 accessions of *Brassica juncea. B. napus* and *B. carinata,* respectively, on winter weeds of north-west India under field conditions. All accessions of these crops significantly decreased the weeds population and biomass compared with non-crop control. The suppression effect was greater in broadleaved weeds than in grassy weeds. The smothering effect of crops on weeds followed the order *B. juncea* > *B. napus* >*B. carinata.* In *B. juncea* `RH-8605 and `RH-8693' accessions decreased the weeds density and biomass 80% over the control. These accessions reduced the population of *Chenopodium album* by 78% but had no effect on *Cyperus rotundus* than control. The accessions of *B. napus* and *B. carinata* had less smothering effect on the weeds compared with *B. juncea.* The effect of these *Brassica* species accessions on weeds biomass was similar. In *B. napus,* `HNS-8902' and `HNS 11-1' and in *B. carinata* `BCCN-5' genotypes proved most smothering to weeds over the other genotypes. Considerable differences were observed amongst accessions and cultivars in these collections, therefore, more research is required in this area. The interfering ability of plants at later growth stages, besides allelopathic potential in younger plants needs to be studied.

It seems that selection and development of competitive crop varieties could greatly help in weed control. However, breeding efforts in this direction are very limited and more attention need to be paid in future. However, obvious stumbling blocks are that (a) autotoxicity must be avoided, (b) allelopathic effects are generally due to a complex of compounds, (c) classical breeding programmes involved multiple generations and (d) crop breeders may have numerous other goals of high priority (Einhellig, 1985 b). In order to bring out practical conclusive results, an integrated approach involving plant breeders, crop physiologists, biochemists and agronomists is imperative.

The genome of crops to enhance herbicidal activity could be altered through classical breeding programmes or using biotechnology. Major problems for the latter are identifying allelochemicals that may effect a desired change, learning how their production is regulated and development of a genetic engineering approach for transferring this controlling gene into an crop variety. An alternative to genome change is to develop chemicals to manipulate crop metabolism to achieve an increase in allelochemicals production. One limitation to enhancing crop originated weed control is that autotoxicity must be avoided. A second concern is that changes in the chemical balance or chemical constituent of a crop must not have adverse effects on humans, livestock or other non-target organisms (Einhellig and Leather, 1988). ARCO Plant cell Research Institute, Dublin, California, USA is looking for plants with desired properties, locate the gene(s) responsible and develop an approach for transferring the controlling gene into a desired plant using an appropriate vehicle (Burke, 1987).

4.1.1.2. Tree litter for weed control

Very little information is available on this aspect but it offers a great scope owing to availability of large quantity of tree litter during leaf fall. It may play a major role, particularly in agroforestry systems.

I. *Leucaena leucocephalla* : Its 3-4 years old plantations do not allow growth of understorey spp. owing to presence of phytotoxins viz., mimosine, quercetin and 8 phytotoxic phenolics in its leaves (Chou and Kuo, 1987). Mimosine @ 50 to 300 ppm killed seedlings of *Ageratum conyzoides*. Therefore, mimosine and 3,4-dihydroxypridine produced by this tree could be used as natural herbicides.

II. *Vitex negundo* : It is a costal vegetation in Taiwan, its natural leachates significantly retarded the growth of *Digitaria decumbens*. In pots, its 1% aqueous extract retarded *D. decumbens* growth. Its leaves contained various allelochemicals viz., phenolic compounds and 10 flavonoids (Chou and Yao, 1983) and offer potential for weed control.

III. Miscellaneous tree spp.: Jobidon (1986) studied the allelopathic potential of fresh leaf, leaf litter extract of balsam fir (*Abies balsamea* L.), black spruce (*Picea mariana*), Norway pine (*Pinus resina* Ait), jack pine (*Pinus divaricota* Ait) and white cedar (*Thuja occidentalis* L.) on seed germination and growth of four weed spp. viz., timothy (*Phleum pratense* L.), Kentucky blue grass (*Poa pratensis* L.), couch grass (*Agropyron repens* L. Beauv.) and fire weed (*Epilobium angustifolium* L.). The extracts significantly inhibited the germination of the test weed spp. Fresh leaf leachates and leaf litter extract of the coniferous spp. reduced the seed germination of fire weed and couch grass to 26 to 51% and 21 to 55% of control, respectively. The extracts inhibited the root elongation and seedling height of Kentucky bluegrass and timothy. Black spruce fresh leaf leachates were most inhibitory to height, growth and root elongation of test weed species.

4.1.2. Allelochemicals

As traditional methods of discovering and developing new herbicides become more difficult and expensive, the interest in natural products as sources of herbicide chemistries have increased. Besides, public awareness and demand for environmentally safer herbicides, with less persistence, more specific targets and less potential for contaminating groundwater makes searches for new weed control strategies using natural products more attractive. Plants and microorganisms produce hundreds of thousands of secondary compounds. Many of these compounds are phytotoxic and have potential as herbicides or as templates for new herbicide classes. It has been estimated that only about 3% of a possible 400,000 secondary metabolites from plants and microorganisms have been isolated and identified. Only a fraction of those identified have been evaluated for herbicidal or bioregulator activity (Dodge, 1987; Einhellig and Leather, 1988). These allelochemicals offer great potential for pesticides because they are free from problems associated with present pesticides. Therefore, allelochemicals are current areas of research for development of new pesticides (herbicides, insecticides, nematicides, fungicides). These could be used for weed control directly or their chemistry could be used to develop new herbicides.

4.1.2.1. Herbicides from Natural Products

A naturally occurring compound is one that is biosynthesized or is a breakdown product of a natural compound found in the natural environment. Growing awareness of the environmental and public health consequences of agricultural chemicals use in agriculture, has stimulated interest in the search for new selective, easily degradable and environmentally safe herbicides. The ability of some natural plant compounds to effectively inhibit the development of other plants has suggested, that they may be used as herbicides. Besides, recent advances in the microbial and plant biochemistry have stimulated scientific interest into the possible role of these natural products (secondary plant metabolites and microbial toxins) as herbicides.

Use of natural compounds as herbicides or as the chemical basis for the development of new herbicides offers several advantages. (a) The wide array of phytotoxic compounds produced by plants provide many complex chemical structures that are unlikely to be discovered in the traditional synthetic strategies used by pesticides companies. (b) Degradation of natural compounds in the environment proceeds faster than that of synthetic compounds and thus reduces the environmental pollution and groundwater contamination etc. (c) The halogenated hydrocarbons which constitute about 60% of the registered herbicides are of environmental concern, while vast majority of natural compounds from plants pose little health hazards and therefore, are environmentally safe. (d) Purified natural compounds have also several other advantages over the bioherbicides (living organisms as weed control agents) viz., longer shelf-life, wider range of storage conditions, broader environmental range for application, less storage space required and greater ease of application. Presently, two major areas of research appear attractive owing to their commercial applications (i) isolation and characterization of microbial toxins or secondary plant metabolities that may be used as herbicides and (ii) evaluation of secondary plant metabolites and microbial toxins with novel chemistries for the chemical synthesis of new herbicides. Since the phytotoxic natural products are primary products of plants or microbial origin, therefore, these have been reviewed separately.

4.1.2.1.1. Plant products as herbicides

Green plants produce hundreds of compounds that are not involved in primary metabolism of the plants and hence, called the secondary products. The compounds (allelochemicals) involved in interspecific chemical interactions (allelopathy) with higher plants are often phytotoxic or herbicidal to other species or even to the species producing them (autotoxicity). The potential use of plant allelochemicals have been reviewed (Duke, 1990, 1991, Lydon and Duke, 1989). In crop plants, most of the allelochemicals have been removed in breeding programmes, because they reduce the palatability and yield of the crops. This has provided weed species with comparatively high levels of allelochemicals, a competitive advantage over crops. Most of the allelochemicals involved in plant -plant interactions are not as phytotoxic as commercial herbicides. Therefore, they are toxic only at very high concentrations (Duke, 1986 b), are sometimes antagonistic in combination and are inactivated quickly in the soil. Such allelochemicals have been discussed briefly according to their chemical classes (Table 69).

I. Shikimate-derived aromatic compounds: A large proportion of secondary plant products and aromatic compounds viz., flavonoids, tannins, alkaloids, coumarins and various aromatic acids are usually derived via the shikimate pathway and are stored in the vacuoles. They may be derived from precursors of aromatic amino acids, beginning with deamination (flavonoids, tannins etc.) or decarboxylation (alkaloids) of an aromatic amino acid. This group represents the largest category of known allelochemicals and studied extensively (Table 70).

The coumarin (umbelliferone, scopoletin and coumarin) at 1-2 mM concentration reduced the germination and seedling (root and shoot) growth of barnyard grass prickly sida and redroot pigweed. Coumarins in general are about 100 times more phytotoxic than most phenolic acids (Duke, 1986b). Quinones have also greatest potential as pesticides.

Table 69. Inhibition of seed germination and seedling growth by natural plant products

Chemical class	Compound	Concentration (mM)	Weed species	Inhibition (%)			Reference
				Germination	Shoot growth	Root growth	
1	2	3	4	5	6	7	8
A. Phenolic compounds							
a. Benzoates							
	Shikimate	2.0	Barnyard grass	ND	18	ND	Shilling and Yoshikawa (1987)
	Benzoate	2.0	Barnyard grass	ND	13	17	Shilling and Yoshikawa (1987)
	P-hydroxybenzoate	2.0	Barnyard grass	ND	ND	15	Shilling and Yoshikawa (1987)
	o-hydroxycinnamate	2.0	Barnyard grass	ND	ND	80	Shilling and Yoshikawa (1987)
	Protocatechuate	2.0	Barnyard grass	ND	20	18	Shilling and Yoshikawa (1987)
	2, 3-dihydroxybenzoate	2.0	Barnyard grass	ND	44	45	Shilling and Yoshikawa (1987)
	p-aminobenzoate	2.0	Barnyard grass	ND	ND	37	Shilling and Yoshikawa (1987)
	Vanillin	2.0	Barnyard grass	ND	44	ND	Shilling and Yoshikawa (1987)
b. Cinnamate derivatives							
	t-cinnamate	2.0	Barnyard grass	ND	37	97	Shilling and Yoshikawa (1987)
	o-hydroxycinnamate	2.0	Barnyard grass	ND	27	80	Shilling and Yoshikawa (1987)
	Caffeate	2.0	Barnyard grass	ND	14	ND	Shilling and Yoshikawa (1987)
	Ferulate	2.0	Barnyard grass	ND	27	12	Shilling and Yoshikawa (1987)
c. Coumarins							
	Coumarin	1.0	Prickly sida	21	ND	ND	Williams and Hoagland (1982)

Contd......

1	2	3	4	5	6	7	8
		1.0	Redroot pigweed	40	ND	ND	Williams and Hoagland (1982)
	Scopoletin	2.0	Barnyard grass	ND	20	89	Shilling and Yohikawa (1987)
	Umbelliferone	2.0	Barnyard grass	ND	52	94	Shilling and Yoshikawa (1987)
	4-hydroxycoumarin	2.0	Barnyard grass	ND	17	89	Shilling and Yoshikawa (1987)
	d. Quinones						
	Juglone	1.0	Velvetleaf	66	ND	ND	Williams and Hoagland (1982)
		3.0	Velvetleaf	94	ND	ND	Williams and Hoagland (1982)
		1.0	Redroot pigweed	99	ND	ND	Williams and Hoagland (1982)
		1.0	Prickly sida	89	ND	ND	Williams and Hoagland (1982)
		2.0	Barnyard grass	ND	68	53	Shilling and Yoshikawa (1987)
	Arbutin	1.1	Leafy spurge	ND	ND	51	Manners and Galitz (1986)
	Hydroquinone	1.1	Leafy spurge	ND	ND	93	Manners and Galitz (1986)
	Menadione	1.2	Velvet leaf	97	ND	ND	Spencer (1986) *et al.*
	Lawsone	2.5	Velvet leaf	90	ND	ND	Spencer (1986) *et al.*
	Plumbagin	0.8	Velvet leaf	88	ND	ND	Spencer (1986) *et al.*
	Phithocol	6.0	Velvet leaf	100	ND	ND	Spencer (1986) *et al.*
	Chimaphillin	4.0	Velvet leaf	92	ND	ND	Spencer (1986) *et al.*
	Lomatiol	8.0	Velvet leaf	45	ND	ND	Spencer (1986) *et al.*
	7-methyljuglone	8.0	Velvet leaf	95	ND	ND	Spencer (1986) *et al.*
	1, 4-nepthoquinone	2.5	Velvet leaf	97	ND	ND	Spencer (1986) *et al.*

Contd.....

1	2	3	4	5	6	7	8
	2-acetoxy-1, 4-napthoquinone	5.0	Velvet leaf	74	ND	ND	Spencer (1986) *et al.*
	2-methoxy-1, 4-apthoquinone	0.6	Velvet leaf	90	ND	ND	Spencer (1986) *et al.*
	5-methoxy-1, 4-napthoquinone	6.0	Velvet leaf	95	ND	ND	Spence (1986) *et al.*
	4-methoxy-1, 2-napthoquinone	2.0	Velvet leaf	95	ND	ND	Spencer (1986) *et al.*
	2-methoxy-3-methy 1-napthoquinone	4.0	Velvet leaf	100	ND	ND	Spencer (1986) *et al.*
	2-methyl-1, 5-methoxyn apthoquinone	5.0	Velvet leaf	100	ND	ND	Spencer (1986) *et al.*
	e. Flavonoids						
	Flavone	2.0	Barnyard grass	ND	55	67	Shilling and Yoshikawa (1987)
	Flavonone	2.0	Barnyard grass	ND	20	ND	Shilling and Yoshikawa (1987)
	Quercetin	2.0	Barnyard grass	ND	11	ND	Shilling and Yoshikawa (1987)
	Tricin	1.0	Cress	ND	ND	66	Nishimura *et. al.* (1984)
		0.4	Cress	ND	ND	50	Weston *et. al.* (1987)
	Dephinidin	1.0	Cress	ND	ND	78	Paszkowski & Kramer (1988)
	Cyanidin	1.0	Cress	ND	ND	53	Paszkowski & Kramer (1988)
	Myricetin	1.0	Cress	ND	ND	54	Paszkowski & Kramer (1988)

Contd......

1	2	3	4	5	6	7	8
	Catechin	1.0	Cress	ND	ND	52	Paszkowski & Kramer (1988)
	Epicatechin	1.0	Cress	ND	ND	59	Paszkowski & Kramer (1988)
B. Prephenate derivatives							
	Phenylacetate	5.0	Common lambsquarter	ND	ND	50	Shilling *et al.* (1985)
		0.5	Redroot pigweed	ND	ND	50	Barnes *et al.* (1986)
C. Terpenoids							
	Artemisinin	0.03	Pigweed	ND	ND	50	Duke *et al.* (1987)
			Purslane	ND	ND	50	Duke *et al.* (1987)
			Velvetleaf	ND	ND	50	Duke *et al.* (1987)
	Axivalin	1.0	Valvetleaf	ND	ND	90	Takiguchi *et al.* (1979)
	Cnicin	0.7	Spotted knapweed	ND	ND	40	Kelsey and Lochan (1987)
	Calein/calein-A	0.05	Palmer amarnath	25	ND	ND	Fisher and Quizanol (1985)
	Conferin	1.0	*Achyranthus aspera*	ND	ND	90	Amodil & Anaya (1978)
			Bidens pilosa	ND	ND	95	Amodil & Anaya (1978)
			Crusea calocephala	ND	ND	95	Amodil & Anaya (1978)
	Cumambrin	1.0	*Ambrosia cumanensis*	ND	ND	70	Amodil & Anaya (1978)
			Bidens pilosa	ND	ND	70	Amodil & Anaya (1978)
			Crubea calocephala	ND	ND	90	Amodil & Anaya (1978)
	Geraniol	2.0	Pigweed	ND	ND	100	Rizvi & Rizvi (1987)

Contd....

1	2	3	4	5	6	7	8
	β-pinene	s.a.	Pigweed	ND	ND	1000	AlSaadawi *et al.* (1985)
	Piquerol-A	0.3	*Lepidium virginicum*	ND	ND	63	Gressel (1985)
			Bidens serrulata	ND	ND	74	Gressel (1985)
	Anhydrovaxillarin	0.8	Velvetleaf	31	ND	ND	Speice and Muselman (1981)
	Illicic acid	4.0	Velvetleaf	73	ND	ND	Takiguchi *et al.* (1979)
D. Alkaloids							
	Primethylformyl-cyclohexadionyl angelate	4.0	Velvetleaf	79	ND	ND	Springer *et al.* (1987)
	DIBOA(2, 4-dihydroxy-1, 4, benzoxazin-3-one)	0.5	Barnyard grass	ND	ND	50	Barnes *et al.* (1986)
			Large crabgrass	ND	ND	100	Barnes *et al.* (1986)
			Prosomillet	ND	ND	75	Barnes *et al.* (1986)
			Redroot pigweed	ND	ND	85	Barres *et. al.* (1986)

ND = Not determined, s.a. = saturated atmosphere.

Table 70. Sources of plant and microbial phytotoxins with promising herbicidal activity

Phytotoxin	Source	Reference
Plant toxins		
Caffein	Coffee plant	Putnam (1985), Duke (1986a)
Dhurrin	Sorghum plant	Putnam (1985)
Gallic acid	Spurge plant	Putnam (1985), Duke (1986a)
Juglone	Black Walnut tree	Reitweld (1983)
Phloridzin	Apple tree root	Putnam (1985)
Trimethylxanthene	Coffee plant	Rizvi *et al.* (1980)
Microbial toxins		
Anisomycin	*Streptomyces* sp.	Yamada *et al.* (1972)
Bialaphos	*S. hygroscopicus*	Mase (1984)
	S. viridochromogene	Springer *et al.* (1981)
Cytochalasins	*Phomopsis* sp.	Cole *et al.* (1981)
Cercosporin	*Cercospora* sp.	
	Pseudocercosporella capsella	Durbin (1981)
Cyanobacterin	*Scytoneme hofmanni*	Gleason *et al.* (1987)
Herbicidins	*Streptomyces saganonensis*	Takiguchi *et al.* (1979)
Mevinolin	*Aspergillus terreus*	Bach and Lichtenhaler (1983)
Moniliformin	*Fusarium moniliforme*	Cole *et al.* (1973)
Patulin	*Penicillium* sp.	Putnam (1985)
Phaseolinone	*Xylaris* sp.	Karl *et al.* (1973)
	Macrophomina phaseolina	Dhar *et al.* (1982)
Phosalacine	*Kitasstosporia phosalascines*	Pickman (1987)
Rhizobitoxine	*Rhizobium japonicum*	Massantini *et al.* (1977)
Stemphytoxin I	*Stemphylium botryozum*	Barash *et al.* (1982)
Tabtoxin	*Pseudomonas tabaci*	Langston-Unkefer *et al.* (1984)
Tentoxin	*Alternaria alternata*	Duke (1986a)
Toyocamycin	*Streptomyces toypcanensis*	Yamada *et al.* (1972)
Viridiol	*Gliocaladium virens*	Howell and Stipanovic *(1984)*
Ziniol	*Alternaria carthami*	Robeson and Strobel (1984)

Source : Hatzios (1987)

Juglone at a concentration of 1-3 mM reduced seed germination in redroot pigweed and velvetleaf and root growth in barnyard grass by more than 91% (Spelce and Muselman, 1981, Shettel and Balke, 1983, Weston *et al.,* 1987). Several other quinones such as lawsone, menadione etc., have also demonstrated phytotoxicity to weeds similar to juglone (Spencer *et al.,* 1986). No data are available on preplant incorporation, pre-emergence or post-emergence application of plant phenolics or on their effect on mature, established weeds or crop plants. In pot studies, pre-plant application of the coumarin, umbelliferone @ 11.2 kg/ha had no significant effect on shoot growth of corn, oat and soybean but reduced the growth of velvetleaf, pigweed and prosomillet by 27, 40 and 34% respectively (Shettel and Balke, 1983). Likewise, salicylate applied @ 11.2 kg/ha reduced the shoot dry weight of velvetleaf and redroot pigweed by 22 and 39%, respectively but had no effect on shoot dry weight of oat, corn, soybean and prosomillet.

Preplant application of p-hydroxybenzoate @ 11.2 kg/ha reduced shoot growth of velvet leaf (22%), corn (17%), and wild prosomillet (53%). Post-emergence application of hydroquinone @ 11.2 kg/ha totally inhibited the shoot growth of redroot pigweed at the 2-3 leaf stage and had no effect on corn and soybean. Redroot pigweed plants at 7-8 leaf stage were less affected (32% reduction) in shoot dry weight. Therefore, the levels of phenolic compounds needed to produce the desired phytotoxic effect may be increased considerably in field conditions, since they are readily metabolised by microbes, adsorbed by soil organic matter and their uptake by plants is reduced under acidic soil conditions (Eagley, 1983; Dale, 1986; Schmidt, 1988). Campbell *et al.* (1982) reported that α-terthinyl @ 0.15 to 1.93 ppm or phenylheptatriyne @ 0.66 to 1.82% caused 50% mortality in the seedlings of common milkweed (*Asclepias syriaca*), common lambsquarter (*Chenopodium album*) and redclover. Common milkweed was most sensitive and redclover was the least sensitive and sunlight or ultraviolet light enhanced their phytotoxicity.

II. Terpenoids : Duke (1991) reviewed the potential of terpenoids as herbicides and reported that these are potent growth inhibitors than as germination inhibitors. The sesquiterpenoids are less phytotoxic than sesquiterpenoid lactones. Strigol, a sesquiterpenoid derivative from cotton roots is a potent germination stimulant of witchweed (*Striga asiatica* (L.) Kuntz), an obligate parasite of maize, sorghum (Cook *et al.,* 1966) and *Orobanche minor* (Spelce and Muselman, 1981). Its application results in germination and death of a large number of striga seeds (Hsiao *et al.,* 1981). Four sesquiterpenoid lactones confertiflorlin, parthenin, desacetycon-fertiflorin and dihydroparthenolide which share common structural features of the lactone ring of strigol stimulated the germination of witchweed (Fischer *et al.,* 1989). Duvatrienediol diterpenoids from tobacco leaf surfaces are phytotoxic to barnyard grass <0.5 mM concentraion reduced its growth > 50% (Lawson *et al.,* 1988). Rizvi *et al.* (1988) tested citral, citronellol and geraniol for herbicidal activity and found that geraniol was most effective. It completely inhibited germination, radicle and hypocotyl growth of *Amaranthus spinosus* at a concentration of 3 mM. Naturally occurring terpenoids are not currently used commercially as herbicides, however, two products viz., toxephene or camphechlor to control sicklepod (*Cassia obtusifolia*) and cinmethylin (close analogue of 1, 8-cineole) similar to naturally occurring terpenoids have been developed. The former has been withdrawn from the market, while the later one is the first plant derived compound which is active against grasses and some dicot weeds. It has shown promise as a pre-emergence herbicide for several crops. Despite the potent activity of a large number of plant terpenoids as phytotoxins, virtually no information is available on selectivity, efficacy or persistence in the field. Heisey and Delwiche (1984) found that monoterpene alcohol terpenin-4-01 from vinegar weed (*Trichodesma lanceolatum* Benth) was twice phytotoxic than cineole but less toxic than camphor. The monoterpenoid *p*-methane-3-8-diol is a potent germination and growth inhibitor of several weed species at concentrations that have no effect on rice (Nishimura *et al.,* 1984). The disappearance of phytotoxic sesquiterpenoid lactone, isolantolactone in the soil was complete within 90 days (Picman, 1987) indicating no residue problem in the soil.

III. Alkaloids : Caffeine showed considerable selectivity, inhibiting germination of *Amaranthus spinosus* L. at a concentration that had no effect on black gram (Rizvi *et al.,* 1980, 1981). Post-emergence application of caffeine @ 11.2 kg/ha was more toxic to redroot pigweed than to corn or soybean (Shettel and Balke, 1983). The application rates of caffeine required for weed control, however, precludes its viability as a commercial herbicide. Lycoridinol (also known as narciclasine), a podolactone-type alkaloid constituent of bulbs of both redspider lily and *Narcissus* species completely inhibited the growth of gardencress roots at a concentration of 1 mM (Katokar and Geissler, 1980), however, its non-selective inhibition of protein synthesis makes it an unlikely candidate as herbicide.

IV. Miscellaneous compounds : AlSaadawi *et al.* (1983) found salts of several fatty acids with 14-22 carbons at 5 ppm concentrations to inhibit growth of bermuda grass (*Cynodon dactylon*). Dihydrohematinic acid, a tricarboxylic acid from oat husks, inhibited germination of *Amaranthus caudatus* by 54% at a concentration of 10 μm (Rudiger and Lohaus, 1987). Isohematinic acid was equally effective in inhibiting seed germination. Benzylisothiocynate, a mustard oil from papaya completely inhibited the germination and growth of velvetleaf at 0.4 mM (Wolf *et al.,* 1984) but had no effect on maize and soybean. Dale (1986) formulated mustard oil into granules that were effective against sicklepod and milo, however, the efficacy of the formulation was reduced to zero after five days in a moist environment. Allylisothiocynate, another mustard oil, has proved to be an effective seed germination inhibitor (Bell and Muller, 1973).

4.1.2.1.2. Herbicides from microbial toxins

Recent advances in microbial chemistry have stimulated scientific interest into the possible use of microbial toxins as natural pesticides (Misato and Yamaguchi, 1984). The major advantages of these compounds compared to those from higher plants, include (a) the ease of isolation, (b) specificity, (c) efficacy at low concentration, and (d) opportunities for fermentative production. Both pathogenic and non-pathogenic bacteria and fungi produce these compounds (Cutler, 1987). The use of phytotoxins from microbes have been reviewed (Cutler, 1991). Although microbial phytotoxin herbicides may be toxic to some livestock or humans, they are less likely to cause environmental damage or alterations compared to synthetic herbicides. The half-life of most natural compounds in the environment is relatively short, since microbes and plants metabolize them for food sources or detoxify them. Many microbially produced phytotoxins have readymade selectivity because the producing microorganisms have been co-evolved with the host plant species. The search for host specific phytotoxins for control of weeds has started (Robeson and Strobel, 1984). All the bacterial and many fugal phytotoxins produced are non-host- specific. In weed control many microbial phytotoxins have been found to be selective (Cutler *et al.*, 1984).

I. Microbial toxins as herbicides : A major problem with most allelochemicals from higher plants is that they are relatively non-selective, often being autotoxic. This is one reason that breeding for crop plants that produce higher levels of allelochemicals is not a worthwhile strategy. Additionally, many of them are not very efficient, requiring uneconomically high application rates to control weeds, while, many microbial phytotoxins are both selective and efficacious at low rates. Microbially produced toxins as herbicides have several potential advantages over those from pathogens : (a) they have a longer shelf-life and require much less storage space, (b) the logistic of formulations and applications are simple, (c) there is no danger of disease spread to non-traget species, (d) results are not as dependent on environmental conditions, (e) the efficacy of particular application rates can be more accurately predicted, and (f) they may be often less expensive to produce. Soil microorganisms have been the focus of several programmes screening for naturally occurring herbicides (Dale, 1986; Gleason *et al.*, 1987) (Table 70). The microbial toxins could be divided into two categories, host selective toxins and non-host selective toxins.

(i). Host-selective toxins : Host-selective toxins are those that affect a particular species or in some cases only one variety of species. All known toxins in this category are of fungal origin and the characterized ones are phytotoxic to commercially valuable plant species. For example, the toxins HC (*Helminthosporium cabonum race* 1), HS (*H. sacchari*). Hmt (*H. maydis* race T.), AK (*Alternaria kikuchiana* Tanaka), A (*A. mall*) AL (*A. alternaria* f. *bycopersici*) affect maize, sugarcane, Japanese pear, apple and tomato, respectively (Scheffer, 1983). Search for weed pathogens produced phytotoxins are presently under way, these are host selective for weed species of high impact such as the velvetleaf, Johnson grass (*Sorghum halepense*) or water hyacinth (*Eichhornia crassipes*) etc. (Robeson and Strobel, 1984).

(ii). Non-host selective toxins : These toxins are toxic to plant species that the producing microorganism does not infect and the producing organism is not pathogenic. Most of these toxin have been characterized and are produced by bacteria or fungi. Although these compounds are not host selective, some such as tentoxin (Durbin and Uchytil, 1977), stemphytoxin I (Barash *et al.,* 1982) and Zinniol (Robeson *et al.,* 1984) are not toxic to all plant species including the commercial crops.

Therefore, these toxins are more important as herbicides than selective toxins which affect only one weed species. Several new herbicides have been developed from these toxins and chemical manipulations of these compounds might further improve their efficacy (Table 70). Duke (1986 a) has listed thirty non-host- selective toxins, their sources and phytotoxic effect on plant growth and few are briefly discussed below.

A. Anisomcin : It is the first commercial herbicide produced from *Streptomyces* sp. 638 and is strongly phytotoxic to *Echinochloa crusgalli* and *Digitaria* spp. with little or no activity towards several horticultural crops including tomato and turnip (Yamada *et al.,* 1972). It provided the chemical basis for the development of the new synthetic herbicide, methoxyphenone or MK-049 (3, 3`-dimethyl-4-methoxy benzophenone), which @ 4 kg/ha causes extreme chlorosis in *Echinochloa* without affecting the rice crop. It is readily degraded in soil and neither the chemical nor its metabolites are toxic to rice (Yamada *et al.,* 1972).

B. Bipoloroxin : It was isolated from the fungus *Bipolaris cynodontis,* a pathogen of bermudagrass (*Cynodon dactylon*). It is phytotoxic to bermudagrass, Johnsongrass and goosegrass (*Eleusine indica*) at a concentration 0.38 mM and also affects sugarcane. AT 0.76 mM it controlled wild oat and affect corn and at higher rate 3.8 mM controlled fescue but did not affect pigweed and sunflower and no symptons were seen on wheat, barley, cotton (Sugawara *et al.,* 1985).

C. Bialaphos : Bialaphos is a microbial product isolated from the fermentation broth of *Streptomyces hygroscopicus* and *S.*

viridochromogenes. Foliar application exhibits strong herbicidal activity against a wide spectrum of grass and broadleaf weeds (Mase, 1984; Misato and Yamaguchi, 1984). It exhibited very little phytotoxicity when applied to the soil due to its remarkable biodegradability by soil microorganisms and its half life is only 18- 20 days (Misato and Yamaguchi, 1984). In sensitive plants it is metabolized to phosphinothricin, a phytotoxic metabolite which inhibits the activity of enzyme glutamine synthetase. Treatments of plants with phosphinothricin causes ammonia to accumulate at levels exceeding those known to uncouple photophosphorylation and as a result CO_2 assimilation is greatly reduced (Lea *et al.,* 1984). Application of bialaphos with nitrogenous fertilizers enhances their efficiency (Nishimura *et al.,* 1982). The LD_{50} of phosphinothricin for sicklepod was about seventy fold than that of couch grass in green house tests of five weed species in which the LD_{50} ranged from <0.125 to 8.5 kg/ha (Putnam, 1985). It is commercially marketed in Japan under the trade name `Herbiaceae' (Springer *et al.,* 1981; Mase, 1984).

Phosalacine is another microbial compound (produced by *Kitasatosporia phosalacinea*) containing phosphinothricin and its herbicidal behavior is similar to that of bialaphos (Omura *et al,* 1984). Hoechst AG, is developing chemically synthesized phosphinothricin as a herbicide (HOE 39866) with the ISO/BSI approved name of glufosinate (Wild and Mandersheid, 1984; Fischer and Quijano, 1985).

D. Tentoxin : It is a cyclic tetrapetide produced by *Alternaria* causes marked chlorosis in many grass and broad leaved weed species (Durbin and Uchytil, 1977; Duke *et al.,* 1980) but not in crops such as corn and soybean. It kills virtually all major weeds in soybean and is the only toxin known to kill Johnson grass in maize (Duke, 1986a). It is efficacious as a pre-emergence herbicide at low application rates in growth chamber tests. It has not been produced on commercial scale owing to difficulties in its synthesis. A number of analogues have been synthesized but they are less effective than tentoxin (Balandrin *et al.,* 1984).

E. Herbicidins or Herbimycin : These selective herbicides are purine analogues and are produced by *Streptomyces segunonensis* (Arai *et al.,* 1976). Herbicidins A and B have been tested in grasses and broadleaf species. Herbicidins E, F and S are awaiting testing for phytotoxicity. Herbicidin A anJ B, when applied to grasses at 37.5 to 300 ppm, controlled barnyard grass, goosegrass (*Eleusine indica*), tufted mannagrass (*Glyceria septentrionalis*) and water hyacinth in rice crop and were phytotoxic to *Portulaca oleracea, Polygonum* spp. *Amaranthus* spp. and *Commelina communis* L., weeds and to radish and tomato (Cutler, 1988 a). These herbicides are more effective as a pre-emergence than post-emergence application.

F. Moniliformin (3-hydroxycyclobut-3-one-2, 3-dinone): This is a potent phytotoxin produced by *Fusarium moniliforme* (Cole *et al.,* 1973) which inhibits the growth and causes chlorosis and necrosis in several weed species. Its several analogues have been synthesized and some have strong, non- selective activity e.g. moniliformin O-dodecyl and 3, 4-dibutoxy analogues are more active than monoliformin.

G. Rhizobitoxine : This is a product from certain strains of *Rhizobium japonicum* and causes chlorosis and other phytotoxicity symptoms in many weed species and has no effect on soybean, the host of the producing bacterium. Its herbicidal analogue 2-amino-4-methoxybut-3-enoic acid (AMB), a product of both *Streptomyces* and *Rhizobium* strains is also a weak herbicide (Fischer and Bellus, 1983).

H. Irpexil : It is produced by the basidiomycete *Irpex pachyodon,* which is structurally similar to the herbicide benzadox (Fischer and Bellus, 1983). Analogues of benzadox closely resembling Irpexil have been patented as herbicides.

Besides the above, several other potential herbicides from microorganisms are under test viz., `Nigerazines' from *Aspergillus niger* (Iwamoto *et al.,* 1985), `Citreoviridin' from *Penicillum* spp. (Cole *et al.,* 1981), `Cladosporin' from *Cladosporium, Aspergillus,* and *Eurotium* (Anke *et al.,* 1978; Springer *et al.,* 1981; Scheffer, 1983), `Recifelolide' from *Cephalosporium recifel* (Vesonder *et al.,* 1991). `Alteichin' from *Alterneria eichorinee* (Robeson and Strobel, 1984) and `Trichothecenes' (Cutler and Jarvis, 1985).

Soil microorganisms have also been studied for the production of phytotoxic compounds (DeFrank and Putnam, 1977). The phytotoxins from non- pathogenic soil microflora are less likely to be extremely selective which could influence their agricultural use. Only some of the potentially useful microbial metaboilites have been described. It is expected that biodegradable, microbially derived pesticides to control insects, nematodes, pathogens and weeds may be available for commercial use in the near future.

4.1.2.2. Synthetic Derivatives of Natural Products as Herbicides

The problems of (i) high phytotoxicity, (ii) limited crop selectivity and (iii) instability under field conditions associated with several naturally occurring compounds, hinder their commercial development as herbicides. These problems may be over come, by a biorational synthesis of more selective and stable analogues of these chemicals.

Besides, chemical alterations of compounds may also improve their absorption and translocation in weeds and improving the efficacy at the site of action. For example, cinmethylin have desirable traits as herbicide than the natural product from which it is derived. Thus, microbial toxins and allelochemicals provides us with novel chemicals that could be manipulated to produce commercial herbicides. Selected examples of commercially developed herbicides, which are chemically similar to naturally occurring phytotoxins are : Cinmethylin, Bezanin, Quinclorae, Methoxyphenone, Benzodox, Picloram, Glufosinate 3, 4-dibtoxy-moniliformin (Table 71). The development of the herbicide methoxyphenone, as a synthetic analogue of the microbial toxin anisomycin, represents an example of biorational synthesis of herbicide based on natural chemistry (Munakata *et al.,* 1973). Methoxyphenone is marketed in Japan, as a selective herbicide for the control of barnyard grass in rice and is easily degraded in the soil (Munakata *et al.,* 1973; Duke, 1986a). The ammonium salt of glufosinate, a synthetic recemic mixture based on the chemistry of phosphinothricin, has been introduced by Hoechst, AG, Germany, as a non- selective herbicide under the trade name `Basta'. It is well known that phosphinothricin is the active ingredient of the microbial herbicide bialaphos (Herbiaceae).

Table 71. Commercially developed herbicides based on natural products chemistry

Natural Product	Source	Herbicide	Manufacturer & Country
Plant products			
Cineole	Widespread in plants	Cinemethylin	Shell, USA
Benzoxazinones (Hydroxamic acids)	Graminae plants	Benzanin	BASF Germany
Quinolinic acid	Tobacco plants	Quinclorae	BASF Germany
Microbial products			
Anisomycin	*Streptomyces* sp.	Methoxyphenone	Nihon Japan
Irpexil	*Irpex pachyton*	Benzodox	Gulf USA
Fusaric acid	*Fusarium* sp.	Picloram	DOW USA
Moniliformin	*Fusarium moniliforme*	3,4-dibutoxy-moniliformin	CIBA-GEIGY Switzerland
Phosphinothricin	*Strepntomyces viridochromogenes*	Glufosinate	Hoechst AG Germany

Source : Hatzios (1987)

Irpexil and its synthetic analogue benzadox, appear to act as proherbicides and after their application, the plants convert these into amino-oxyacetic acid, a potent inhibitor of pyridoxyl phosphate requiring enzyme (Fischer and Bellus 1983; Duke 1986 a).

Fusaric acid is a marasmin produced by many species of *Fusaria* fungi. Its chlorinated analogue `Picloram' is marketed as a herbicide since last many years.

These examples demonstrate, that the structures of naturally occurring phytotoxins can serve as leads for the synthesis of new successful herbicides. Undoubtedly, the use of biorational design for the discovery of new herbicide chemistries will become increasingly important in the future, following the isolation and characterization of additional microbial toxins and allelochemicals from higher plants.

4.1.3. Future Lines of Research

the indiscriminate use of pesticides including herbicides are causing many problems such as environmental pollution and development of tolerance/resistance in pests (insects, nematodes, pathogens and weeds). Therefore to overcome such problems, research may be undertaken on the following aspects of crop interference: (a) In the world about five hundred plant species are grown, these crop plants offer a great scope for selection of weed smothering species. (b) Within the promising crops, their varieties may be screened for weed smothering potential. (c) The genes responsible for production of allelochemicals, which smother the weeds may be identified and possibly transferred in new varieties of the same or other crops using latest techniques of biotechnology, so that plants could produce their own herbicides. (d) Research efforts may be directed to utilize microbial and fungal toxins to control weeds. (e) New synthetic analogues of natural products having grater selectivity, stability and efficacy to control weeds may be developed at lower cost of commercial production. (f) The agronomical practices (sowing time, plant population, row spacing, crop geometry, crop mixture/intercropping) may be further improved to provide competitive edge to the crops over weeds. (g) Crop rotations with allelopathic crops to control weeds may be developed.

4.1.4. Perspectives

In view of numerous problems arising from chemical weed control, questions have been raised about the continuous use of herbicides. Hence, alternative approaches free from such problems needs to be considered. Use of crop allelopathy and allelochemicals for weed control is a sound alternative, provided more information is generated. Crop interference and allelochemicals may be used in future to partially/fully replace herbicides, to save our plant earth and future generations from ill effects of present hazardous herbicides.

4.2. MANAGEMENT OF INSECT PESTS*

The term allelopathy includes all types of chemical interactions among plants (Molish, 1937). The allelopathic effects of the compounds (allelochemicals) may be either inhibitory or stimulatory (Rice, 1979) and investigations have mainly concentrated on the inhibitory effects of allelochemicals on plants and microorganisms. Studies on inhibitory effects of allelochemicals on insects are not related to allelopathy, while insect allelochemical interactions are included in allelopathy (Ahmad, 1983; Dickens and Payne, 1985; Barbosa and LeTourneau, 1988). Hence, allelopathy may be referred to as interactions among plants, microorganisms and insects which are mediated by chemical products (allelochemicals). The allelochemicals are used in pests control as repellents, antifeedents, growth disrupters and toxicants (Table 72).

Table 72. Distribution of plant chemicals affecting insect behaviour

Attractants	%	Epellents	%	Feeding stimulants	%	Feeding deterrents and inhibitors	%
Alcohols	25	Amino acids	10	Amino acids	20	Alkaloids	24
Alkaloids	3	Phenolics	45	Flavonoids	20	Amino acids	5
Amino acids	5	Terpenes	45	Glucosides	15	Glucosides	7
Disulfides	3			Nucleic acids	5	Phenolics	10
Isothiocynates	5			Sterols	15	Sterols	2
Phenolics	28			Sugars	20	Sesquiterpene lactones	16
Terpenes	31					Tannins	1
						Terpenes	35
n = 61		n = 9		n = 54		n = 19	

Source : Smith (1986)

4.2.1. Allelochemicals

The allelochemicals in entomological literature have been broadly grouped under semiochemicals (Law and Regnier, 1971) and have been further divided according to their intraspecific (pheromones) and interspecific (allelochemics) effects (Table 73).

Table 73. Classification of semiochemicals

Type	Function
Allelochemic	Interspecific chemical messenger
Allomone	Interspecific chemical messenger adaptive advantage with emitter
Kairomone	Interspecific chemical messenger adaptive advantage with receiver
Homeochemic	Intraspecific chemical messenger
Pheromone	Intraspecific chemical messenger specific reaction in receiver

Source : Dickens and Payne (1985).

Among semiochemicals, allelochemicals are of paramount importance in the management of inset pests and have been designated as secondary metabolities. The distinction between primary and secondary plant compounds is that, the former (carbohydrates, amino acids, purine and pyrimidine bases, lipids) are ubiquitous and perform vital roles in all living organisms; while, the later (alkaloids, terpenoids, flavonoids, steroids, tannins) are distributed among organisms and have little value outside ecological context (Kogan, 1986). Although the physiological role of many allelochemicals in plants is still obscure, yet, their ecological funtions are well known (Bell and Charlwood, 1980). In view of their ecological significance, they may be defined as non-nutrients produced by one organism that affect behaviour, health or ecological welfare of another organism of the same or other species. If their activity favours the producing organisms They are called allomones and if their activity favours the receiving organisms they are called kairomones (Kogan, 1986) (Table 74).

Table 74. Principal classes of allelochemicals and their behavioural or physiological effects on insects

Allelochemical factors	Behavioural or physiological effects
ALLOMONES	Give adaptive advantage to the producing organism
Antixenotics	Disrupt normal host selection behaviour
Repellents	Orient insect away from plant
Locomotory excitant	Start or speed up movement
Suppressants	Inhibit biting or piercing
Deterrents	Prevent maintenance of feeding or oviposition
Antibiotics	Disrupt normal growth and development of larvae, reduce longevity and fecundity of adults
KAIROMONES	Give adaptive advantage to the receiving organism
Attractants	Orient insect towards host plant
Arrestants	Slow down or stop movement
Feeding or oviposition excitants	Elicit biting, piercing or oviposition, promote continuation of feeding

Source : Kogan (1986).

Significant success in the management of insect pests using allomones, have attracted the attention of Entomoligists but it is not true with kairomones. Among allomones, antixenotics, repellents, locomotary excitants, suppressants and deterrents have been used to alter the behaviour of inset pests to their disadvantage, but their large scale field applications in agroecosystem is still obscure. On the other hand, antibiotics successfully reduces the insect populations in agricultural crops, therefore, their role in the management of insect pests have been discussed.

4.2.1.1. Antibiotic allelochemicals

The term antibiotic denotes adverse effects of phytochemicals on insect growth, survival and reproduction. Antibiotic allelochmeicals commonly found in plants are toxic amino acids, protease inhibitors, alkaloids, cynogenic glycosides, phenols, tannins, lignins, flavonoids, glucosinolates, toxic lipids, terpenoids, saponins, phytohaemagglutinins. Their role in the control of insects have been summarised (Tables 75 and 76) and discussed below.

I. Toxic amino acids : Among the toxic amino acids, α-canavanine found in *Canavalia ensiformis* caused larval mortality and reduced larval development in *Manduca sexta* (Dahlaman and Rosenthal, 1975), while α-azetidine-2-carboxylic acid found in *Convallaria majalis* reduced larval development in *Heliothis zea* (Adeyeye and Blum, 1989).

II. Protease inhibitors : Several protease inhibitors caused antibiotic effect, on the growth and development of different phytophagous insects. Trypsin inhibitor-I found in cowpea caused larval mortality in *Callosobruchus maculatus* F. (Gatehouse and Boulter, 1983) and reduced the larval development in *Epilachna varivestis* Mulsant and *M.sexta* when larvae was fed on soybean (Shukle and Murdock, 1983). While the same trypsin inhibitor when found in soybean and potato, affected the digestive physiology of *Spodoptera exigua* (Hub.) and *H. zea* (Broadway and Duffey, 1986) and reduced the larval development in *Leptinotarsa decemlineata* (Say), when fed on potato and tomato (Green and Ryan, 1972). Similarly, Trypsin inhibitor-II found in potato affected digestive physiology of *H. zea* and *S. exigua* (Broadway and Duffey, 1986).

III. Phytohaemagglutinins : The phytohaemagglutinin found in *Phaseolus vulgaris* caused larval mortality in *C. maculatus* (Janzen *et al.,* 1976), while, lectin in soybean inhibited larval growth in *M. sexta* (Shukle and Murdock, 1983). Similarly the E-type lectin found in *P. vulgaris* adversely affected the digestion of *C. maculatus* (Gatehouse *et al.,* 1984).

Table 75. Antibiotic allelochemicals and their effects on inect biology

Chemical class	Chemical species	Source plant	Insect species	Effecs	Reference
1	2	3	4	5	6
Toxic amino acid	α- canavanine	*Canavalia ensiformis*	*Manduca sexta* (L.)	Larval mortality and reduced development	Dahlman and Rosenthal (1975)
	α - azetidine-2-carboxylic acid	*Convallaria majalis*	*Heliothis zea* (Boddie)	Reduced development	Adyeye and Blum (1989)
Protease inhibitor	Trypsin inhibitor-I	Cowpea	*Callosobruchus maculatus* f.	Larval mortality	Gatehouse and Boulter (1983)
	Trypsin inhibitor-I	Soybean	*Epilechna varivestis Manduca sexta*	Reduced development	Kraemer *et al.* (1987), Shukle and Murdock (1983)
	Trypsin inhibitor-I	Potato, tomato	*Leptinotarsa decemlineata*	Reduced development	Green and Ryan (1972)
	Trypsin inhibitor-I	Soybean	*Heliothis zea, Spodoptera exigua* (Huls.)	Affected digestive physiology	Broadway and Duffey (1986)
	Trypsin inhibitor-II	Potato	*H.zea, Spodoptera exigua* (Huls.)	Affected digestive physiology	Broadway and Duffey (1986)
Phytochaem-agglutinins	Phytohaemagglutinin	*Phaseolus vulgaris*	*C. maculatus*	Larval mortality	Janzen *et al.* (1976)
	Lectin	Soybean	*M. sexta*	Inhibited larval growth	Shukle and Murdock (1983)

Contd....

1	2	3	4	5	6
	E-type lectin	*P. vulgaris*	*C. maculatus*	Adverse effect on digestive physiology	Gatehouse *et. al.* (1984)
Total phenols	Phenols	Tomato	*H. zea*	Inhibited larval growth	Isman and Duffery (1982)
Acetogenins	DIMBOA MBOA	Maize	*Ostrinia nubilalis*	Reduced growth	Klun *et. al.* (1967), Klun and Robinson (1969)
Saponin	Saponin	Soybean	*Callosobruchus chinensis*	Larval mortality	Appelbaum *et al.* (1969)
	Pea	Ca-precipitable saponin	*Callosobruchus chinensis*	Inhibited growth	Applebaum *et al.* (1969)
Alkaloids	L-tomatine	Tomato	*H. zea*	Larval mortality	Juvik and Stevens (1982)
	Solanin, Demissine	*Solanum* spp.	*L. decemlineata*	Inhibited growth	Kuhn and Gauhe (1947)
	2-tridecanone	Tomato	*M. sexta*	Inhibited growth	Pathak and Dale (1983)
Glucosinolate	Sinigrin	*Brassica* spp.	*Papilio polyxenes*, *Myzus persicae*	Inhibited growth	Kogan (1986)
	Sinalbin, AP-isothiocynate		*Drosophila spp.*, *D. melanogaster*	Killed adult	Lichtenstein *et. al.* (1962)
Terpenoids	Gossypol	*Gossypium* sp.	*Heliothis virescens*, *Earias vittella*	Reduced growth and development	Lukefahr and Houghtaling (1969), Singh and Agarwal (1988)

Contd....

1	2	3	4	5	6
	Heliocide, Hemig - ssypolone	*Gossypium hirsutum*	*H. virescens*	Reduced growth	Lukefahr *et al.* (1974, 1977)
Tannins	Condensed tannins, tannins	*Gossypium hirsutum*	*H. virescens, E. vittella*	Reduced growth	Chan *et al.* (1978), Kamboj (1991)

Table 76. Flavonoids and their effects on insect biology

Flavonoid	Source plant	Insect species	Effects (s)	Reference
Quercetin	*Gossypium* spp.	*Pectinophora gossypiella, Heliothis zea, H. virescens, Schizaphis graminum*	Reduced development	Hedin *et al.* (1974)
	Quercus macrocarpa	*Scolytus multistriatus*	Reduced development	Norris (1977)
Myristicin	*Quercus macrocarpa*	*Bombyx mori*	Growth inhibitor	Isogai *et al.* (1973)
Morin	*Quercus macrocarpa*	*H. virescens*	Reduced development	Hamamura (1970)
Sesamin, Kobusin	*Magnolia kobus*	*B. mori*	Growth inhibitor	Kaminkado *et al.* (1975)

IV. Phenolics, acetogenins and tannins : The phenols present in tomato inhibited the larval growth in *H. zea* (Isman and Duffey, 1982), while DIMBOA (2, 4-dihydroxy-7-methoxy (2H)-1, 4-benzoxazin-3-(4H)-one) and MBOA (6-methoxy-2-2-benzoxazolinone) among acetogenins found in maize reduced larval development of *Ostrinia nubilalis* (Hubner) (Klun *et al.,* 1967). Tannins occurring in cotton adversely affected the larval development of *Heliothis virescens* (Chan *et al.,* 1978) and *Earias vitella* (F.) (Kamboj, 1991).

V. Saponins : Saponin found in soybean and cal-precipitable saponin in *Pisum sativum* caused larval mortalities and inhibited larval growth, respectively in *Callosobruchus chinensis* (L.) (Applebaum *et al.,* 1965).

VI. Alkaloids : Several alkaloids present in different plant species exert antibiosis effects in insects feeding on them. For example, α-tomatine found in tomato resulted in heavy larval mortality of *H. zea,* demissine present in *Solanum* spp. inhibited the larval growth of *L. decemlineata* (Kuhn and Gauhe, 1947) and 2-tridecanone occurring in wild tomato (*L. esculentum*) caused high larval mortality of *M. sexta* (Pathak and Dale, 1983).

VII. Glucosinolates : The sinigrin and sinalbin found in *Brassica* spp. inhibited the growth of *Myzus persicae* S. (Kogan, 1986) and the AP-isothiocynate found in *Brassica* spp. killed the adults of *Drosophilla melanogaster* Meigen (Lichtenstein *et al.,* 1962).

VIII. Terpenoids : Several terpenoids present in various species of cotton, caused anibiosis effects in different species of their insect pests. Gossypol found in some cotton spp. reduced larval growth and development of *H. virescens* (Lukefahr and Houghtaling, 1969) and *E. vittella* (Singh and Agarwal, 1988). While heliocide and hemigossypolone found in *Gossypium hirsutum* reduced the larval growth of *H. virescens* (Lukerfahr *et al.,* 1974).

IX. Flavonoids : Of the several flavonoids reported from different plant species (Table 76), quercetin found in *Gossypium* spp. inhibited the growth of *P. gossypiella, H. zea. H. virescens* and *Schizaphis graminum* Rond. (Hedin *et al.,* 1974), while the same flavanoids when found in *Quercus macrocarpa* inhibited the growth of *Scolytus multistriatus* (Marsham) (Norris, 1977). Similarly myristicin and morin, found in *Q. macrocarpa* inhibited the growth of *Bombyx mori* L. (Isogai *et al.,* 1973) and *H. virescens* (Hamamura, 1970) and sesamin and kobusin found in *Magnolia kobus,* adversely affected insect growth (Keminkado *et al.,* 1975).

4.2.2. Insect-Pest Resistant Varieties

The antibiotics have led to the development of several resistant genotypes in different crops. Some of the insect-pest resistant varieties/cultivars identified in rice, wheat, maize, sorghum, cotton and alfalfa are being used for breeding commercial varieties (Table 77). These studies showed that antibiotic metabolities responsible for antibiosis in different varieties were : saponins, phenolic acids, 6-MBOA, DIMBOA, gossypol, quercetin and 2-tridecanone (Pathak and Dale, 1983). There are several other secondary metabolities responsible for producing antibiosis, these may be utilized to breed resistant varieties or strains against insect-pests of field, fruit and vegetable crops.

Table 77. Insect resistant cultivars/lines used for breeding commercial varieties (Based on antibiotic factors)

Crop	Insects	Varieties
Rice	Brown planthopper Rice gall midge	Mudgo, ASDT, IR 26, Eswarakora MTUIS, HR 42, HR 63, Ptb 18. Ptb 21, TBS 446, JBS 673, Siam 29.
	Rice whorl maggot	IR 26, OR94-13, IR 40
	Striped stem borer	TKM6, Tailing 16
	White backed plant hopper	N 22, ARC 6003, Dharia
Alfalfa	Pea aphid, Spotted alfalfa aphid	Apex, Washoe, Dawson Meşilla, Mesa-Sirsa, Zia
Cotton	Boll weevil	AK Dajura NC Margin
Maize	Corn earworm	Zapalote chies
	European corn borer	B 52
Sorghum	Chinch bug	`Atlas' sorgo
	Green bug	IS 809, KS 30, SA 7536-1
Wheat	Hessianfly	W 38, Pl 94587, Ribetro

4.2.3. Conclusions

Allelopathy may be referred to as interactions among plants, microorganisms and insects, which are mediated by chemical products (allelochemicals). Allelochemicals and homeochemicals constitute semiochemicals in entomological context. Among allelochemicals, allomones are most promising in the management of insect-pests of agricultural importance owing to their inhibitory effects on insect growth, survival and reproduction. Direct adverse effects of antibiotic allomones on insect biology make them potential sources for inbuilt defence mechanisms in plants against insects. Efforts are continuing to exploit antibiotic compounds, for developing insect-pests resistant plants. Some of the important antibiotic allelochemicals, viz, toxic amino acids, flavonoids, phenols, lignins, tannins, proteinase inhibitors, alkaloids, cynogenicglycosides, glucosinolates, terpenoids, saponins and phytohaemagglutinins have been listed and discussed.

4.3. MANAGEMENT OF PHYTONEMATODES

In nematology, allelopathy word is not used commonly. For the control of plant-parasitic nematodes, allelopathic effects may be implicit in some organic amendments of soil, use of phytotherapeutic substances, toxic plants/their parts/extracts and antagonistic crops. Allelopathy is defined as chemical warfare between plants. It represents the plant against plant aspect in a broader arena of chemical ecology, within which all the organisms tend to respond to, or regulate one another by producing a variety of toxins, chemical inhibitors, stimulators, or attractants. Allelochemicals affect the interactions between different plants and include effects mediated through associated organisms. Plants release them in several ways viz., volatilization, root exudates, leachates and during decomposition of crop residues. Nematoxicity of plant extracts and crop residues also reflects allelopathy. The field of allelopathy offers challenges for scientists of several disciplines, such as botanists, chemists, weed scientists, entomologists, plant pathlologists and nemotologists. Nematologists can hopefully exploit allelopathy for the management of phytonematodes very usefully. However, tremendous research inputs are needed to evolve economical nematode management practices, which are closely in tune with nature, ensuring environmental safety and sustainability. Using plant products or some plants *per se,* being available near the farmers` fields, is likely to be handy for nematode control. Such materials may replace chemical pesticides, which cause environmental hazards. Also, initially, characteristic patchy occurrence of phytonematodes warrants effective use of such materials of low cost or no cost.

4.3.1. Nematicidal Plants

Bhatti and Nandal (1994) have reviewed the information on nematode management through phytotherapeutic substances. In agroecosystem, there are naturally occurring plants like weeds, shrubs and cultivated plants which are nematicidal or nematostatic in action. Plants with nematicidal compounds have been discovered, owing to their marked suppression of certain nematode populations in screenhouse or in field experiments. Toxic compounds are not uniformly distributed throughout the plants tissues/cells and variation in nematoxicity depends upon the plants parts used (Rhode, 1972). For example, the nematicidal compounds in the roots of marigolds (Uhlenbroek and Bijloo, 1959) are located mainly in the endodermis. Plant shoots and leave are often toxic to nematodes, whereas, roots are parasitised by phytonematodes (Bhatti, 1988). Therefore, it is often difficult to relate the *in vitro* mode of action to that occurring *in vivo.*

Research programmes on screening nematicidal plants and isolation and characteristation of toxic principles may yield new groups of nematicides. However, Bhatti and his associates have projected the potential of nematicidal plants and their products for management of nematodes in different economic crops under field conditions. Other examples are (a) the inhibitory effects of isothiocynates in root diffusates of some cruciferous non-hosts of potato-cyst nematodes on the emergence of *Globodera rostochiensis* larvae (Triffit, 1930; Ellenby, 1945b) and (b) inhibition of *Angunia tritici* in wheat by root diffusates of *Polygonum hydropiper* (Sukul, 1970).

Juices from numerous plants or their parts and extracts with organic solvents or root exudates contain nematicidal or nematostatic compounds (Gommers, 1973; Rice, 1984 and Devakumar, 1994). Chemical structures of naturally occurring nematicides have been described. Bhatti and Nandal (1994) have reviewed and listed effects of phytotherapeutic substances on nematode mortality and hatching.

4.3.2. Crop Residues

Mishra and Mojumder (1994) have reviewed the management of plant- parasitic nematodes through organic amendments of soil including crop residues. Soil incorporation of mature crop residues such as wheat straw, rice straw, paddy husk, etc. is not a common practice but many workers have reported their nematicidal value. It is the proper decomposition of such materials, that suppresses the nematode populations. Johnson (1963) used every type of dried crop residues in pots containing soil infested with root-knot nematode and observed reduction in the number of root-knot galls per plant. In 1967, he confirmed that chopped mature dried residues of lespedeza, alfalfa, oats and flax when incorporated @ 10 tons per acre into field infested with *Meloidogyne incoginta,* significantly reduced the incidence of root-knot in tomato. The ploughing of rye crop in the infested soil effectively controls the *Pratylenchus penetrans* identical to D.D. fumigation (Mountain and Elliott, 1962). The soil amendment with cotton waste, lucerne pellets and lucerne hay reduces the *Tylenchulus semipenetrans* (Mankau and Minteer, 1962). Rice straw @ 9.0 or 17.9 tonnes per acre reduced the populations of *Belonolaimus longicaudatus* and other plant-parasitic nematodes (Tomerlin and Smart, 1969). Johnson (1972) studied the effects of organic matter on root-knot incidence in potted tomato plants and observed 75-90% reduction in galling on plants grown in soil mulched with flax, lucerne or orchard grass residues. Prasad *et al.* (1972) found that wheat straw and *neem* cake + NPK gave maximum reduction in the plant-parasitic nematodes associated with wheat and mungbean. Mishra and Prasad (1978) reported good reduction in *M. incoginta* incidence in tomato, through application of wheat straw and paddy husk. In fact, nematodes are killed during the decomposition of organic matters. Therefore, the residues of cultivated crops, weeds, grasses etc. if properly decomposed at the right time may suppress nematode population. Only appropriate research efforts can develop easy pest control practices in this potential area.

4.3.3 Phytotherapeutic substances affecting rhizosphere microbes

Bhatti (1988) observed that leaf extracts of *Azadirachta indica, Bougainvillea spectabilies, Calotropis procera, Cannabis sativa, Chenopodium album, Cirsium arvense, Datura stramonium, Eucalyptus* spp., *Leucaena leucocephala, Mentha piperita* and *Ricinus communis* either increased or decreased the mycelial dry weight or *Paecilomyces lilacinus* (oviparasitic fungus), *Aspergillus nidulans, Alternaria alternata, Cephalophora irregularis, Rhizoctonia solani* and *R. bataticola* (pathogens). All the plant extracts except *B. Spectabilis* encouraged the growth of *P. lilacinus* (biocontrol agent of nematode pests), while against pathogens. The effects varied from beneficial to harmful, indicating necessity of a detailed study on all the related micro- organisms inhibiting the rhizosphere.

The extracts from *Convolvulus pluricaulis* and *Evolvulus alsinoides* almost completely inhibited spore germination and mycelial growth of *Alernaria brassicae, A. brassicola* and *Fusarium oxysprum* (Kapoor *et al.,* 1981). Bhowmik and Choudhary (1982) studied the antifungal activity of leaf extracts of 10 medicinal plant species against *Alternaria alternata* and greatest inhibiton *in vitro* was obtained from *Acalypah indica* followed by camphor, *Vitex negudo* and *Azadirachta indica.* The extract of *Azdirachta indica* at 2.5 and 5.0% concentration inhibited the growth of *Fusarium moniliforme, Gibrella fujikurol* by 57.8- 66.7%, *Aspergillus niger* by 77.8%, *Drechslera rostratum* and *Setosphaera rostrata* by 83.3-88.9% and *Macrophomina phaseolina* by 61.1-75.6% and at 10% concentration gave 100% inhibition in radicle growth of all the test pathogens (Vir and Sharma, 1985). Nath and Bordoli (1986) screened 80 plant species for fungitoxic properties, 12 of them gave promising results against *Curvalaria eragrostides* and *Heminthosporium oryzae* and the maximum inhibition was obtained with the extract of

Ranunclus scleratum containing protoanemonin. Thus it is evident that different plant species have varying reducing effect against beneficial/pathogienic fungi, which is attributed to their chemical composition. Since the information on the qualitiative chemical composition of different plant is meagre, it is difficult to speculate the role of any particular compound. However, the mertis and mertis of any particular plant species and its role after addition to field soil should be thoroughly assessed before making any recommendation for field application.

4.3.4. Field Utilization of Nematicidal Plants

Numerous workers have attempted to determine the potential of nematicidal plants for control of nematode pests under field conditions. For this purpose, chopped plant materials were mixed in potted/field soil in varying proportions before sowing. Bhatti (1988) tested 8 plant species against *Heterodera cajani* and 7 plants against *H. avenae.* Chopped leaves mixed in potted soil @ 10, 20, 40 and 80 g/kg soil were highly superior. *Azadirachta indica, Datura stramonium, Leucaena leucocephala* and *Ricinus communis* were most effective against *H. cajani, B. spectabilis* and *H. avenae.* Similar observations with different plants against nematode pests of economic importance have been reported (Hameed, 1970; Yuhara, 1971; Bhatti and Dhawan, 1980; Gupta and Ram 1981; Nandal and Bhatti, 1983; Dutt and Bhatti, 1986; Nandal and Bhatti, 1986; Goswami and Vijayalakshmi 1986 b; Dahiya *et al.,* 1986; Jain and Bhatti, 1988 and Akhtar and Alam, 1990).

In fields studies, Bhatti (1988) obtained maximum wheat yield (520 g/m^2) by mixing chopped leaves of castor @ 40 g/kg soil, which was identical to *Calotropis procerea* @ 40 g/kg (410 g) and carbofuran @ 2 kg a.i./ha (435 g) and significantly superior from other treatments including check. However, the differences in reduced cyst population were non-significant. Similar observations with plant species against different nematode pests were reported by Patel and Desai (1964), Gommers (1972) and Egunjobi and Afolami (1976).

4.3.5. Allelochemicals as Nematicides

Devakumar (1994) has reviewed the nematicidal principles of plant origin. Control of plant parasitic nematodes using antagonistic plants and phytochemicals offers an alternate strategy to the prevalent use of synthetic nematicides. Some of the advantages of natural pest control chemicals compared to synthetic are their facile biodegradability, selective toxicity only to target pests, safety to non-target organisms, cleaner environment and their renewable nature. Earlier, Gommers (1981) reviewed the biochemical interactions between nematodes and plants and their relevance to control.

The compounds isolated and identified as nematicidal principles from plants are chemically diverse viz., simple straight chain organic acids, phenolic compounds, phenolic phytoalexins, ployacetylenes, organosulphurous compounds, alkaloids and other N-containing compounds.

4.3.6. Microbial Products as Nematicides

Of the large number of organisms that are parasites and predators of soil nematodes, till now none has proved very effective for biological control of economically important plant-parasitic nematodes and disease agents (Mankua, 1975).

The avermectins represent the allelochemicals and are macrocyclic lactones derived from the mycelia of *Streptomyces avermitilis* (Stretton *et al.,* 1987). They have potent anthelimntic and insecticidal activities and are widely used as agents affecting parasitic nematodes. In microplot studies, avermectins were compared against *Meloidogyne* spp. and *Tylenchulus semipene-trans* with ethoprop, aldicarb, phenamiphos, oxamyl and carbo-furan. Yields of crops with avermectins were reported better than with chemical pesticides at the doses used.

4.3.7. Conclusions

The total number of plants screened is much less than number of insecticidal plants known (Parmar and Devakumar, 1993). Some of the nematicidal plants such as *Calotropis procera, Datura stramonium* and *Xanthium strumarium* are available in abundance close to farmer's fields and can be used advantageously for nematode control. Some cultivable plants like *Ricinus communis* could be mass produced. With the disillusionment and fears of health hazards and environmental pollution from synthetics, people have become more attracted towards the concept of organic farming. Generally, the leaf extracts of nematicidal plants are thermostrable. The nematicidal action seems to be a composite effect of several compounds. Further, there appears to be nematicidal specificity in some plants. Efficacy of phytotherapeutic substances in the soil may persist for 6-8 weeks. Chopped plant material mixed in the soil needs decomposition period of 2-4 weeks before sowing (Bhatti and Nandal, 1994).

Nematicidal compounds in numerous plants have been described, efforts are needed to characterise nematicidal principles, study their mode of action and test structural analogues for potentiation of nematicidal activity. Allelopathic management of plant- parasitic nematodes offers potential in view of cost-effectiveness, environmental safety and socio-economic viability.

4.4. BIOREGULATORS

These are exogenous non-nutrient substances that manipulate growth, development and composition of plants and function by interaction with the endogenous phytohormone groups. Their actions include growth retardation, flower induction, hastening maturity or senescene, enlarged biomass production etc. Allelochemicals provide a promising source for new growth regulating compounds. Three compounds agrostemin, triacontanol and brassinolide have received maximum attention, Bioprodukt (1984) summarized Yugoslavian work showing that 100 g agrostemin per hectare through seed treatment or foliar spray hastened germination and increased yields of wheat, maize, sunflower and sugarbeet by 10, 15, 15 and 10%, respectively. It also enhanced the oil consent of sunflower by 4%. It has proved beneficial to vegetables, flowers, fruits, pastures and forests species. Triacontanol, a 30-carbon primary alcohol, was isolated as a growth promoting factor from alfalfa. Its foliar applications increased the yields in cucumber, carrot, rice, corn, soybean and others. Inconsistant results, perhaps due to formulation problems and to method, rate and time of application, reduced its efficacy (Laughlin *et al.,* 1983). Extensive work has been done on evaluation of brassinolide, a sterioid isolated from rape (*Brassica napus* L.) pollen as a yield stimulant. Brassionlide and several analogues have been synthesized (Maugh, 1981) but they are too expensive for use in field crops. All the bioregulators have shown increase in yields of major crops, but inconsistency between the locations, genotypes and spraying dates, plus difficulties in formulations have hindered their commercial use.

CHAPTER - 5

REMOVAL OF PHYTOTOXINS FROM AGRICULTURAL LANDS

The possibility of accumulation of toxic compounds from crop plants in agricultural soils was recognized early in this century (Schreiner and Reed, 1907 a,b). As indicated in text earlier, a large variety of organic compounds are released by numerous plant species, each having an active life in the soil. These toxins may be used to control weeds, insects, nematodes and disease pathogens as already outlined but these may be also harmful to the crops. Therefore, it is necessary to minimize their toxic effects on crop plants in following ways :

I. Improvement of water drainage

In waterlogged areas, where two or three rice crops are grown continuously in one calendar year, the yields of second and third rice crops is reduced partly due to the phytotoxic substances produced during the decomposition of rice residues in soil, besides denitrifying organism *Pseudomonas putida* was dominant microbe during the decomposition process. Chou (1987) reported that drainage of waterlogged fields increased the rice yields by 30%, because higher amounts of phytotoxins in the poorly drained soil caused greater phytotoxicity compared with the well drained soil. This indicated that drainage removed the phytotoxins from the poorly drained soil and this increased the soil productivity and rice yields. Wu *et al.* (1976 b) found that total quantity of phytotoxic phenolics and fatty acids (volatile and non-volatile) in the well drained soil were lower (0.52 x 10^{-2} mole and 7.61 μmole per 100 g soil) compared to poorly drained soil (1.92 x 10^{-2} mole and 37.61 μmole per 100 g soil, respectively). Therefore, rice yields were lower in poorly drained soil.

II. Water flooding

After several years of sugarcane rationing, a significant reduction in the yield generally occur (Section 3.1.1.1.4.I). This was attributed to phytotoxins produced during the decomposition of sugarcane residues left in the soil and imbalance of the microbial population (Wu *et al.,* 1976 a,b). The population of fungi *Fusarum oxysporum* and concentration of its product fusaric acid was exceedingly high in the rhizospere of the low yielding sugarcane fields (Wu *et al.,* 1976 b). The water flooding of these fields considerably decreased the *Fusarium* population as well as the quantity of fusaric acid in the rhizospere and thereby, increased sugarcane yields (Chou, 1987).

III. Crop rotation

In many monoculture fields, often soil sickness problem is reported, which is assumed due to the imbalance of soil microorganisms, accumulation of soil toxins, mineral deficiency or abnormal soil pH, resulting in decreased crop yields and soil productivity. Crop rotations have been reported to avoid or eliminate such problems but only a few studies are concerned with allelopathic effects. In Taiwan, pangola grass produces a highly productive pasture and constitutes a dominant specie, however, after several years of growth its productivity declines. Chou and his associates found that the pangola grass produced phytotoxins which suppressed its own growth (Chou, 1983). Therefore, a crop rotation of pangola grass-watermelon was recommended. The yield of pangola grass after watermelon was 40% higher compared with its monoculture, owing to disappearance of phytotoxins from the pangola grass fields. In 8-year study at IRRI, Phillipines, continuous monoculture reduced the growth and yield of rice, maize, mungbean and cowpea. The growing of these corps in rotations such as sorghum/maize/rice - mungbean/cowpea or rice- sorghum/maze increased their growth and yields because crop rotation overcome the soil sickness of monocultures (Ventura *et al.,* 1984).

IV. Application of nutrients

In field studies on poorly drained rice soil, Wu *et al.* (1976 a) applied ammonium sulphate, lime and green manure to decrease the phytotoxic effects of decomposing rice residues in soil. The liming increased the rice yields than other treatments, because calcium adsorbed and detoxified some phytotoxins. Chou and Chou (1979) reported that application of ammonium sulphate fertilizer produced higher rice yields compared with nitrate fertilizer, indicating that ammonium ions in former fertilizer bound some phytotoxins and thus eliminated the phytotoxicity of decomposing rice residues (Chandrasekaran and Yoshida, 1973). The root system of rice plants was healthy and well-developed under the ammonium sulphate dressing than under the nitrate fertilizer dressing (Chou, 1987). It is concluded that some nutrients may play a chelating role in detoxifying the phytotoxins in the soil and consequently increased the yield of crops.

V. Application of phytotoxin adsorbents

The adsorbents may effectively detoxity the contaminated soil. Breazeale (1924) reported that extracts of certain soils were toxic to wheat seedlings in water culture and this toxicity was removed wholly or partly by carbon black, $CaCO_3$ and ferric hydrate etc. The soil toxicity from asparagus residues was eliminated with addition of activated charcoal to pots (Putnam, 1983 b). Soils rich in organic matter have sufficient adsorption sites and microbial population to inactivate the allelochemicals rapidly. Adsorbents are very useful in perennial crops, where toxins tend to accumulate. The use of companion plants that contribute organic matter or inoculation with microganisms which readily metabolized the toxins might prove useful in perennial crop ecosystem to overcome soil toxicity (Putnam and Duke, 1978).

The binding of many phytotoxic substances to clay minerals or other organic compounds decreased the phytotoxicity (Wang *et al.*, 1971; Rice, 1984). Wang *et al.* (1978) found that protocatechuie acid could be polymerized with humic acid using clay minerals as heterogeneous catalysts. Humic acid may polymerize amino acids, flavonoids, terpenoids, aliphatic acids and other nitrogen containing compounds, thus maintaining the soil productivity.

CHAPTER - 6

FUTURE PROSPECTS

Most of the research in allelopathy has been done during the last three decades (1960 onwards) and quantum of research is increasing. It is hoped that till the end of Twentieth Century, allelopathy may be used in various forms to increase crop productivity. Recent developments (a) in isolation, characterization and identification techniques of allelochemicals, (b) new techniques in biotechnology, (c) use of allelochemicals to control pests (weeds insects, nematodes, diseases) and to stimulate crop growth and yield and (d) new weed control practices using allelopathic crops may help in achieving this goal.

6.1 PROSPECTS OF RESEARCH

Allelopathy is a very young field of science, therefore, research may be continued in all areas investigated in the past but certain areas need special emphasis.

I. Weed control : To achieve complete weed control in field crops with minimum use of present herbicides, research may be pursued in following fields :

(a). *Biological weed control :* Research may be done to use allelopathic crops in (i) mulches, (ii) cover crops, (iii) crop rotations and (iv) in crop mixtures/intercropping systems, underplanting in orchards or agroforestry systems to supprss or control weeds and to develop allelopathic crop cultivars through bio-technology for control of major weeds.

(b). *Reducing weed seed Bank:* To decrease weeds seed banks in soil, either through their decay using microbial toxins as adsorbents or stimulation of their seed germination.

(c). *Allelochemicals :* Possible use of certain allelochemicals from plants or microbes (bacteria, fungi etc.) as herbicides or as structural models for herbicide development.

II. Develop Biotechnology : Techniques for incorporating a controlling gene into crops for their own production of allelochemicals and to control their production. To breed pests (insects, nematodes, disease) resistant varieties of crops based on their high allelochemical content.

III. Biological Nitrosen Fixation : Development of legume crops with high Biological Nitrogen fixation potential and least inhibitory effects on component crops in intercropping systems and on succeeding crops in crop rotations.

IV. Weed Interference : To quantity effect of weed interference on crop yields and relative contribution of allelopathy and competition for each weed species.

V. Crop to Crop Relationships : These need to be investigated in great depth, to determine which crops can follow others with the least inhibitory or stimulatory effects. The autotoxicty of crops also needs in-depth investigations.

VI. Antagonistic Effects : To study antagonistic effects of plants on soil-borne plant pathogens and the effects of allelochemicals in the predisposition of plants to infection by pathogens. To understand the production of microbal allelochemicals in soil, that affect plant growth.

VII. Isolation and identification of allelochemicals : and to screen them for biological activity. Bacterial and fungal metabolities have great potential for use as herbicides insectricides, nematicides, fungicides or as biorgulators particularly those which stimulate the crop growth and yield.

VIII. Threshold Concentrations: After identification, concentration of allelochemicals in substrate may be calculated and their threshold concentrations for activity should be determined against test plants, using combinations of compounds present in the substrate, in addition to individual ones. This is suggested because, many important allelopathic effects have been overlooked owing to use of single allelochemical in determining threshold concentrations for activity.

IX. Concentration of Allelochemicals: Little information is available on factors affecting concentration of Allelochemicals (except phenolics) in soil.

X. Activation and Effectiveness of Allelochemicals : Studies are needed on factors affecting in activation and effectiveness of Allelochemicals after their release from donor plants. Very little is known concerning the binding of these chemicals in soil and the effects of the binding on their activity. Virtually nothing is known, concerning the role of soil texture in the accumulation of allelochemicals to physiologically active concentrations. Temperature stress markedly accentruates the allelopathic effects of some allelochemicals on growth of crops, detailed studies are needed on this aspect.

XI. Mechanism of Action: The mechamism of action of different allelochemicals also need in-depth studies.

These suggestions for future research in allelopathy related to crop production are only a few of the large numbers that could be listed. It is hoped that these areas of future research may provide new impetus to progress in some vital areas of allelopathy.

6.2 PROBLEMS IN ALLELOPATHIC RESEARCH

Mandava (1985) listed some of the following problems in allelopathy research : (a) Inability to transfer laboratory results into field situations where the phenomenon is observed, (b) complexity of allelopathic chemical mixtures and improper assessment of their concentrations and (c) failure to detect chemicals that might be biologically active because it may be impossible to detect them in the presence of other chemicals.

Once we know the chemical nature of the allelopathic agents and their effects on plant growth dynamics, as well as on health and environment, we can apply genetic manipulation and biotechnology to develop phytoxin-resistant plants and to reduce phytotoxin levels from the donor plants. These approaches have a dual purpose, because they may contribute to increased agricultural productivity and help to minimise the potential risks on health and environment.

CHAPTER - 7

GLOSSARY OF CROPS AND WEEDS

Allelopathy being a multidisciplinary field of research, hence, the readers of this book may be from diffirent disciplines like agricultural sciences and biosciences. Therefore, to improve their understanding of the book, common English names of Crops and weed species have been used; however, at the first mention of common name, its scientific name has also been given. For ready reference, common and scientific names of all the crops and weed species included in this book have been in Tables I and II of this chapter.

Table I. Common and scientific names of crops

Common Name	Scientific Name
* Alfalfa	*Medicago sativa* L.
* Asparagus	*Asparagus officinalis* L.
* Barley	*Hordeum vulgare* L.
* Beans (Cuba, dry, kidney, navy, pinto, snap)	*Phaseolus vulgaris* L.
Broad/field bean	*Vicia faba* L.
Horse bean	*V. faba* var. *equina* L.
Mothbean	*Vigna aconitifolius* (Jack) Mar.
* Mung bean	*V. radiata* (L.) Wilczek
Urd bean	*V. mungo* (L.) Hepper
* Birdsfoot trefoil	*Lotus corniculatus* L.
* Broccoli	*Brassica oleracea* L. var. *italica* L.
* Buckwheat Annual	*Fagopyrum esculentum* Moench.
* Perennial	*F. cymosum* Meissn
* Cabbage/Chinese cabbage	*Brassica oleracea* L. Var. *capitata* L.
Cauliflower	*B. oleracea* L. var. *botrytis* L.
Chickpea	*Cicer arietinum* L.
Chillies/red pepper	*Capsicum annuum* L.

Clover Alsike	*Trifolium hybridum* L.
* Crimson	*T. incarnatum* L.
Egyptian	*T. alexandrinum* L.
* Ladino/white	*T. repens* L.
Persian	*T. resupinatum* L.
Red	*T. pratense* L.
Subterranean	*T. subterranean*
Clusterbean	*Cyamopsis tetragonoloba* (L.) Taub.
Cogon grass/Alang Alang	*Imperata cylindrica* (L) Raeusch
*Corn/maize	*Zea mays* L.
Sweet	*Z. mays* Subsp. *rugosa* Bonf.
Cotton	*Gossypium hirsutum* L.
Cowpea	*Vigna unguiculata* (L.) Walp.
* Cucumber	*Cucumis sativus* L.
Cumulative cress	*Lepidium sativum* L.
Egg plant	*Solanum melongena* L.
Fenugreek	*Trigonella foenum-graecum* L.
Finger millet	*Eleusine corocana* (L.) Gaertn
Flax/linseed	*Linum usitatissimum* L.
Foxtail Italian millet	*Setaria italica* (L.) Beauv.
Garlic	*Allium sativa* L.
Groundnut	*Arachis hypogaea* L.
Hemp	*Cannabis sativa* L.
Indian colza	*Brassica campestris* var. *sarson* Prain.
Rape	*B. campestris* var. *toria* Dutt & Full.
Indigo	*Indigofera articulata* Gouan.
* Jute	*Corchorus capsularis L.*
Kentucky bluegrass	*Poa pratensis* L.
Lentil	*Lens culinaris* Medik.
Lettuce	*Lactuca sativa* L.
* Lupine white	*Lupinus albus* L.
Yellow	*L. luteus* L.
* Mustard	*Brassica campestris* L.
* Black	*B. nigra* (L.) Koch
* Indian	*B. juncea* (L.) Czern & Cosson
* White	*B. hirta* (L.) Moench
* Oat	*Avena sativa* L.
Okra	*Abelmoschus esculentus* (L.) Moench
Onion	*Allium cepa* L.

Pangola grass	*Digitaria decumbens Stent.*
* Pea/field pea	*Pisum sativum* Subsp. *arvense* Poir (L.)
* Pearlmillet	*Pennisetum glaucum* Br. emend Stuntz (L.)
* Pigeonpea	*Cajanus cajan* (L.) Millsp.
Poppy	*Papaver* sp.
* Potato	*Solanum tuberosum* L.
Prosomillet	*Panicum miliaceum* L.
* Radish	*Raphanus sativus* L.
* Rape/rapeseed	*Brassica napus* L.
* Oilseed	*B. napus* L. var. oleifera Metz.
* Rice	*Oryza sativa* L.
Rye	*Secale cereale* L.
Rye grass	*Lolium multiflorum* Lam.
Safflower	*Carthamus tinctorius* L.
* *Sarson* brown	*Brassica campestris* var *brown sarson* Prain
* Yellow	*B. campestris var yellow sarson* Prain
* Sesbania	*Sesbania* spp.
* Sorghum	*Sorghum bicolor (L.) Moench.*
* Sugarbeet	*Beta vulgaris* L.
* Sugarcane	*Saccharum officinarum* L.
* Sunflower	*Helianthus annuus* L.
Sweet clover	*Melilotus indica* (L.) All.
White	*M. alba* Medik.
* Sweet potato	*Ipomoea batatas* (L.) Poir.
* Taro	*Colocasia esculenta* (L.) Schott.
Timothy grass	*Phleum pratense* L.
* Tobacco	*Nicotiana tabacum* L.
* Tomato	*Lycopersicon lycopersicum* (L.).
Turnip	*Brassica rapa* (L.)
Vetch	*Vicia atropurpurea* L.
Bitter	*V. ervilia* (L.) Willd.
* Hairy	*V. villosa* Roth.
* Watermelon	*Citrullus vulgaris* Schrad.
* Wheat	*Triticum aestivum* L.

* Allelopathic crops

Table II. Common and scientific names of weeds

Scientific name	Common name
* *Abutilon theophrasti* Medic.	Velvet leaf
* *Ageratum conyzoides* L.	Goat weed
* *Agropyron repens* (L.) Beauv.	Quack grass
* *Agrostemma githago* L.	Corn cockle
* *Agrostis tenuis* Sibth.	Colonial bentgrass
* *Alhagi pseudoalhagi* (Bieb.) Desv.	
* *A sparsifolia*	
Alopecurus myosuroides Huds.	Slender foxtail
Alternaria sessiles	
Alysicarpus vaginalis (L.) DC	
Amaranthus dubius Mart. ex Thell.	
A. hybridus L.	Smooth pigweed
A. leucocarpus Wats.	
* *A. palmeri* (L.) Wats.	Palmer amaranth
* *A.retroflexus* L.	Red root pigweed
* *A.spinosus* L.	
* *A. tricolor* L.	
* *A. viridis* L.	
* *Ambrosia artemisiifolia* L.	Common ragweed
* *A. psilostachaya* DC.	Western ragweed
* *A. trifida* L.	Giant ragweed
Ammania baccifera L.	
* *Anagallis arvensis* L.	
* *Antennaria microphylla*	Pussy toes
Anthemis arvensis L.	Dog fennel
* *Argemone mexicana* L.	Mexican poppy
* *Aristida adcensionis*	
* *Artemisia annua* L.	
* *Asclepias syriaca* L.	Common milkweed
* *Avena fatua* L.	Wild oat
* *A. sterilis* subsp. *ludoviciana*	
* *Bidens bipinnata* L.	
* *B. pilosa* L.	
Bilder dykia	
Borreria articularis (L.) Will.	
* *Bothriochloa pertusa* (L.) Camus.	

Brassica campestris L.	Mustard
B. hirta Moench	White mustard
B. kaber (DC) Wheeler	Wild mustard
B. nigra (L.) Koch	Black mustard
Bromus japonicus Thumb. ex. Murr.	Japanese brome
Cachia maritima	
* *Calendula arvensis L.*	
* *Camelina alyssum*	Flax weed
* *C. sativa* (L.) Crantz	Largseed false flax
Capsella bursa-pastoris (L.) Medic	Shepherd purse
* *Cassia auriculata* L.	
C. jalapensis L.	
* *C. obtusifolia* L.	Sicklepod
* *C. sophera* L.	
* *C. tora* L.	
* *Celosia argentea* L.	
* *Centaurea diffusa* Lam.	Diffuse knapweed
* *C. maculosa* Lam.	Spotted knapweed
* *C. nigra* L.	Black knapweed
* *C. repens* L.	Russian knapweed
* *Chenopodium album* L.	Common lambsquarter
* *C. murale* L.	
* *Cirsium arvense* (L.) Scop.	Canada thistle
* *Citrulus colocynthis* (L.) Kuntz	
* *C. lanatus* (Thumb.) Mansf.	
Cleome barchycarpa Vahl.	
* *Commelina benghalensis* L.	
Convolvulus arvensis L.	Hedge bindweed
* *Crotolaria medicaginea* Lamk	
* *C. pallida*	
C. verucosa L.	
* *Croton bonplandianum* Baill.	
* *Cucurbita pepo* L.	Pumpkin/squash
Cuscuta sp.	Dodder
* *Cynodon dactylon* (L.) Pers.	Bermuda grass
* *Cyperus esculentus* L.	Yellow nutsedge
* *C. iria* L.	
* *C.rotundus* L.	Purple nutsedge
C. serotinus L.	

* *Datura stramonium* L.	Thorn apple/Jimson weed
Daucus carota L.	Wild carrot
* *Digera alternifolia* (L.) Ascher	
* *D. arvensis* Forsk.	
* *D. muricata* (L.) Mart.	
Digitaria adscendens (HBK) Henr.	Henry grass
D. ischaemum (Scherb.) Muhl.	Smooth crabgrass
* *D. sanguinalis* (L.) Scop.	Large crabgrass
Echinochloa colonum (L.) Link	
**E. crusgalli* (L.) Beauv.	Barnyard grass
E. Crusgalli var *frumentaceae* (Roxb.) Lamk.	Japanese barnyard grass
* *Echinops echinatus* Roxb.	
E. leusine indica (L.) Gaertn.	Goose grass/wild finger millet
* *Erica arborea* L.	
* *E. australis* L.	
Erigeron canadensis L.	
* *Eupatorium adenophorum* Spreng.	
* *E. odoratum* L.	
* *Euphorbia esula* L.	Leafy spurge
* *E. geniculata Ort.*	
* *E. granulata*	
* *E.heterophylla* L	
* *Fagopyrum cymosum* Meisen.	Wild perrenial buckwheat
Festuca arundinacea Shreb.	
* *Fumaria indica* Pugsly.	
* *Galium aparine* L.	
* *G.mollungo* L.	Smooth bedstraw
Gisekia pharnaceoides L.	
* *Gloriosa superba* L.	
* *Gomphrena decumbens* Jacq.	
Halopappus ciliatus	
Helianthus annuus L.	Sunflower
H. rigidus L.	
* *Heliotropium eichwaldi* (DC) Stent.	
H. europeum	Wild heliotrope
Helminthia echiodes	
* *Heracleum laciniatum*	
Hibiscus trionum L.	
Hirschfeldia incana (L.) Largreze-Fossat.	

Holcus mollis L.	Velvet grass
* *Hydrocotyle sibthorpioides* Lam.	Lawn pennywort/twitch grass
* *Imperata cylindrica* (L.) Beauv.	Cogonn grass/blady grass/ alang alang
* *Indigofera cordifolia* Heyne ex. Roth	Wild Indigo
* *Ipomoea aquatica* (L.) Forsk	
I. hedracea (L.) Jacq.	Ivy morning glory
I. hederifolia L.	
I. lacunosa L.	Pitted morning glory
I. pandurata	Bigroot morning glory
I. pestigridis L.	
I. purpurea (L.) Roth.	Tall morning glory
I. sindica (L.) Stapf	
* *Kochia scoparia* (L.) Schreb.	Kochia
Lamium amplexicaule L.	
* *Lantana camara* L.	
* *Leersia hexandra* SW.	Rice grass
Lepidium sativum L.	Garden cress
L. virginicum L.	
* *Lolium multiflorum* Lam.	Italian ryegrass
L. perrene L.	Perrenial ryegrass
* *L. temulentum* L.	
* *Ludwigia adscendens* (L.) Hara	
L. octovalis (Jacq.) Raven.	
* *Madia glomerata* Hook.	
Matricaria inodora L.	
M. maritima L.	
Ochthodium aegyptiacum	
* *Oryza perrenis* Moench	
* *Panicum dichotomiflorum* Mischx.	Fall panicum
P. miliaceum L.	Proso millet
Papaver rhoeas L.	
* *Parthenium hysterophorus* L.	
* *Paspalum dilatatum* Poir.	Dallis grass
* *Peganum harmala* L.	
Phalaris lanceolata L.	
P. major	
**P. minor* Retz.	
P. paradoxa L.	Hood canarygrass

* *Plantago lanceolata* L.	
P. major L.	
Polycarpaea corymbosa (L.) Lam.	
* *Polygonum aviculare* L.	Prostrate knotweed
* *P. lapathifolium* L.	
* *P. orientale* L.	Prince's feather
* *P. pennsylvanicum* L.	Pennsylvania smartweed
* *Portulaca grandiflora* Hook.	
* *P. oleracea* L.	Common purslane
* *Pulicaria crispa* (Cass.) Benth & Hook.	
* *Rottboellia exaltata* (L.) L.f.	Itch grass
Rudbekia hirta L.	
* *Rumex acetosella* L.	Sheep sorrel
R. crispus L.	Curlydock/red sorrel
R. dentatus L.	
* *Saccharum spontaneum* L.	Thatch grass/wild cane
Sagittaris pygmaea L.	
Salsola kali L.	Russian thistle
Salvia austriaca	
S.leucocephalla	
* *S.reflexa* Hornem.	Mint weed
S. syriaca	
* *Schinus molle* L.	Piru
* *Scoparia dulcis* L.	
* *Sesbania exaltata* (Raf.) Rydb.	Sesbania
* *Setaria faberii* Herm.	Giant foxtail
* *S.glauca* (L.) Beavu.	Yellow foxtail
* *S.italica* (L.) Beavu.	Foxtail millet
S. viridis (L.) Beavu.	Green foxtail
* *Sida spinosa* L.	Prickly sida
* *Silybum marianum* (L.) Gaertn.	
* *Sinapis arvensis* L.	Wild mustard
Sisymbrium polyceratum	
Solanum nigrum L.	Black nightshade
* *S. surattense* Burm.	
* *Sonchus arvensis* L.	
* *Sorghum halepense* (L.) Pers.	Johnson grass
* *Spergula arvensis* L.	Spurrey weed
* *Stachaya parviflora*	

Stellaria media (L.) Villars.	Common chickweed
* *Steuia eupatoria*	
* *Stizolobium deeringianum* Bort.	
Striga asiatica (L.) Kuntz.	Witch weed
* *Tagetes minuta* L.	Marigold
* *Tephrosia purpurea* (L.) Pers	
Thalpi arvense L.	Field pennygrass
* *Trianthema portulacastrum* L.	
* *Tribulus terrestris* L.	Caltrop/Puncture vine
* *Trichodesma amplexicaule* Roth.	
T. sedgwickianum Banerji	
* *Urginea indica*	
Vulpia myuros	
* *Xanthium pennsylvanicum* Wall.	Common cocklebur
* *X. strumarium* L.	Cocklebur.

* Allelopathic weeds

APPENDIX

Appendix. Inhibitory effects of weed residues/extracts on the seed germination, growth and yield of crops

Donor weed			Recipient crop		Reference
Weed species	Weed residues / extract	Phytotoxin	Crops affected	Inhibitory effect	
1	2	3	4	5	6
Agropyron repens (L.) Beauv.	Root, rhizome extract	ND	Oat, pea, alfalfa, soybean, rapeseed	Reduced seed germination and seedling growth	Osvald (1948), Ohman and Kommedahl (1959, 1960, 1964), Welbank (1960), Weston and Putnam (1986), Touchette *et al.* (1988)
	Root rhizome extract	ND	Wheat, rapeseed, pea, alfalfa, snapbean	Inhibited seedling growth	Kommedahl *et. al.* (1959), Welbank (1963), LeTourneau and Heggeness (1957), Weston and Putnam (1986), Toai and Linscott (1979)
	Shoot, rhizome extract	SH group	Red clover	Blocked xylem vessels	Bogdan and Grodzinsky (1973)
	Rhizome extract	Glucoside	Maize, oat, cucumber, alfalfa	Inhibited root growth	Gabor and Veatch (1981)
	Whole plant extract	5-hydroxy-tryptophan, 5-hydroxy indole-3-acetic acid	Maize, soybean	Decrease growth at higher concentrations	Hagin (1989)

Contd....

1	2	3	4	5	6
	Rhizomes	Acetic acid, butyric acid	Barley	Inhibited seedling growth	Lynch and Penn (1980)
	Root rhizomes	Ethylene, P-hydroxybenzoic acid, vanillic acid, agropyrene			Grummer (1961), Grummer and Beyer (1970), Harvey and Linscott (1978)
	Leaves	Tricin	—	—	Weston *et al.* (1987)
	Plant residues added to soil	ND	Soybean, snapbean, navybean	Decreased nodule number, nodule weight nitrogen fixation	Weston and Putnam (1985)
	Plant residues added to soil	ND	Wheat, oat, maize, soybean	Decreased seed yield	Buccholtz (1971), Kommedahl *et al.* (1970), Minar (1974)
	Live plants	ND	Wheat, maize, potato, *Trifolium* spp.	Decreased growth and yield	Ahlgren and Aamodt (1939), Islieb (1960), Bandeen and Buchholtz (1967)
Cyperus rotundus L.	Tubers extract	ND	Barley, oat, rice, pearlmillet, sorghum, mustard, sesame, sannhemp, cowpea, blackgram, groundnut, soybean, tomato, onion, radish	Inhibited seed germination and seedling growth and seedling survival of tomato, onion and radish	Singh (1968), Friedman and Horowitz (1971), Lucena (1974), Lucena and Doll (1975), Meissner *et al.* (1979), Komai *et al.* (1981), Castro *et al.* (1983)

Contd....

1	2	3	4	5	6
	Tubers extract	ND	Rice	Completely inhibited root development and reduced shoot growth	Castro *et al.* (1984)
	Tubers added to soil	ND	Barley	Inhibited seedling growth	Horowitz and Friedman (1971)
	Tubers added to soil	ND	Sorghum, pepper, tomato	Reduced seed germination and seedling growth	Mohmed-Saleem and Fawusi (1983)
	Infested soil	ND	Sorghum, barley, tomato, cucumber	Inhibited growth	Meissner *et. al.* (1979)
	Live plants	ND	Rice, tomato, cabbage, cucumber, carrot, okra, garlic	Reduced yields	Williams and Warren (1975), Okafor and DeDutta (1976)
	Plant parts	Polyphenols, sequiterpenes	—	Inhibited seedling growth	Komai and Ueki (1975, 1980), Komai *et al.* (1977)
C. esculentus L.	Shoot extract	ND	Wheat, soybean	Inhibited seed germination	Bansal *et. al.* (1981), Mallik and Tesfai (1988), Singh *et al.* (1988)
	Tubers extract	ND	Maize, sorghum, soybean	Decreased seedling growth	Drost and Doll (1980)

Contd.....

1	2	3	4	5	6
	Tubers extract	p-hydroxybenzoic acid, vanillic acid, syringic acid, ferulic acid, *p*-coumaric acid	Sugarbeet, pea, lettuce, tomato, white clover	Inhibited seed germination and seedling growth	Tames *et al.* (1973)
C. iria L.	Inflorescence, seed extract	ND	Blackgram	Inhibited 32P uptake	Bansal and Kalia (1984)
Parthenium hysterophorus L.	Shoot extract	ND	Wheat, maize, soybean, sunnhemp, blackgram, groundnut	Inhibited seedling growth	Sarma *et al.* (1976), Mersie and Singh (1987a b)
	Leaf, inflorescence extract	ND	Mustard	Inhibited seedling growth due to decrease in cell survival and chlorosis	Kumari *et. al.* (1985)
	Leaf stem root extract	Alkaloids, parthenin, caffeic acid, *p*-coumaric acid	Wheat, barley, maize, sorghum, pearlmillet, sesame, mustard, pea, pigeon pea, blackgram, greegram, fenugreek, radish, cabbage, cauliflower	Inhibited seed germination and seedling growth	Rajan (1973), Kanchan (1975), Dube *et al.* (1979) Mall and Dagar (1979), Ambika and Jayanchandra (1980), Nath (1981), Dharmaraj and Ali (1985), Kumari *et al.* (1985), Srivastava *et al.* (1985)

Contd.....

1	2	3	4	5	6
	Whole plant extract	Parthenin, Coronopilin	—	—	Jarvis *et al.* (1985)
	Leaf leachates	ND	Wheat, bean	Reduced drymatter and chlorophyll content	Kanchan and Jayachandra (1980 b)
	Pollens	ND	Bean, brinjal, tomato, chillies	Inhibited fruit setting and cause chlorosis	Kanchan and Jayanchandra (1980a)
	Weed residues	ND	Wheat, rice, maize, sorghum, pearl-millet, fingermillet, cowpea, groundnut, greengram, black-gram, soybean, frenchbean	Reduced seed germination and seedling growth	Muniyappa and Krishnamurthy (1981)
	Weed residues	ND	Cowpea, bean, tomato, fingermillet, oat, barley	Inhibited growth of all test crops, besides decreased nodulation and yields of legumes	Kanchan and Jayanchandra (1979 a b), Muniyappa and Krishnamurthy (1981)
Sorghum halepense (L.) Pers.	Rhizome extract	ND	Tomato	Completely inhibited germination and seedling growth	Castro *et al.* (1983)

Contd.....

1	2	3	4	5	6
	Rhizome extract	Dhurrin, taxiphyllin, p-hydroxybenzaldehyde p-hydroxybenzoic acid, ferulic, p-coumaric acid, phloroglu-cinol, p-hydroxybenzyl alcohol	Wheat, barley, rice, maize, mustard, soybean, radish, tomato	Inhibited seedling growth	Friedman and Horowitz (1970), Horowitz and Friedman (1971), Lolas and Coble (1982), Mainx and Tauscher (1983), Castro *et al.* (1984), Mikulas (1984), Nicollier *et al.* (1983, 1985)
	Leaf, rhizome extract	Chlorogenic acid, *p*-coumaric acid, *p*-hydroxybenzaldehyd	Sunflower, tomato	Inhibited seed germination and seedling growth	Abdul-Wahab and Rice (1967)
	Plant extract	ND	Wheat	Inhibited seed germination and seedling growth	Singh *et al.* (1988)
	Weed residues	ND	Sunflower, onion, cabbage,	Inhibited growth	Menges (1987)
	Live plants	ND	Soybean, maize, sugarcane, cotton	Decreased growth and grain yields	Kleifeld (1970), Millhollon (1970), McWhorter and Hartwig (1972), Roeth (1973)
	Infested soil	ND	Rapessed	Inhibited germination and growth	Osvald (1947)

Contd.....

1	2	3	4	5	6
Amaranthus palmeri (L.) Wats.	Weed residues	2,6-dimethoxy-benzoquinone, Vanillin, phytol, chondri-llasterol, 3-met-hoxy-4-hydroxy nitrobenzene	Sorghum, onion, carrot, cabbage	Inhibited growth	Bradow (1985), Fischer and Quijano (1985), Menges (1987, 1988)
A. retroflexus L.	Weed residue extract	ND	Maize, soybean	Inhibited radicle elongation of maize and hypocotyl elongation of soybean	Bhowmik (1982), Bhowmik and Doll (1982)
	Weed residues soil added	ND	Maize, soybean, tobacco	Reduced growth and yield	Bhowmik and Doll (1982, 1983)
A. spinosus L.	Leaf, inflorescence extract spray	ND	Wheat, mustard	Severely decreased growth	Datta and Bandyopadhay (1981)
A. tricolor L.	Leaf extract	ND	Sorghum	Delayed seed germination and severely inhibited seedling growth	Chatterjee (1975), Mohnot and Soni (1979)
A. viridis L.	Leaf extract	ND	Wheat	Inhibited seedling growth	Bhatia *et al.* (1982)

Contd.....

1	2	3	4	5	6
	Plant parts extract	ND	Pearlmillet, sesame	Inhibited seed germination of pearlmillet and seedling growth of both crops	Singh and Sen (1981), Singhal and Sen (1982)
	Weed residues added to soil	ND	Pearlmillet, sesame	Decreased growth and yield	Singh and Sen (1981)
Centaurea maculosa Lam.	Shoot extract	Cnicin	Lettuce	Inhibited germination and root growth	Kelsey and Locken (1987)
C. nigra L.	Leaf extract	ND	Wheat, tomato, mustard	Partially inhibited germination and seedling growth	Vezina and Doyon (1983)
C. repens, *C. diffusa* Lam	Leaf, root extract	Polycetylenes VIII-XIV	Barley, alfalfa, tomato, lettuce	Decreased root elongation	Fletcher and Renny (1963), Stevens (1986)
Polygonum aviculare L.	Underneath soil	Phenolic glycosides	Sorghum, cotton, lettuce, *Medicago* spp.	Reduced seed germination, growth, cell division and *Rhizobium* and *Azotobacter* bacteria	AlSaadawi and Rice (1982a b), Kloot and Boyce (1982)
P. lapathifolium L.	Live plant	Apigenin, lutolin	inseed	Decreased seed yield	Grummer and Beyer (1960)

Contd....

1	2	3	4	5	6
P. orientale L.	Plant parts extract	ND	Rice, mustard, blackgram, pea, lettuce	Reduced seed germination, growth	Datta and Chatterjee (1978, 1980)
P. pennsyl-vanicum L.	Whole plant extract	ND	Wheat, alfalfa, cabbage, carrot	Reduced seed germination of all test species and seedling growth of alfalfa and wheat	LeTourneau et al. (1956) Gressel and Holm (1964)
	Whole plant extract	ND	Soybean	Reduced seed germination, stimulated nodulation at lower concentration but inhibited at higher ones.	Mallik and Tesfai (1988)
Setaria faberii Herm.	Weed residue, live plant	ND	Maize, soybean	Reduced growth and yield	Schreiber and Williams (1967), Bell and Koeppe (1972) Bhowmik and Doll (1979)
S. glauca (L.) Beauv.	Seed extract	ND	Maize, soybean, cabbage, radish, turnip	Inhibited germination and seedling growth	Gressel and Holm (1964), Schreiber and Williams (1967) Bhowmik and Doll (1979)
	Live plant	ND	Sorghum, maize, soybean	Decreased seed yields	Staniforth (1961), Feltner et al. (1969)

Contd.....

1	2	3	4	5	6
S. italica (L.) Beauv.	Plant parts	ND	Sorghum, pearlmillet, rapeseed	Inhibited germination and seedling growth	Hussain et al. (1981)
S. viridis (L.) Beauv.	Shoot extract	ND	Soybean	Reduced growth and nodulation	Mallik and Tesfai (1988)
Cirsium arvense (L.) Scop.	Shoot extract	ND	Wheat, rye, soybean	Inhibited root growth, reduced the mitotic activity of meristmetic cells and damaged the epidermis, cortex and vascular bundles of roots. Reduced nodulation in soybean	Bukolova (1971), Bhandari *et al.* (1982)
	Shoot, root extract	ND	Wheat, barley, maize, soybean, alfalfa, turnip, cucumber, linseed	Inhibited germination of soybean and turnip and growth of all test crops	Hegelson and Konzak (1950), Woodward and Glenn (1983), Kovacs *et al.* (1988)
	Crop residues	ND	Pearlmillet, sesame	Inhibited germination and growth	Stachon and Zimdahl (1980), Wilson Jr. (1981)

Contd....

1	2	3	4	5	6
	Live plant	ND	Wheat	Decreased seed yields	Hodgson (1958)
Digera arvensis Forsk.	Root, shoot, seed extract	ND	Pearlmillet, sesame	Inhibited germination and seedling growth. Root extract most toxic to root growth	Sarma (1974), Rao *et al.* (1979) Ashraf and Sen (1980)
D. alternifolia (L.) Ascher	Whole plant extract	ND	Pearlmillet, sorghum, sesame, cotton, pigeonpea, clusterbean, tomato, brinjal, onion, chillies	Severely inhibited seed germination of all crops except cotton, pigeonpea, tomato	Dubey (1973), Ashraf and Sen (1980)
	Crop residues	ND	Pearlmillet, sesame	Reduced corp growth and yield	Ashraf and Sen (1980)
D. muricata (L.) Mart.	Leaf, root, stem extract	ND	Pearlmillet	Inhibited seedling growth	Sen *et al.* (1969)
Abutilon theophrastii Medic.	Leaf extract	Phenols	Soybean, radish	Inhibited germination of radish and growth of soybean, interfered with water balance and chlorophyll content	Colton and Einhellig (1980)

Contd....

1	2	3	4	5	6
	Seed extract	Amino acids, flavonoids, delphindin, cynidin, quercetin, myricitin(+) - catechin, (-)-epicatechin		Inhibited germination and growth of test crops. Inhibit growth of seed rotting fungi	Paszkowski and Kremer (1988)
	Weed residue	ND	Maize, soybean	Reduced growth and yield	Bhowmik and Doll (1980, 1982), Bhowmik (1982)
	Live plant	ND	Soybean, cotton	Decreased seed yield	Hagwood *et al.* (1980)
Bidens bipinnata L.	Soil extract	Phenylhepatriyne	Maize, sunflower, onion, carrot, radish, cucumber, lettuce, squash, tomato	Delayed seed germination	Ruth *et al.* (1986)
Celosia argentea L.	Leaf, shoot, root extract	ND	Pearlmillet	Inhibited seedling growth	Pandya (1975, 1977), Rao *et al.* (1979)
	Leaf extract spray	ND	Sorghum	Inhibited chlorophyll synthesis	Soni and Mohnot (1988)
	Weed residues	ND	Pearlmillet, sesame	Severely decreased seed germination	Ashraf and Sen (1978)
	Live plants	ND	Sorghum, maize, pearlmillet, blackgram, mothbean	Decreased seed yields, reduced nodulation in legumes	Pandya and Dash (1982), Vyas (1982), Pandya and Vyas (1984), Pandya and Sidha (1987)

Contd....

1	2	3	4	5	6
Euphorbia esula L.	Whole plant extract	ND	Wheat, pea	Inhibited seed germination of wheat and growth of both test crops	LeTourneau and Heggeness (1957)
	Soil extract, weed residues	ND	Tomato	Inhibited seedling growth	Steenhagen and Zimdahl (1979)
E. geniculata Ort.	Root, seed extract	ND	Wheat	Reduced seed germination	Sugha (1979, 1980)
E. granulata	Whole plant extract	ND	Pearlmillet	Inhibited seed germination and root growth of seedlings	Hussain (1980)
E. hetero-phylla L.	Weed residues	ND	Sorghum, tomato, pepper	Slightly reduced seed germination and growth	Mohmed-Saleem and Fawusi (1983)
Xanthium strumarium L.	Whole plant extract	ND	Wheat, chickpea	Decreased seed germination	Ambasht and Rai (1981)
	Leaf, stem, flower extract, weed residues	Caffeic acid, p-coumaric acid, p-hydroxy-benzoic acid, chlorogenic acid, ellagic acid	Maize, pearlmillet, rapeseed, lettuce	Inhibited seed germination and seedling growth	Bushra *et al.* (1987)

Contd....

1	2	3	4	5	6
	Allelochemical	Carboxyatractyloside	Wheat, maize, tobacco	Etiolated wheat coleoptiles, stunts maize seedlings, deformed tobacco leaves	Cutler (1985)
Imperata cylindrica (L.) Beauv.	Leaf, root, rhizome extracts	ND	Rice, sorghum, maize, tomato, cucumer	Inhibited root elongation	Eussen (1978), Eussen and Soerjani (1978)
	Plant extract	ND	Corn, soybean	Inhibited germination at higher concentration	Singh *et al.* (1989)
	Live plants	ND	Maize, sorghum	Decreased seed yield	Essen (1979)
	Leaf, culm, rhizome	P-&-o-coumaric, gentisic, vanillic, benzoic, ferulic, syringic, p-hydroxychlori c, chlorogenic, iso-chlorogenic acids, p- hydroxybenzaldehyde vanillin, scopolin, scopoletin,	—	—	Abdul-Wahab and Al-Naib (1972), Eussen and Niemann (1981), Amritlingam (1984)

Contd....

1	2	3	4	5	6
Avena fatua L.	Hull extract	Vanillin	Lettuce	Inhibited seed germination	Chem *et al.* (1982)
	Live plants	—	Wheat, barley, linseed	Reduced yields	Bell and Nalewaja (1968 a b)
A. Sterilis sub sp. *ludoviciana*	Straw, glumes extract	ND	Wheat, greengram	Inhibited seedling growth and nodulation in greengram	Bhandari *et al* (1982), Bhatia *et al.* (1982)
Alhagi sparsifolia	Leaf, stem, seed extract	ND	Chickpea, greengram	Reduced seed germination and seedling growth	Mavlanov (1982)
A.psuedo-alhagi (Bieb.) Desv.	Leaf, stem, seed extraek	ND	Pearlimillet	Reduced seed germination and seedling growth	Singhvi and Sharma (1985)
Asclepias syriaca L.	Leaf extract	ND	Sorghum	Inhibited seedling growth	Rasmussen and Einhellig (1975)
	Shoot extract	ND	Wheat	Inhibited germination and growth	LeTourneau *et al.* (1956)
	Live plants	ND	Sorghum, maize, soybean	Decreased seed yields	Cramer (1981)

Contd....

1	2	3	4	5	6
Argemone mexicana L.	Leaf extract	Phenolics	Wheat, sorghum, fingermillet, mustard, fenugreek, tomato, cucumber	Complete inhibition of seed germination in wheat, mustard, fenugreek and reduced germination and growth of remaining crops	Leela (1981)
	Whole plant extract	ND	Wheat, pearlmillet, mustard, radish	Inhibited seed germination and seedling growth	Sharma and Nathawat (1987)
	Leaf litter added to soil	ND	Wheat	Inhibited seedling growth	Leela (1981)
Artemisia annua L.	Allelochemicals	Artemisinin	Lettuce	Reduced seed germination and seedling growth	Duke et al. (1987)
Chenopodium album L.	Shoot extract	ND	Soybean, chickpea	Reduced seed germination, growth and nodulation of soybean and seedling growth of chickpea	Goel (1987), Mallik and Tesfai (1988)

Contd....

1	2	3	4	5	6
	Plant extract	ND	Wheat, rye, maize	Inhibited growth, reduced mitotic activity of meristmetic cells, damaged epidermis, cortex and vascular bundles of roots	Bukolova (1971), Bhatia *et al.* (1982), Caussanel (1979), LeTourneau *et al.* (1956)
	Plant extract	ND	Wheat	Reduced germination and growth	LeTourneau *et al.* (1956)
	Plant leachates, weed residues	ND	Tomato	Reduced shoot growth and uptake of N PK Ca Mg	Qasem and Hill (1989)
	Weed residues and their extract	ND	Soybean, maize	Reduced growth and seed yield	Bhowmik and Doll (1980, 1982, 1984)
	Live plants	ND	Oilseed rape, sugarbeet	- do -	Holm *et al.* (1979)
C. murale L.	Leaf extract	Oxalic acid	Mustard	Decreased seed germination	Datta and Ghosh (1982)
Datura stramonium L.	Leaf leachates	Scopolamine, hyoscyamine	Linseed	Depressed seed germination and root elongation	Lovett *et al.* (1981)

Contd....

1	2	3	4	5	6
	Seed washings	- do -	Sunflower	Decreased seed germination	Harris (1984)
Fagopyrum cymosum Meisen.	Plant extract	ND	Rice, rapeseed	Inhibited seedling growth	Tsuzuki *et al.* (1977)
	Allelochemicals	Ferulic, caffeic, chlorolgenic acids	Rice	Inhibited sedling growth	Tsuzuki and Yamamoto (1987)
Galium mollungo L.	Shoot extract	ND	Wheat, sunflower, radish, onion	Inhibited seed germination of wheat and radish, seedling growth of sunflower and bulb of onion	Kohlmuenzer (1965)
	Seed extract	Asperulosidic acid, Deacetyl asperulosidic acid	Alfalfa	Inhibited seed germination and growth	Komai *et al.* (1986)
Indigofera cordifolia Heyne ex Roth	Leaf, stem, root extract & weed residues	ND	Pearlmillet, sesame	Inhibited germination and seedling growth	Bhandari (1977), Singh and Sen (1982)
Lantana camara L.	Weed residues	ND	Wheat, maize, soybean	Decreased seedling growth	Mersie and Singh (1987a, b)

Contd.....

1	2	3	4	5	6
	Plant extract	ND	Wheat	Inhibited germination and seedling growth	Singh *et al.* (1988)
Peganum harmala L.	Plant extract	Harmaline	Rapeseed, pearlmillet	Decreased seed germination and seedling growth	Sharma *et al.* (1982), Yang *et al.* (1987).
	Shoot & soil extract, shoot litter, volatiles	Caffeic, ferulic *p*-Coumaric and *p*-oH-benzoic acids	Pearlmillet, rapeseed, lettuce, persian clover	Decreased seed germination and seedling growth	Hussain *et al.* (1988)
Pulicaria crispa (Cass.) Benth, Hook	Leaf, stem, root leachates	ND	Pearlmillet	Decreased seed germination and seedling growth	Bhandari *et al.* (1982)
Salvia reflexa Hornem	leaf washings	alpha-pinene, beta-pinene, cineole	Wheat	Decreased root length	Lovett (1983b)
S. syriaca	Shoot, rhizome extract	ND	Wheat	Delayed seed germination and decreased growth	Qasem and Abu-Irmaileh (1985)
Ambrosia artemisifolia L.	Weed residues added to soil	ND	Maize, soybean	Inhibited growth	Bhowmik and Doll (1984)

Contd....

1	2	3	4	5	6
A. psilostachaya DC.	Shoot, rhizome extract	ND	Wheat, oat, rye, alfalfa, sunflower, tomato	Delayed seed germination and decreased growth	Dalrymple and Rogers (1983)
A. trifida L.	Plant extract	ND	Wheat, sorghum, radish	Reduced seed germination and growth	LeTourneau and Heggeness (1957), Rasmussen and Einhellig (1979)
Ageratum conyzoides L.	Leaf, root extract	ND	Wheat	Inhibited seed germination and seedling growth	Sugha (1979, 1980), Bansal *et al.* (1981), Singh *et al.* (1988)
	Seed, inflorescence extract	ND	Kidneybean	Reduced 32P uptake	Bansal and Kalia (1984)
Agrostis tenuis Sibth.	Plant extract	ND	Rapeseed	Reduced radicle elongation	Welbank (1963)
Anagallis arvensis L.	Shoot extract	ND	Chickpea	Reduced shoot elongation	Goel (1987)
Antenarria microphylla	Plant extract	Caffeic, acid, arbutin, hydroquinone	Lettuce	Inhibited root elongation of seedlings	Manners and Galitz (1986)
Aristida adscensionis	Live plants	ND	Indigo	Inhibited nodulation	Sarma (1983)
Calendula arvensis L.	Plant extract	ND	Wheat	Inhibited germiantion and growth	Chaghtai *et al.* (1987)

Contd.....

1	2	3	4	5	6
Camelina alyssum	Rain leachates of live plants	Vanillic, ferulic, p-hydroxybenzoic acids	Linseed	Decreased growth and seed yield	Grummer and Beyer (1960)
Cassia sophera L.	Root, stem, leaf, flower extract	Scopoletin, quercetin triterpene	Wheat, rice, rapeseed, blackgram, lettuce	Inhibited germination	Datta and Dasmahapatra (1984)
C. tora L.	Seed coat extract	ND	Sesame	Inhibited seedling growth	Bhatia and Chawan (1976)
C. auriculata L.	Root extract	ND	Pearlmillet	Inhibited root elongation	Rao *et al.* (1979)
Citrullus colocynthis (L.) Kurtz.	Shoot, root extract, rain drip, weed residues	ND	Sorghum, pearlmillet, turnip, alfalfa, lettuce	Decreased germination and growth	Bhandari (1974), Shaukat *et al.* (1985)
C. lanatus (Thumb.) Mansf.	Fruit pulp extract	ND	Pearlmillet	Delayed germination and suppressed growth	Bhandari and Sen (1972)
Commelina benghalensis L.	Plant extract	ND	Corn, soybean	Inhibited germination and growth	Singh *et al.* (1989)
Crotolaria medicagenia Lamk.	Shoot extract	ND	Sesame	Inhibited seedling growth	Singh and Sen (1981, 1982)

Contd....

1	2	3	4	5	6
C. pallida	Root, stem, leaf, flower extract	Quercetin-3-glycoside	Wheat, rice, rapeseed, blackgram, lettuce	Inhibited seed germination	Datta and Dasmahapatra (1984)
Croton bonplandiaum Baill.	Leaf leachates	ND	Rice, mustard, lettuce	Inhibited seed germination	Datta and Sinha-Roy (1983)
Cucurbita pepo L.	Plant extracts	ND	Maize, cowpea, cabbage	Reduced germination and seedling growth	Gliessman (1983)
Cynodon dactylon (L.) Pers.	Root, rhizome extracts	ND	Rice, tomato, pea, cucumber	Reduced germination and seedling growth	Sapankevitch *et al.* (1965) Castro *et al.* (1983, 1984)
	Soil extract	ND	Barley	Inhibited sedling growth	Horowitz and Friedman (1971)
Digitaria sanguinalis (L.) Scop.	Whole plant extract	Chlorogenic, isochlorogenic, sulfosalicylic acids	Sunflower, wheat	Inhibited seed germination of both species and growth of sunflower	Parenti and Rice (1969), Bansal *et al.* (1981)
	Whole plant extract	ND	Kidneybean	Decreased phosphorus uptake	Bansal and Kalia (1984)
	Decaying roots	ND	Maize	Decreased root growth	Schreiber and Wiliams (1967)

Contd....

1	2	3	4	5	6
	Weed residues	ND	Sorghum, tomato, pepper	Decreased germination and seedling growth	Mohmed-saleem and Fawusi (1983)
	Live plants	ND	Cotton, groundnut	Decreased yields	Hill and Santelmann (1969), Robinson (1976)
Echinochloa crusgalli (L.) Beauv.	Live plants	ND	Rice, maize, soy-bean, fieldbean, sugarbeet	Decreased growth and yield	Smith Jr. (1974), Dawson (1977), Bhowmik and Doll (1982)
	Plant extract	ND	Wheat	Inhibited germination and seedling growth	Singh *et al.* (1988)
Echinops echinatus Roxb.	Leaf, stem, root extract, weed residues	ND	Pearlmillet, sesame	Decreased germination and growth	Jha and Sen (1981)
Erica arborea L. *E. australis* L.	Aerial parts extract	ND	Red clover	Reduced seedling growth	Vietiz and Vietiz (1979)
Eupatorium adenophorum Spreng.	Plant extract	ND	Wheat	Reduced seed germination and seedling growth	Tripathi *et al.* (1981), Singh *et al.* (1988)
E. odoratum L.	Leaf extract	ND	Wheat, mustard	Reduced seedling growth	Datta and Bandyopadhyay (1981)

Contd......

1	2	3	4	5	6
Festuca arundinacea Schreb.	Leaf, root extract	Lactic, succinic, citric, shikimic, glyceric, fumaric and quinic acids	Rapeseed, red clover	Reduced seedling growth	Peters and Luu (1985)
Fumarica indica Pugsly	Plant parts extract	ND	Wheat	Inhibited germination and growth	Chagtai *et al.* (1986)
Gloriosa superba L.	Tuber extract	ND	Pea	Inhibited germination and growth	Sugha (1978)
Gomphrena decumbens Jacq.	Leaf, stem, root extract	ND	Sorghum, sesame	Inhibited seedling growth	Solomon and Bhandari (1981)
Hydrocotyle sibthopoides Lam	Leaf, soil extract	ND	Rice, rapeseed, tomato	Inhibited germination and seedling growth	Tsuzuki *et al.* (1978)
Ipomea aquatica (L.) Forsk.	Leaf, stem, root extract	ND	Pearlmillet	Inhibited seedling growth	Singhvi and Sharma (1984)
Kochia scoparia (L.) Schred.	Plant extract, weed residues	Caffeic, chloro genic, ferulic acids, myrice-tin, quercetin	Sorghum, soybean	Decreased seedling growth and water content	Lodhi (1979b), Einhellig and Schon (1982)

Contd....

1	2	3	4	5	6
Lolium multiflorum Lam.	Shoot, soil leachates	ND	Oat, lettuce	Inhibited seedling growth	Naqvi and Muller (1975)
L. temu-lentum L.	Root, stem, leaf, inflorescence extract	ND	Rice	Inhibited seedling growth	Bansal and Singh (1986)
Leersia hexandra SW.	Plant extract	ND	Rice, maize, soybean, greengram	Inhibited seed germination	Mercado and Manuel (1978)
Madia glomerata Hook.	Whole plant leachates, weed residues	ND	Radish	Decreased seed germination and growth	Carnahan and Hull (1962)
Oryza perrenis Moench.	Plant leachates, weed residues	ND	Rice, lettuce	Phytotoxicity to roots	Chou *et al.* (1984)
Panicum dichotom-iflorum Michx.	Weed residue extract	ND	Wheat, maize, soybean, kidneybean	Decreased growth of wheat, maize and soybean and P uptake of kidneybean	Bansal *et al.* (1981), Bhowmik and Doll (1982), Bhansal and Kalia (1984)
Paspalum dilatatum Pair.	Plaint extracts	ND	Wheat, Kidneybean	Decreased growth of wheat and P uptake of kidneybean	Bansal *et al.* (1981), Bansal and Kalia (1984)
Phalaris minor Retz.	Plant parts extract	ND	Wheat, rice, greengram	Decreased seedling growth	Bhandari *et al.* (1982), Bhatia *et al.* (1982), Bansal and Singh (1986)

Contd.....

1	2	3	4	5	6
Plantago lanceolata L.	Seed, leaf, stem, root extract	ND	Wheat	Inhibited seed germination	LeTourneau *et al.* (1956), Sugha (1979, 1980)
Portulaca oleracea L.	Plant extract	ND	Wheat	Inhibited seed germination	LeTourneau *et al.* (1956)
P. grandiflora Hook.	Stem extract	ND	Chickpea	Inhibited germination and growth	Afaq and Sinha (1970)
Rotoboellia exaltata (L.) L.f.	Seed, plant extract	ND	Rice, miaze, soybean, greengram	Inhibited seed germination	Mercado and Manuel (1978)
Rumex crispus L.	Leaf extract	ND	Sorghum, maize	Inhibited growth	Einhellig and Rasmussen (1973)
Saccharum spontaneum L.	Rhizome extract	ND	Wheat	Inhibited growth	Anritphalle and Mall (1978)
Salsola kali L.	Leaf extract	Phenolics	Radish	Inhibited germination and growth	Lodhi (1979b)
Schinus molle L.	Leaf, fruit, root extract	ND	Wheat, pearlmillet, cucumber	Inhibited germination and growth	Anaya and Gomez-Pompa (1971), Chatterjee (1975)
Scoparia dulcis L.	Leaf extract	ND	Wheat, rice, pigeon-pea, greengram	Inhibited germination	Trivedi and Trivedi (1987)

Contd....

1	2	3	4	5	6
Silybun marianum (L.) Gaertn.	Plant parts extracts, weed residues	ND	Wheat	Inhibited germination and growth	Chagtai *et al.* (1988)
Solanum surratense Burm.	Stem, fruit pulp extract	ND	Sorghum, pearlmillet, sesame	Inhibited germination and growth	Mohnot and Soni (1971), Sharma and Sen (1971)
Sonchus arvensis L.	Plant extract	ND	Wheat, rye	Inhibited root growth	Bukolova (1971)
Spergula arvensis L.	Plant extract, weed residues	ND	Rice maize, oat, alfalfa, lettuce, radish, turnip, tomato	Inhibited germination and growth	Gliessman (1983), Tsuzuki and Araki (1984)
Stizolobium deeringianum Bort.	Plant extract	ND	Cabbage	Inhibited germination	Gliessman (1983)
Wild Sunflower	Shoot extract, weed residues	ND	Maize, soybean	Inhibited germination of soybean and growth, yield of soybean and maize	Bhowmik (1982), Mallik and Tesfai (1988)
Stcchaya parviflora	Root, shoot extract, volatiles, weed residues	ND	Lentil, greengram, chickpea, Egyptian clover, rapeseed, radish, chinese mustard	Inhibited germination and seedling growth	Farrukh *et al.* (1986)

Contd.....

1	2	3	4	5	6
Stevia eupatoria	Plant parts leachates	ND	White clover	Inhibited germination and growth	Lovett (1982a)
Tagetes minuta L.	Soil extract	a-tertheinyl	Maize, sunflower, squash, tomato, onion, cucumber, lettuce, carrot	Inhibited seedling growth	Ruth *et al.* (1986)
Tephrosia purpurea (L.) Pers.	Shoot extract	ND	Sesame, greengram	Inhibited seed germination and growth	Mohnot and Choudhary (1981), Singh and Sen (1981, 1982)
Trianthema portulacastrum L.	Leaf extract	ND	Sorghum, mothbean	Inhibited seed germination and growth	Chatterjee (1975), Sethi and Mohnot (1988)
Tribulus terrestris L.	Leaf, shoot, root extract	ND	Pearlmillet	Inhibited seedling growth	Sen *et al.* (1969)
Trichodesma sedgwickanium Banerji	Plant extract, weed residues added to soil	ND	Pearlmillet, sesame	Inhibited growth and yield	Singh and Sen (1981, 1982), Bansal and Sen (1982)
Urginea indica	Seed, leaf, bulb extract	ND	Sorghum, maize, greengram, pigeonpea	Inhibited germination	Khare (1980)

ND = : Not determined

REFERENCES

Abdullah, M.M.F., Elmore, C.L. and Khalaf, K.A. (1989). Allelopathic effects of tomato root extracts on the germination of some weeds seeds *in-vitro*. *Egyptian Journal of Horticulture* **16:** 25-34.

Abdul-Rahman, A.A. and Habib, S.A. (1989). Allelopathic effects of alfalfa (*Medicago sativa*) on bladygrass (*Imperata cylindrica*). *Journal of Chemical Ecology* **15**: 2289-2300.

Abdul-Rahman, A.A.S. and Habib, S.A. (1986). *JAWRR* 5, 53-64.

Abdul-Wahab, A.S. and Al-Naib, F.A.G. (1972). Inhibitional effects of *Imperata cylindrica* (L.) P.B. *Bulletin Iraq Natural History Museum* **5(2):** 17-23.

Abdul-Wahab, A.S. and Rice, E.L. (1967). Plant inhibition by johnsongrass and its possible significance in old field succession. *Bulletin of the Torrey Botanical Club* **94:** 486-497.

Adeyeye, O.A. and Blum, M.S. (1989). *Environ. Entomol.* 18, 608-611.

Afaq, S.H. and Sinha, A.S. (1970). Effect of stem extracts of *Portulaca grandiflora* on the germination and growth of *Cicer arietinum* seeds. *Science and Culture* **36:** 402-404.

Ahlgren, H.L. and Aamodt, O.S. (1939). Harmful root interactions as a possible explanation for effects noted between various species of grasses and legumes. *Journal of American Society Agronomy* **31:** 982-985.

Ahmad, S. (1983). Herbivorous Insects-host seeking behaviour and mechanisms. Academic Press, New York. 257 pp.

Ahrens, J.F. (1989). Meeting the challenge. *Weed Technology* **3:** 531-536.

Akram, M. and Hussain, F. (1987). The possible role of allelopathy exhibited by root extracts and exudates of Chinese cabbage in hydroponics. *Pakistan Journal of Science and Industrial Research* **30:** 918- 920.

Al Saadavi, I.S. Arif, M.B. an AlRubea, A.J. (1985). *J. Chem. Ecol.* **11:** 1527-1534.

Alderman, W.H. and Middleton, J.A. (1925). Toxic relations of other crops to tomatoes. *Proceeding, American Society of Horticultural Science* **22:** 307-308.

Al-Saadawi, I.S. and Rice, E.L. (1982). Allelopathic effects of *Polygonum aviculare* L. 1. Vegetational pattering. *Journal of Chemical Ecology* **8:** 993-1010.

Al-Saadawi, I.S., Al-Ugaili, J.K., Al-Hadithy, S.A.M. and Al-Rubeaa, J. (1985). Effect of gamma irradiation of allelopathic potential of *Sorghum bicolor* against weds and nitrification. *Journal of Chemical Ecology* **11:** 1732-1745.

Altieri, M.A. and Doll, J.D. (1978). The potential of allelopathy as a tool for weed management in crop fields. *PANS* **24:** 495-502

Altieri, M.A. and Liebman, M. (1988). Weed management: Ecological guidelines. In *Weed Management in Agroecosystems : Ecological Approaches* (Eds., M.A. Altieri and M. Liebman). pp. 331-337. CRC Press, Boca Raton, Florida, USA.

Amador, M.F. and Gliessman, S.R. (1982). Response of three species (corn, beans and squash) in polyculture, in the Chontalpa, Tabasco, Mexico, Unpublished Manuscript, pp. 20.

Ambasht, R.S. and Rai, Y. (1981). Abstracts of papers, *Annu. Conf. Indian Soc. Weed Sci.* p. 2. Indian Society of Weed Science, Hyderabad, India.

Ambika, S.R. and Jayachandra (1980). Suppression of plantation crops by *Eupatorium* weed. *Current Science* **49:** 874-875.

Amo, S.D. and Anaya, A.L. (1978). Effect of some sesquiterpenic lactones on the growth of certain secondary tropical species. *Journal of Chemical Ecology* **4:** 305-313.

Amritlingam, R. 1984. Cited by Lovett, J.V. (1986). In "The Science of Allelopathy" (A.R. Putnam and C.S. Tang, eds.) pp. 75-99. Wiley Interscience, New York.

Amritphale, D. and Mall, L.P. (1978). Allelopathic influence of *Saccharum spontaneum* L. on the growth of three varieties of wheat. *Science and Culture* **44:** 28-30.

Anaya, A.L. (1989). Recent advances in allelopathy research in Mexico. In: *Phytochemical Ecology: Allelochemicals, Mycotoxins and Insect Pheromones and Allomones* (Eds. C.H. Chou and G.R. Waller). Institute of Botany, Academia Sinica Monograph Series No. **9:** 167-182. Taipei, China.

Anaya, A.L. 1989a. In Phytochemical Ecology : Allelochemicals, mycotoxins insect pheromones and allomones (C.H. Chou and G.R. Waller, eds). Monograph No. 9, 69-80. Institute of Botany, Academia Sinica, Taipei, Taiwan.

Anaya, A.L. and Gomez-Pompa, A. (1971). Inhibition of growth produced by 'piru' *Schinus molle* L. *Rev. Soc. Mex Hist Nat.* **32:** 99-109.

Anaya, A.L., Calera, M.R., Mata, R. and Pereda-Miranda, R. (1990). Allelopathic potential of compounds isolated from *Ipomoea tricolor* Cav. (Convolvulaceae). *Journal of Chemical Ecology* **16:** 2145- 2152.

Anaya, A.L., Ramos, L., Cruz, R., Harnandez, J.C. and Nava, V. (1997). Perspectives in allelopathy in Mexican traditional agroecosystems: A case study in Tlaxcala. *Journal of Chemical Ecology* **13:** 2083-2101.

Anaya, A.L., Ramos, L., Harnandez, J.G. and Cruz, R. (1987). Allelopathy in Mexico. In: *Allelochemicals : Role in Agriculture and Forestry* (Ed. G.R. Waller). ACS Symposium Series No. 330: 89-101. American Chemical Society, Washington DC.

Andrews, F.W. (1945). *Ann. Appl. Biol.* 32, 193-206.

Angiras, N.N., Singh, S.D. and Singh, C.M. (1988). Allelopathic effects of some weeds on germination and growth of chickpea. *Indian Journal of Weed Science* **20:** 85-87.

Anguilar-Erazo, H. (1984). Evaluation of Velvetbean allelopathy. M. Sc. Thesis, Univ. Mayaguez, Puerto Rico.

Anke, H., Zachuer, H. Koening, W.A. (1978). *Arch. Microbiol.* 116, 253-259.

Applebaum, S.W., Gestetner, B. and Birk, Y. (1965). *J. Insect. Physiol.* 11, 611-616.

Applebaum, S.W.,Macro, S. and Birk, Y. (1969). *Agri. Food. Sci.* 17, 618-22.

Arai, M., Hancisi, T., Kitahara, N., Enokilla, R., Kawakubo, K. and Kondo, Y. (1976). *J. Antibiot.* 29, 863-868.

Ashraf, N. and Sen, D.N. (1978). Allelopathic potential of *Celosia argentea* in arid land crop fields. *Ecologia Plantarum* **13:** 331-338.

Ashraf, N. and Sen, D.N. (1980). Allelopathic influence of *Digera alternifolia* on cultivated crops in Indian arid zone. *Indian Journal of Weed Science* **12:** 69-74.

Ashraf, N., Sharmah, M.M. and Sen, D.N. (1977). *Geobios*, 4, 212-213.

Asian Vegetables and Research Development Centre (1978). *Soybean Report* 1976. AVRDC, Shanhua, Taiwan, China.

Bach, T.J. and Lichtenhaler, H.K. (1983). *Physiol. Plant.* 59, 50-60.

Badillo-Feliciano, J. and Lugo-Lopez, M.A. (1983). Exploratory tests on possible injurious aftereffects of pigeon peas on subsequent crops. *Journal of Agricultural University Puerto Rico* **67:** 171-173.

Baker, K.F. and Cook, J.R. (1974). "Biological Control of Plant Pathogens" Freeman, San Francisco.

Baladrin, M.F., Klocke, J.A., Wurteles, E.J. and Bollinger, W.H. (1985). Natural plant-chemicals-source of industrial and medicinal chemicals. *Science* **228:** 1154-1160.

Bandeen, J.D. and Buchholtz, K.P. (1967). Competitive effects of quackgrass upon corn as modified by fertilisation. *Weeds* **15:** 220-224.

Bansal, G.C. and Kalia, A.K. (1984). Allelopathic effects of aqueous extracts of some weeds on the ^{32}P uptake and distribution in *Phaseolus mungo* L. Cv T 9 seedlings. *Journal of Nuclear and Agricultural Biology* **13:** 89-90.

Bansal, G.L. and Singh, C.M. (1986). Allelopathic effect of different plant parts on grassy weeds of wheat (*Triticum aestivum*) on the germination and growth of rice (*Oryza sativa*). *Indian Journal of Weed Science* **18:** 108-110.

Bansal, G.L., Modgal, S.C. and Singh, K.K. (1981). Allelopathic effect of aqueous extracts of some weeds on two wheat cultivars. In: *Proceedings, 8th Asian-Pacific Weed Sciences Society Conference.* Pp. 825- 827.

Bansal, R.P. and Sen, D.N. (1982). Effect of arid zone weed *Trichoderma sedgwickianum* on growth and yield of pearlmillet and sesame crops. *Indian Journal of Botany* **5:** 45-49.

Barash, I., Pupkin, T., Natzer, D. and Kashman, Y. (1982). *Plant Physiol.* **69:** 23-27.

Barbosa, P. and LeTourneau, D.K. (1988). Novel Aspects of Insect- Plant Interactions. John Wiley and Sons, New York.

Barnes, J.P. and Putnam, A.R. (1987). Role of benzoxazinones in allelopathy by rye (*Secale cereale* L.). *Journal of Chemical Ecology* **13:** 889-906.

Barnes, J.P. and Putnam, A.R. (1986). Evidence for allelopathy by residues and aqueous extracts of rye (*Secale cereale*). *Weed Science* **34:** 384-390.

Barnes, J.P. and Putnam, A.R. (1981). *Proc. North Central Weed Control Conf.* pp. 36-64.

Barnes, J.P. and Putnam, A.R. (1983a). *Proc. North Central Weed Control Conf.* pp. 38-39.

Barnes, J.P., Putnam, A.R. and Burke, B.A. (1986). Allelopathic activity of rye (*Secale cereale* L.). In: *The Science of Allelopathy* (Eds., A.R. Putnam and C.S. Tang). pp. 271-286. John Wiley & Sons Inc., New York.

Barnes, J.P., Putnam, A.R., Burke, B. and Assen, A.J. (1987). Isolation and characterization of allelochemicals in rye herbage. *Phytochemistry* **26:** 1305-1390.

Becker, Y., Guyot, L., Massenot, M. and Montegut, J. (1950). Sur la presence d'excretats racinaires toxiques dans les sols de certains groupement vegetaux spontanes. *C R hebd seances Acd Agric Fr* **36:** 689-697.

Behmer, D.E. and McCalla, T.M. (1963). The inhibition of seedling growth by crop residues in soil inoculated with *Penicillium urticae* Bainer. *Plant and Soil* **18:** 199-206.

Bell, A.R. and Nalewaja, J.D. (1968). Competition of wild oat in wheat and barley. *Weed Science* **16:** 505- 508.

Bell, A.R. and Nalewaja, J.D. (1968). Competitive effects of wild oat in flax. *Weed Science* **16:** 501-504.

Bell, A.R. and Nalewaja, J.D. (1963). Effect of duration of wild oat competition in flax. *Weed Science* **16:** 509-512.

Bell, D.T. (1970). Allelopathic effects of *Brassica nigra* on annual grassland. Ph.D Thesis, University of California, Santa Barbara, USA.

Bell, D.T. and Muller, C.H. (1973). *Amer. Midl. Nat.* 90, 227-229.

Bell, E.A. and Charlwood, B.V. (1980). Secondary plant products. Springer and Verlag, Berlin. 674pp.

Beltrano, J. and Montaldi, E.R. (1982). Allelopathic effects of subterranean organs of sorghum (*Sorghum halepense*) on the growth of corn seedlings. *Latinoamericana de Fisiologia Vegetal*. P. 67

Beltrano, J. and Montaldi, E.R. (1982). Mutual allelopathy between sorghum and corn. *Rev Fac. Agronomia Universidad de Buenos Aires* **3:** 265-269.

Bendall, G.M. (1975). The allelopathic activity of California thistle (*Cirsium arvense* (L.) Scop.) in Tasmania. *Weed Research* **15:** 77-81.

Berestetskiy, O.A. and Kalmykova, N.A. (1974). *Mikrobiologi Cheskiy Zhurnal* **36:** 726-732.

Bhandari, D.C. (1977). Studies on arid zone legumes with special reference to *Indigofera* spp. - their ecology, nitrogen metabolism and role in crop ecosystems. Ph.D. thesis, University of Jodhpur, Jodhpur (Raj.).

Bhandari, D.C. and Sen, D.N. (1983). Allelopathy an insight into Indian work. *Indian Review of Life Science* **3:** 207-245.

Bhandari, D.C. and Sen, D.N. (1979). *Agro-ecosystems*, **5:** 257-262.

Bhandari, M. C. (1974). A report of growth regulators in the seed coat of old and fresh seeds of *Citrillus colocynthis* (L). Schrad. *Geobios* **1:** 145-146.

Bhandari, M.C. and Sen, D.N. (1971). *Z. Pflanzenphysiol.* **64:** 466-469.

Bhandari, M.C. and Sen, D.N. (1972). *Z. Naturforch.* **27:** 72- 75.

Bhandari, S.C., Khurana, A.S. and Bhatia, R.K. (1982). *Geobios* **9:** 261-262.

Bhaskar, A. and Vyas, K.G. (1987). Effect of aqueous extracts of *Celosia argentea* on *Rhizobium* sp. *Geobios* **14:** 42-44.

Bhatia, R.K. and Chawan, D.D. (1976). Occurrence of phenolic substances in seed coat of *Cassia* species and their effect on early seedling growth. *Geobios* **3:** 214-216.

Bhatia, R.K., Gill, H.S. and Mehra, S.P. (1982). Allelopathic potential of some weeds on wheat. *Indian Journal of Weed Science* **14:** 108-114.

Bhatti, D.S. (1988). Final Technical Report. Project IN-ARS-497 : FG-IN-608. Haryana Agricultural University, Hisar, India. 231 pp.

Bhatti, D.S. and Dhawan, S.C. (1980). *Haryana Agril Univ. J. Res.* **10:** 419-421.

Bhatti, D.S. and Nandal, S.N. (1994). In Nematode Pest Management in Crops. (D.S. Bhatti and R.K. Walia, eds). pp. 148-164 CBS Publishers, New Delhi.

Bhowmick, B.N. and Choudhary, B.K. (1982). *Indian Bot. Rep.* **1:** 164-165.

Bhowmik, P. C. and Doll, J. D. (1983). Growth analysis of corn and soybean response to allopathic effects of weed residue at various temperatures and photosynthetic photon flux densities. *Journal of Chemical Ecology* **9:** 1263-1280.

Bhowmik, P.C. (1982). *Diss. Abst. Int. B.* **42:** 2637.

Bhowmik, P.C. and Doll, J.D. (1980). Physiology and ecology of weeds. *Research Report, North Central Weed Control Conference* **37:** 64-65.

Bhowmik, P.C. and Doll, J.D. (1979). *Proc. North Central Weed Control Conf.* 34, 43-45.

Bhowmik, P.C. and Doll, J.D. 1983b). *J. Chem. Ecol.* 10, 161-170.

Bhowmik, P.C. and Doll. J.D. (1982). Corn and soybean response to allelopathic effects of weed and crop residues. *Agronomy Journal* **74:** 601-606.

Bioprodukt, (1984). Agrostemin-Gift of Nature Novi Dani, Beograd, Yugoslavia.

Bogdan, G.P. and Grodzinsky, A.M. (1973). *Ukr. Bot. Zh.* 30, 771-778.

Bogdan, G.P., Shpilevoi, B.E., Maistrenko, O.V. and Sidorova, N.V. (1982). Ultrastructural alterations in the clover root xylem under soil fatigue. *In Rol' Allelopatii v Rastenievodstve* (Ed., A.M. Grodzinski). pp. 33-42. Naukova Dumka, Kiev.

Bogdan, G.P., Shpilevoy, B.E., Maistrenko, O.U. and Sidorova, N.V. (1982). In "Rol allelopatii Vrastenie Vcdstve". pp. 33-42. Naukova Kumka Kiev.

Bogdon, G.P. and Grodzinski, A.M. (1973). Role of sulfhydryl groups in protective reactions of plants during allelopathic damage. *Ukr. Bot. Zh.* **30:** 771-778.

Bonner, J. (1950). The role of toxic substances in the interactions of higher plants. *Botanical Review* **16:** 51-65.

Bonner, J. (1960). *Bot. Rev.* 26, 546-569.

Boucher, D. and Espinosa, J. (1982). *Trop. Agric.* 59, 279- 282.

Bradow, J.M. (1985). Germination regulation by *Amaranthus palmeri* and *Ambrosia artemisifolia.* In: *Chemistry of Allelopathy: Biochemical Interactions Among Plants* (Ed., A.C. Thompson). ACS Symposium Series No. **268:** 285-299. American Chemical Society, Washington, D.C.

Bradow, J.M. (1985). Germination regulation by *Amaranthus palmeri* and *Ambrosia artemisiifolia. The Chemistry of Allelopathy : Biochemical Interactions Among Plants* (Ed., A.C. Thompson). ACS Symposium Series 268: 285-300. Washington D.C.

Bradow, J.M. (1985). In "The Chemistry of Allelochemicals" (A.C. Thompson, ed.) pp. 285-289. ACS Symposium Series No. 268. American Chemical Society, Washington, D.C.

Bradow, J.M. and Connick Jr., W.J. (1987). Allelochemicals from palmer amaranth, (*Amarnathus palmeri* S. Wats). *Journal of Chemical Ecology* **13:** 185-202.

Bradow, J.M. and Connick Jr., W.J. (1988). Seed-germination inhibition by volatile alcohols and other compounds associated with *Amaranthus palmeri* residues. *Journal of Chemical Ecology* **14:** 1633-1648.

Bradow, J.M. and Connick Jr., W.J. (1988). Volatile methyl ketone seed germination inhibitors from *Amaranthus palmeri* S. Wats. *Journal of Chemical Ecology* **14:** 1617-1630.

Breazeale, J.F. (1924). The injurious after-effects of sorghum. *Journal of American Society of Agronomy* **16:** 689-700.

Breazeale, T.F. 1924. *J. Am. Soc. Agron.* **16:** 689-700.

Broadway, R.M. Duffey, S.S. 1986. *J. Insect Physiol.* **32:** 827-833.

Brown, S.A. 1969. *Bioscience* **19:** 115.

Buchholtz, K.P. (1971). The influence of allelopathy on mineral nutrition. In: *Biochemical Interactions Among Plants.* pp. 86-89. Environmental Physiology Subcommittee for the International Biological Programme, Division of Biology and Agriculture, National Research Council, National Academy of Sciences; Washington DC.

Bukolova, T.P. (1971). A study of the mechanism of acation of water-solbile substances of weeds on cultivated plants. In: *Fiziologo-Biokhimicheskie Osnovy Vzaimodeistviya Rastenii v Fitotsenozash* (Ed., A.M. Grodzinskii). Pp. 66-69. Naukova Dumka, Kiev.

Bukolova, T.P. 1971. In "Physiological-Biochemical Basis of Plant Interactions in Phytocenosis" (A.M. Grodzinsky, ed.) 2, 66-69. Naukova Dumka, Kiev.

Burgeos-Leon, W., Ganry, F., Nicon, R., Chopart, J.L. and Dommergues, Y. 1980. *Agron. Trop.* 35, 319-334.

Burgos, Leon, W., Ganry, F., Nicou, R., Chopart, J.L. and Dommergues, Y. (1980). Un cas de fatique des sols induite par la culture du sorgho. *Agron. Trop.* **35:** 319-334.

Burhan, H.O. and Mansi, M.G. 1967. *J. Agric. Sci.* Camb. 68, 255- 261.

Burke, B.A. (1987). Allelopathy : A biotechnological-agrochemical approach. In: *Allelochemicals : Role in Agriculture and Forestry* (Ed., G.R. Waller). ACS Symposium Series No. **330:** 147-155. American Chemical Society, Washington DC.

Bushra, I., Farrukh, H. and Forhat, B. 1987. *Pakistan J. Sci. Ind. Res.* 30, 530-33.

Bye, R.A. 1979. *The Kiva* 44, 237-56.

Caamal, A. and del Amo, R.S. 1987. *Turrialba* 37, 195-210.

Campbell, G., Lambert, J.D.H., Arnasson, T. and Towers, G.H.N. (1982). Allelopathic properties of α-terthinyl and phenylheptatriyne, naturally occurring compounds from species of Asteraceae. *Journal of Chemical Ecology* **8:** 961-973.

Carley, H.E. and Watson, R.D. 1967. *Phytopathology* 57, 401-4.

Carnahan, G. and Hull, A.C. (1962). The inhibition of seeded plants by tarweed. *Weeds* **10:** 87-90.

Castro, P.R.C., Rodrigues, J.D. and Maimoni-Rodella, R.C.S. (1984). Allelopathic action of some extracts of plants on the germination of rice (*Oryza sativa* L.). *An E S A Luiz de Querioz* **41:** 369-381.

Castro, P.R.C., Rodrigues, J.D., Maimoni-Rodella, R.C.S., Rabelo, J.C., Veiga, R.F.A., Lima, G.P.P., Jureidini, P. and Denadai, I.A.M. 1984. *Anais. Esc. Sup. Agri. Luiz. Queiroz.* 41, 369-381.

Castro, P.R.C., Rodrigues, J.D., Moraes, M.A. and Carvalho, V.L.M. (1983). Allelopathic effects of some plant extracts on the germination of some tomato (*Lycopersicum estulentum* Mill. Cv. Santa Cruz). *Planta Daninha* **6:** 79-85.

Castro, P.R.C., Rodrigues, J.D., Moreas, M.A. and Carvalho, V.L.M. 1983. *Plant Daniha* 6, 79-85.

Caussanel, J.P. (1979). Non-competitive effects between lambs quarters (*Chenopodium album* L.) and maize. *Weed Research* **19:** 123-135.

Caussanel, J.P. 1979. *Weed Res.* 19, 123-135.

Caussanel, J.P. and Kunesch, G. (1979). Qualitative and quantitative study of growth inhibitors in root exudates of common lambsquarters (*Chenopodium album* L.) at the beginning of its flowering in hydroponic culture and under controlled conditions. *Z Pflazenphysiol* **93:** 229-243.

Caussanel, J.P. and Kunesch, G. 1979. *Z. Pflanzenphysiol.* 93, 229-243.

Cernusko, K. and Boreky, V. 1992. *Rostlinna Vyroba* 38, 608-9.

Chacon, J.C. and Gliessman, S.R. (1982). Use of the 'Non-weed' concept in traditional tropical agro-ecosystems of south eastern Mexico. *Agro-Ecosystems* **8:** 1-11.

Chagtai, S.M., Sadiq , A and Ibrar, M. 1988. Pakistan J. Bot. 20, 213-220.

Chagtai, S.M., Sadiq, A. and Ibrar, M. 1987. Biologia (Lahore) 33, 27-32.

Chagtai, S.M., Sadiq, A. and Shah, J. (1988). Phytotoxicity of *Silybum marianum* Gaertn. on wheat. *Pakistan Journal of Botany* **220:** 231-240.

Chagtai, S.N., Sadiq, A. and Ibrar, M. (1986). Phytotoxicity of *Fumaria indica* on wheat. *Pakistan Journal of Botany* **18:** 59-64.

Chan, B.G., Waiss Jr., Binder, R.G. and Elliger, C.A. 1978. *Entomol Exp. Appl.* 24, 94-100.

Chandler, J.M. (1977). Competition of spurred anoda, velvetleaf, prickly sida and venice mallow in cotton. *Weed Science* **25:** 151-158.

Chandramohan, D. Purushothaman, D. and Kothandataman (1973). Soil phenolics and plant growth inhibition. *Plant and Soil* **39:** 303-308.

Chandrasekaran, S. and Yoshida, T. (1973). Effect of organic acid transformation in submerged soils on growth of rice plant. *Soil Science and Plant Nutrients* **19:** 39-45.

Chandrasekaran, S. and Yoshida, T. 1973. Soil Sci. Plant Nutri. 19, 39-45.

Chang, M.D., Netzly, D.H., Butler, L.G. and Lynn. D.G. 1986. *J. An. Chem. Soc.* 108, 7858-7860.

Chase, W.R., Nair, M.G. and Putnam, A.R. (1991). 2,2'-Oxo-l, 1 '-Azobenzene. II. Selective toxicity of rye (*Secale cereale* L.) allelochemicals to weed and crop species. *Journal of Chemical Ecology* **17:** 9-19.

Chatterjee, V.N. 1975. *Indian J. Plant Physiol.* 18, 91-96.

Chatterji, U.N. (1975). Some aspects of plant-plant chemical interactions. *Indian Journal Plant Physiology* **18:** 91-96.

Chen, F.S., Taggart, J.M. and Elofron, R.M. (1982). Chemical constituents in wild oat (*Avena fatua*) hulls and their effects on seed germination. *Canadian Journal of Plant Science* **62:** 155-161.

Chivinge, O.A. 1985. *Zimbabwe. Agric. J.* 82, 151-152.

Chou, C.H. (1983). Allelopathy in agroecosystem. In: *Allelochemicals and Pheromones* (Eds., C. H. Chou and G. R. Waller). Pp. 27-64. Institute of Botany, Academia, Sinica, Taipei, Taiwan.

Chou, C.H. (1986). The role of allelopathy in subtropical agroecosystems in Taiwan. In: *The Science of Allelopathy* (Eds., A.R. Putnam and C.S. Tang). pp. 57- 73. John Wiley & Sons Inc., New York.

Chou, C.H. and Lin, H.J. (1976). Autointoxication mechanism of *Oryza sativa.* I. Phytotoxic effects of decomposing rice residue in soil. *Journal of Chemical Ecology* **2:** 353-367.

Chou, C.H., Chaing, Y.C. and Cheng, H.H. (1981). Autointoxication mechanism of *Oryza sativa*. III. Effect of temperature on phytotoxins production during rice straw decomposition in soil. *Journal of Chemical Ecology* **7:** 741-752.

Chou, C.H. (1987). Allelopathy in subtropical vegetation and soils in Taiwan. In: *Allelochemicals : Role in Agriculture and Forestry* (Ed., G.R. Waller). ACS Symposium Series No. **330:** 102-117. American Chemical Society, Washington DC.

Chou, C.H. and Chiou, S.J. (1979). Autointoxication mechanism of *Oryza sativa*. II. Effects of culture treatments on the chemical nature of paddy soil and on rice productivity. *Journal of Chemical Ecology* **5:** 839-859.

Chou, C.H. and Kuo, Y.L. 1987. *J. Chem. Ecol.* 12, 1431- 1448.

Chou, C.H. and Yao, C. (1983). Phytochemical adaptation of coastal vegetation in Taiwan. I. Isolation, identification and biological activities of compounds in *Vitex negundo* L. *Botany Bulletin of Academia Sinica* **24:** 155-168.

Chou, C.H., Lee, M.L. and Oka, H.I. (1984). Possible allelopathic interactions between *Oryza perennis* and *Leersia hexandra. Botanical Bulletin of Academia Sinica* **25:** 1-19.

Cochran, V.L., Elliott, L.F. and Papendick, R.I. (1977). The production of phytotoxins from surface crop residues. *Soil Science Society of American Journal* **41:** 903-908.

Cochrane, V.W. 1948. *Phytopathology* 38, 185-196.

Cole, J.S. and Halstead, A.L. 1922. Methods of Winter Wheat Production at Fort Hyes Branch Station. USDA *Bull.* No. 1094.

Cole, R.J., Dorner, J.W., Cow, R.H., Hill, R.A., Cutler, H.G. and Well, J.M. 1981. *Appl. Environ. Microbiol.* 42, 677-681.

Cole, R.J., Kirksey, J.W., Cutler, H.G., Doupnik, B.L. and Peckham J.C. 1973. *Science* 179, 1324-1326.

Cole, R.J., Wilson, D.M., Harper, J.L., Cox, R.H., Cochran, T.W., Cutler, H.G. and Bell, D.K. 1981. *J. Agric. Food Chem.* 30, 301-303.

Cole, S. 1991. M.A. Thesis, University of South Dakota, Vermillon, South Dakota, USA.

Collison, R.C. (1925). The presence of certain organic compounds in plants and their relation to the growth of other plants. *Agronomy Journal* **17:** 58-68.

Collison, R.C. and Conn, H.J. (1925). The effect of straw on plant growth. New York. *Agriculture Experiment Station Technical Bulletin* **114:** 1-25.

Colton, C.E. and Einhellig, F.A. (1980). Allelopathic mechanisms of velvetleaf (*Abutilon theophrasti* Medic (Malvaceae) on soybean. *American Journal of Botany* **67:** 1407-1413.

Connick Jr., W.J., Bradow, J.M., Legendre, M.G., Vail, S.L. and Menges, R.M. (1987). Identification of volatile allelochemicals from *Amaranthus palmeri* S. Wats. *Journal of Chemical Ecology* **13:** 463-472.

Conrad, J.P. (127). Soome causes of the injurious after-effects of sorghum and suggested remedies. *Journal of American Society of Agronomy* **19:** 1091-1110.

Conrad, J.P. 1927. *American Soc. Agron. J.* 19, 1091 (Abstract).

Cook, C.E., Whichard, L.P., Turner, B., Well, M.E. and Egley, G.H. 1966. *Science* 154, 1189-1190.

Cooke, T. 1908 In The Flora of the Presidency of Bombay, vol. 2 (Reprinted 1958). Botanical Survey of India, Calcutta.

Cramer, G.L. 1981. *Diss. Abst. Int. B.* 41, 3969-3970.

Cruz, O.R., Anaya, A.L. and Ramos, L. 1988. *J. Chem. Ecol.,* 14, 71-86.

Cubbon, M.H. 1925. *Agron. J.* 17, 568-577.

Culepreper, N. 1633. English Physitian and Complete Herball Foulsham, London (Reprinted, 1955).

Currie, J.A. 1962 *J. Sci. Food Agric.* 13, 380-385.

Curtis, J.T. and Cottam, G. (1950). Antibiotic and autotoxic effects in Prairie sunflower. *Bulletin of the Torrey Botanical Club* **77:** 187-191.

Curtis, J.T. and Cottam, G. 1950. *Bull. Torrey Bot. Club.* 77, 187-191.

Cutler, H. G. (1984). Biologically active natural products from fungi: templates for tomorrow's pesticides. In: *Bioregulators : Chemistry and Uses* (Eds., R.L. Ory and F.R. Ritting). ACS Symposium Series No. **257:** 153-170. American Chemical Society, Washington, D.C.

Cutler, H.G. (1985). Secondary metabolites from plants and their allelochemic effects. In: *Bioregulators for Pest Control* (Ed., P.A. Hedin). ACS Symposium Series No. **276:** 445-468. American Chemical Society, Washington. D.C.

Cutler, H.G. (1987). Japanese contributions to development of allelochemicals. In: *Allelochemicals : Role in Agriculture and Forestry* (Ed. G.R. Waller). ACS Symposium Series No. **330:** 23-38. American Chemical Society, Washington DC.

Cutler, H.G. (1988). Natural products and their potential in Agriculture: A personal overview. In *Biologically Active Natural Products: Potential Uses in Agriculture* (Ed. H.G. Cutler). ACS Symposium Series No. **380:** 1-21. American Chemical Society. Washington DC.

Cutler, H.G. (1988). Perspectives on discovery of microbial phytotoxins with herbicidal activity. *Weed Technology* **2:** 525-532.

Cutler, H.G. (1991). Phototoxins of microbial origin. In: *Hand Book of Natural Toxins - Toxicology of Plant and Fungal Compounds* (Eds., R.F. Keeler and A.T. Tu). **6:** 411-438. Marcel Dekker Inc., New York.

Cutler, H.G. and Jarvis, B.B. (1985). Preliminary observations on the effect of macrocylcic tricothecenes on plant growth. *Environmental and Experimental Botany* **25:** 115-128.

Cutler, H.G., Crumley, F.G., Cox, R.H., Wells, J.M. and Cole, R.J. (1984). The biological properties of cyclopenin and cyclopenol. *Plant and Cell Physiology* **25:** 251-263.

Dadykin, V.P., Stepanou, L.N. and Ryzhkova, B.E. (1970). On importance of volatile plant secretions under the development of closed systems. In: *Physiological and Biochemical Interactions Between the Plants in Phytocenosis* (Ed. A.M. Grodzinsky) **1:** 118-123. Naukova Dumka, Kiev, USSR.

Dagley, S. 1983. *Residue Review* 85, 127-137.

Dahiya, R.S., Dutt, R. and Gupta, D.C. (1986). Effect of leucaena and sunhemp leaves on cereal cyst nematode *Heterodera avenae* in wheat. *Haryana Agricultural University Journal of Research* **16:** 389-391.

Dahlman, D.L. and Rosenthal, G.A. 1975. In "CRC Handbook of natural pesticides : Methods". Vol. 1 : Theory, practice and detection (N.B. Mandawa, ed). Pp. 201-230. CRC, Boca Raton, Florida, USA.

Dailey Jr., O.D., Severson, R.E. and Arrendale, R.F. (1989). Nonpolar lipids of *Amaranthus palmeri* S. Wats. I. Fatty alcohols and wax esters (saturated). *Journal of Agricultural and Food Chemistry* **37:** 1317- 1321.

Dale, J. E. (1986). Decline in phytotoxicity of benzyl isothyiocyanate formulated in granules. *Weed Science* **34:** 325-327

Dalrymple, R.L. and Rogers, J.L. (1983). Allelopathic effects of western ragweed on seed germination and seedling growth of selected plants. *Journal of Chemical Ecology* **9:** 1073-1078.

Das, N.R. and Ghosh, N. (1978). Ramification of wheat root as influenced by root excretions of jute (*Corchorus capsularis*). *Science and Culture* **44:** 41-42.

Datta, S.C and DasMahaptra, S. (1984). Allelopathy in two legumes: Inhibition of mustard (*Brassica*) germination and seedling growth. *Comparative Physiology and Ecology* **9:** 286-289.

Datta, S.C. and Chatterjee, A.K. (1978). Some characteristics of an inhibitory factor in *Polygonum. Indian Journal of Weed Science* **10:** 23-33.

Datta, S.C. and Chatterjee, A.K. (1980). Pollution by plants. In: *Symposium on Environmental Pollution and Toxicology* (Ed., S.P. Raychaudhuri). pp 195-234. Today and Tomorrow, New Delhi.

Datta, S.C. and Ghosh, K.N. (1982). Effect of pre-sowing treatment of mustard seeds with leaf and inflorescence extract of *Chenopodium murale. Indian Journal of Weed Science* **14:** 1-6. !C

Datta, S.C. and Sinha-Roy, S.P. (1975). Phytotoxic effects of *Croton bonplandianum* Baill on weedy associates. *Vegetatio* **30:** 157-163.

Datta, S.C. and Sinha-Roy, S.P. (1983). Characteristics of an inhibitory factor from *Croton bonplandianum* Baill. *Acta Agronomica Academia Scientiarum Hungaricae* **32:** 124-129.

Datta, S.C., Bandyopadhyay, A.K. 1981. *Proc.* 8th *Asian Pacific Weed Sci. Soc. Conf.* vol. 1, pp. 391-399.

Dawson, J. H. (1977). Competition of late-emerging weeds with sugarbeets. *Weed Science* **25:** 168- 170.

Dawson, J.H. (1965). Competition between irrigated sugarbeets and annual weeds. *Weeds* **13:** 245- 249.

Dawson, R.F. 1948. *Adv. Enzymol.* 8, 203-257.

de Candolle, M.A.P. (1832). Physiologie Vegetale. *Tome III, Bechet Jeune*, Paris. Pp. 1474-1475.

DeFrank, J. and Putnam, A.R. 1977. *Proc. North Central Weed Control Conf.* 32, 23-24.

DeFrank, J. and Putnam, A.R. 1983. *Crop Protection* 2, 173- 81.

Dekker, J. and Meggitt, W.F. 1982. *Weed Res.* 23, 91-101.

DelMoral, R. and Cates, R.G. 1971. *Ecology* 52, 1030-1037.

Deperalta, A. and Estioko, R.P. 1923. *Philippines Agric.* 11, 205-211.

Devakumar, C. and Dev, S. (1994). Chemistry of neem. In : *Abstracts* International Symposium *Allelopathy in Sustainable Agriculture, Forestry and Environment* (Eds. S.S. Narwal, P. Tauro, G.S. Dhaliwal and J. Parkash). pp. 119. Indian Society of Allelopathy, Haryana Agricultural University, Hisar, India.

Dhar, T.K., Siddigqui, K.A.T. and Ali, E. 1982 *Tetradhedron Lett.* 23, 5459-5462.

Dharmaraj, G. and Ali, A.M. 1985. Abstracts of Papers, *Annu. Conf. Indian Soc. Weed Sci.* pp. 46. Indian Society of Weed Science, Hyderabad, India.

Dilday, R.H. (1992). Victims no one mourns. *Agricultural Research* **40**: 10-11.

Dilday, R.H., Frans, R.E., Semidey, N., Smith, R.J. and Oliver, L.R. (1992). Weed control with crop allelopathy. *Arkansas Farm Research* **41:** 14-15.

Dock Gustovsson, A.M. (1989). Growth of dicotyledonous weeds. *Crop Production Science Publication* No. 5. Department of Crop Production, Swedish University of Agricultural Science, Uppsala, Sweden.

Dodge, A.D. (1987). Potential new targets for herbicides. *Pesticide Science* **20:** 301-313.

Drost, D.C., Doll, J.D. and Moody, K. 1980. *Philippines J. Weed Sci.* 7, 34-39.

Drost, S.C. and Doll, J.D. (1980). The allelopathic effect of yellow nutsedge (*Cyperus esculentus*) on corn (*Zea mays*) and soybean (*Glycine max*). *Weed Science* **28:** 229-233.

Dube, V.P., Singhal, V.P. and Tyagi, S. (1979). *Parthenium hysterophorus* : Allelopathic effects on vegetable crops. *Botanical Progress* **2** : 62-69.

Dubey, P.S. (1973). Phytotoxicity of weeds to crops. I. Effect on germination. *Science and Culture* **39:** 556-558.

Duke, S.O. (1986). Microbially produced phytotoxins as herbicides - A perspective. In: *The Science of Allelopathy* (Eds., A. R. Putnam and C. S. Tang). pp. 287-304. John Wiley & Sons Inc., New York.

Duke, S.O. (1990). Natural pesticides from plants. In: *Advances in New Crops* (Eds., J. Janick and J.E. Simon) pp 511-517. Timber Press, Portland, OR, U.S.A.

Duke, S.O. (1991). Plant terpenoids and pesticides. In: *Hand Book of Natural Toxins -Toxicology of Plant and Fungal Compounds* (Eds., R.F. Keeler and A.T. Tu). **6:** 269-296. Marcel Dekker Inc., New York.

Duke, S.O. 1986a. In Chemistry of Allelopathy (A.C. Thompson, ed.) pp. 113-131. ACS Symposium Series No. 268, American Chemicla Society, Washington, D.C.

Duke, S.O. 1986C. In ``The Science of Allelopathy" (A.R. Putnam and C.S. Tang, eds) PP. 287-294. Wiley Interscience, New York.

Duke, S.O. and Lydon, J. (1987). Herbicides from natural compounds. *Weed Technology* **1:** 122- 127.

Duke, S.O. Paul, R.N. and Wickliff, J.L. 1980. *Physiol. Plant.* 49, 27-36.

Dukes, P. Jones, A., Schalk, J.M. and Peterson, J. 1989. *Agric. Res.* 37 (11), 24-25.

Durbin, R.D. 1981. ``Toxins in plant disease". Academic Press, New York.

Durbin, R.D. and Uchytil 1977. *Phytophthology* 67, 602-3.

Dutt, R. and Bhatti, D.S. (1986). Determination of effective doses and time of application of nematodes and castor leaves for controlling *Meloidogyne javanica* in tomato. *Indian Journal of Nematology* **16:** 8-11.

Dzyubenko, N.N. and Krupa, L.I. (1974). On interactions of cultivated plants vegetation and weeds in agrophytocenoses. In: *Physiological and Biochemical Interactions Between the Plants in Phytocenosis* (Ed., A.M. Grodzinsky) Vol. **5:** 55-56. Naukova Dumka, Kiev, USSR.

Dzyubenko, N.N. and Petrenko, N.I. (1971). On biochemical interactions of cultivated plants and weeds. In: *Physiological and Biochemical Interactions Between the Plants in Phytocenosis* (Ed., A.M. Grodzinsky) Vol. **2:** 60-66. Naukova Dumka, Kiev, USSR

Eaton, B.J., Russ, O.G. and Feltner, K.C. (1976). Competition of velvetleaf, prickly sida and venice mallow in soybeans. *Weed Science* **24 :** 225-228.

Egley, G.H. (1986). Stimulation of weed seed germination in soil. *Review of Weed Science* **2:** 67-89.

Egunjobi, O.A. and Afolami, S.O. 1976. *Nematologica* 22, 125-132.

Einhellig, F.A. (1985). Allelopathy-a natural protection allelochemics. In: *Handbook of Natural Pesticides Methods. Theory, Practice and Detection* (Ed., N.B. Mandava) **1:** 161-200. CRC Press, Boca Raton, Florida.

Einhellig, F.A. (1986). Mechanisms and modes of action of allelochemicals. In: *The Science of Allelopathy* (Eds., A.R. Putnam and C.S. Tang). pp. 171-188. John Wiley & Sons Inc., New York.

Einhellig, F.A. and Eckrich, D.C. (1984). Interactions of temperature and ferulic acid stress on grain sorghum and soybean. *Journal of Chemical Ecology* **10:** 161-170.

Einhellig, F.A. and Leather, G.R. (1988). Potentials for exploiting allelopathy to enhance crop production. *Journal of Chemical Ecology* **14**: 1829-1844.

Einhellig, F.A. and Rasmussen, J.A. (1973). Allelopathic effects of *Rumex crispus* on *Amaranthus retroflexus,* grain sorghum and corn in field. *American Midland Naturalist* **90:** 79-86.

Einhellig, F.A. and Rasmussen, J.A. (1989). Prior cropping with grain sorghum inhibits weeds. *Journal of Chemical Ecology* **15:** 951-959.

Einhellig, F.A. and Schon, M.K. (1982). Non-competitive effects of *Kochia scoparia* on grain sorghum and soybeans. *Canadian Journal of Botany* **60:** 2923-2930.

Einhellig, F.A. and Souza, I.F. (1992). Phytotoxicity of sorgoleone found in grain sorghum root exudates. *Journal of Chemical Ecology* **18**: 1-11.

Einhellig, F.A., Rasmussen, J.A., Hejl, A. and Souza, I.F. (1993). Effects of root exudate sorgoleone on photosynthesis. *Journal of Chemical Ecology* **19:** 369-375.

El-Habbasha, K.M. and Behairy, A.G. 1975. *Zeit.Acker. Pflanzen.* 145, 66-74.

Elkan, G.H. 1961. *Canadian J. Microbiol.* 7, 851-856.

Ellenby, C. 1945. *Ann. Appl. Biol.* 32, 237-239.

Elliott, L.F., Cochran, V.L. and Papendick, R.I. 1981. *Soil Sci.* 131, 48-52.

Elliott, L.F., Miller, D.E. and Richards, A.W. 1979, *Plant Diss. Reptr.* 63, 882-886.

Ellis, J.R. and McCalla, T.M. 1973. *Appl. Microbiol.* 25, 562-566.

Ellison, L. 1959. *K. Lantbr. Hogsk, Annlr.* 25, 285.

Elmore, C.D. (1980). Inhibition of turnip germination by velvetleaf (*Abutilon theophrasti*) seed. *Weed Science* **28:** 658-660.

Else, M.J. and Ilnicki, R.D. (1989). Weed Science Society of America *Abstracts* No. 157.

Enache, A. and Ilnicki, R.D. 1988. *Proc. N.E. Weed Sci. Soc.* 42, 34.

Escarzaga, G.E. 1987. M.Sc. Thesis. Inst. Technol. Estudios Superior de Monterrey, Campus Queretaro, Mexico.

Eussen, J.H.H. 1978. *Proc.* 6th *Asian-Pacific Weed Sci. Soc. Conf. Indonesia* 1, 138-151.

Eussen, J.H.H. 1979. *Oecologia* (Berlin), 40, 351-356.

Evenari, M. (1949). Germination inhibitors. *The Botanical Review* **15:** 153-194.

Evetkovic, R. 1980. *Fragm. Herbol. Jugosl.* 9, 47-51.

Evetts, L.L. and Burnside, O.C. (1972). Germination and seedling development of common milkweed and other species. *Weed Science* **20:** 371-378.

FAO. 1970. Proc. FAO *Int. Conf. Weed Control.* University of California, Davis, USA.

Farrukh, H., Ahmed, N. and Akram, M. 1986. *Pakistan J. Sci. Ind. Res.* 29, 458-460.

Fay, P.K. and Duke, W.B. (1977). An Assessment of allelopathic potential of *Avena* germplasm. *Weed Science* **25:** 224-228.

Feltner, K.C., Hurst, H.R. and Anderson, L.E. (1969). Yellow foxtail competition in grain sorghum. *Weed Science* **17:** 211-213.

Fischer, H.P. and Bellus, D. 1983. *Pestic. Sci.* 14, 334- 346.

Fischer, N.H. and Quijano, L. (1985). Allelopathic agents from common weeds. In: *The Chemistry of Allelopathy: Biochemical Interactions Among Plants* (Ed., A.C. Thompson). ACS Symposium Series No. **268:** 133-147 American Chemical Society, Washington DC.

Fischer, N.H., Weidenhamer, J.D. and Bradow, J.M. (1989). Inhibition and promotion of germination by several sesquiterpenes. *Journal of Chemical Ecology* **15:** 1785-1793.

Fischer, N.H., Weidenhamer, J.D., Riopel, J.L., Quijano, L. and Menelaou, M.A. (1990). Stimulation of witchweed germination by sesquiterpene lactones: A structure activity study. *Phytochemistry* **29:** 2479-2483.

Fisher, R.F. 1979. In Plant Disease, An Advance Treatise (F.G. Horrfall and E.B. Cawling, Eds.) pp. 313 Academic Press, New York.

Fletcher, F. 1912. *J. Agric. Sci. Camb.* 4, 245-247.

Fletcher, R.A. and Renee, A.J. (1963). A growth inhibitor found in *Centaurea* spp. *Canadian Journal of Plant Science* **43:** 475-481.

Forney, R.D. and Foy, C.L. (1985). Phytotoxicity of products from rhizospheres of a sorghum- sudangrass hybrid (*Sorghum bicolor* x *Sorghum sudanense*). *Weed Science* **33:** 597-604.

Forney, R.D., Roy, C.L. and Wolf, D.D. 1983. *Proc. South Weed Sci. Soc.* 36, 358.

Forney, R.D., Roy, C.L. and Wolf, D.D. 1985. *Weed Sci.* 33, 490-497.

Francis, C.A. 1989. *Adv. Agron.* 42, 1-43.

Friedman, T. and Horowitz, M. (1971). Biologically active substances in subterranean parts of purple nutsedge. *Weed Science* **19:** 398-401

Friesen, H.A. (1961). Some factors affecting the control of wild oats with barban. *Weeds* **9:** 185-194.

Fuerst, E.P. and Putnam, A.R. (1983). Sepaı ating the competitive and allelopathic components of interference: Theoretical principles. *Journal of Chemical Ecology* **9:** 937-944.

Fujii, Y. 1992. The potential biological control of paddy weeds with allelopathy-allelopathic effects, of some rice varieties. In: *Proceeding Biological Control and Integrated Management of Paddy and Aquatic Weeds in Asia.* pp. 305-320. National Agriculture Research Centre, Tsukuba, Japan.

Fujii, Y., Shibuya, T. and Yasuda, T. 1991b. *Agric. Biol. Chem.* 55, 617-618.

Fujii, Y., Shibuya, T. and Usami, Y. (1991). Allclopathic effect of *Mucuna pruriens* on the appearance of weeds. *Weed Research* (Japan) **36:** 43-49.

Fujii, Y., Shibuya, T. and Usami, Y. 1988. *Weed Research (Japan)* 33 (Suppl), 107-114.

Funke, G.L. (1943). The influence of *Artemisia absinthium* on neighbouring plants. *Blumea* **5:** 281-293.

Gabor, W.A. and Veatch, C. (1981). Isolation of a phytotoxins from quackgrass (*Agropyron repens*) rhizomes. *Weed Science* **29:** 155-164.

Gaidamak, V.M. (1971). Biologically active substances in nutrient solutions after cucumbers and tomatoes grown on pure and multiple used broken bricks. In: *Physiological and Biochemical Interactions Between the Plants in Phytocenosis* (Ed. A.M. Grodzinsky) Vol. **2:** 56-60. Naukova Dumka, Kiev, USSR.

Gajic, D. 1966, *J. Agric, Res.* 19, 63-96.

Gajic, D. and Nikocevic, G. (1973). Chemical allelopathic effects of *Agrostemma githago* upon wheat. *Fragmenta Hebologica Jugoslavica* **18:** 1-5.

Gajic, D., Malencic, S., Varbaski, M. and Varbaski, S. (1976). Study of possible quantitative and qualitative improvement of wheat yield through agrostemin as an allelopathic factor. *Fragmenta Herbologica Jugoslavica* **63:** 121-141 (In Serbian).

Garb, S. (1961). Differential growth inhibitors produced by plants. *Botanical Review* **27:** 422-443.

Garcia, A.G. 1983. Seasonal variation in allelopathic effects of corn residues on corn and cress seedling, Ph. D. Dissertation, Iowa State University, Ames, USA.

Gatehouse, A.M.R. and Boulter, D. 1983. *J. Sci. Food Agric.* 34, 345-350.

Gatehouse, A.M.R., Dewey, F.M., Dove, J., Fenton, K.A. and Pusztal, A. 1984. *J. Sci. Food Agric.* 35, 373-380.

Geller, I.A., Kalmykova, N.A. and Petrusha, L.K. 1977. *Sakharnarya Sbekla* 1, 31-32.

Gilmore, A.R. (1977). Effect of soil moisture on monoterpenes in loblolly pine. *Journal of Chemical Ecology* **3:** 667-676.

Gleason, F.K., Thoma, W.J. and Carlson, J.L. 1987. In Progress in photosynthetic research (J. Biggins, ed). Vol. 3, 763-766 Martinus Nijhoff, Dodrecht.

Glick, J.L., Pierce, M.V., Anderson, D.M., Vaslet, C.A. and Hsiao, H.Y. 1984. In Genetic Engineering, BARC Symposium 7 (L.D. Owens, ed.), pp. 67-67. Rowman and Allanheld, Totolova, New Jersey.

Gliessman, S. R. (1982). Allelopathy and biological weed control in agroecosystems. In : *Proceedings of the Seminar on Allelochemicals and Pheromones* (Eds., C.H. Chou and G.R. Waller). pp. 77-86. Institute of Botany, Academia Sinica, Taipei, Taiwan, China.

Gliessman, S.R. (1983). Allelopathic interaction in crop weed mixtures: Application for weed management. *Journal of Chemical Ecology* **9**: 991-999.

Gliessman, S.R. (1989). Allelopathy and agricultural sustainability. In: *Allelochemicals, Mycotoxins and Insect Pheromones and Allomones* (Eds., C.H. Chou and G.R. Waller). Institute of Botany, Academia Sinica Monograph Series No. **9:** 69-80. Taipei, RCC.

Gliessman, S.R. 1982. *Plant Soil* 67, 105-117.

Gliessman, S.R. and Altieri, M.A. (1982). Polyculture cropping has advantage. *California Agriculture* **36:** 14-16.

Gliessman, S.R. and Garcia, E.R. 1979. In Tropical legumes : Resources for the future Pub. No. 27, 292-293. National Academy of Science, Washington, D.C.

Gliessman, S.R., Garcia, E.R. and Amador, M.A. 1981. *Agro- Ecosyst.* 7, 173-185.

Goel, U. 1987. *Acta Bot. Indica* 15, 129-130.

Golovko, E.A., Elaneskaya, I.A. and Kostroma, E.Y. (1981). Allelopathic soil fatigue and phytotoxic properties of soil microscopic fungi. In: *Allelopathy in Natural and Artificial Phytocenosis: Collection of Scientific Papers* (Ed., A.M. Grodzinsky). Pp. 86-95. Naukova Dumka, Kiev, USSR.

Gommers, F.J. 1972. *Acta. Bot. Nederlandica* 21, 111-112.

Gommers, F.J. 1973. M.Sc. Dissertain. Agricultural University, Wageningen, Netherlands, 73 Pp.

Gommers, F.J. 1981. *Helminthol Abot* 50 B, 9-24.

Gorobets, S.A. and Serdyuk, L.S. (1981). Dynamics of organic substances in perlite when cultivating khibinskaya cabbage in monoculture under artificial condition. 2. Plant substances. In: *Allelopathy in Natural and Artificial Phytocenosis: Collection of Scientific Papers* (Ed., A.M. Grodzinsky). Pp. 159-164. Naukova Dumka, Kiev, USSR.

Goswami, B. K. and Vijaylakshmi, K. (1986). Efficacy of some indigenous plant materials and non- edible oilseed cakes against *Meloidogyne incognita* on tomato. *Indian Journal of Nematology* **16:** 280-281.

Grechkanev, O.M. and Rodionov, V.I. (1971). Interactions of the components of the nector fodder mixture with wild heliotrope. In: *Physiological and Biochemical Interactions Between the Plants in Phytocenosis* (Ed., A.M.Grodzinsky). Vol. **2:** 88-94. Naukova Dumka, Kiev, USSR.

Green, T.R. and Ryan, C.A. 1972. *Science* 175, 776-777.

Greenwood, D.J. 1961. *Plant Soil* 14, 360-367.

Greenwood, D.J. and Lee, H. 1960. *Plant Soil* 12, 69-80.

Gressel, J. 1985. In "Weed Physiology : Vol. II. Herbicide physiology" (S.O. Duke, ed) PP. 159-189. CRC Press, Boca Raton, Florida, USA

Gressel, J.B. and Holm, L.G. (1964). Chemical inhibition of crop germination by weed seeds and the nature of inhibition by *Abutilon theophrasti. Weed Research* **4:** 44-53.

Grodzinsky, A.M. (1974). Role of root systems in chemical interaction of plants. In: *Physiological and Biochemical Interactions Between the Plants in Phytocenosis* (Ed., A.M. Grodzinsky) Vo1. **5:** 10-14. Naukova Dumka, Kiev, USSR.

Grodzinsky, A.M. (1982). Evolutionary problems of the chemical interactions among plants. In: *Evolution and Environment* (Eds., V.J.A. Nevak and J. Mlikovsky). Pp. 133-143. CSAV, Praha.

Grodzinsky, A.M. 1969. *Qual. Plant Mat. Veg.* 17, 93-100.

Grodzinsky, A.M. and Golovko, E.A. (1983). Allelopathic problems in soil fatigue. *Soviet Soil Science* **15:** 54-62.

Grodzinsky, A.M., Bogdan, G.P. and Golovko, E.A. 1979, Allelopathic Soil Fatigue pp. 248. Naukova Dumka, Kiev.

Grodzinsky, A.M., Seredyuk, L.S. and Krupa, L.I. (1982). Organic acids in field and steppe cenosis soils. In: *Role of Allelopathy in Plants, Science Colloquim* (Ed., A.M. Grodzinsky). pp. 46-50. Naukova Dumka, Kiev, USSR.

Grummer, G. 1955. Die gegenseitige bonerer Pflanzen- Allelopathic, Fischer, Jena, pp. 162.

Grummer, G. 1961. *Symp. Soc. Exp. Biol.* 15, 219-228.

Grummer, G. and Beyer, N. 1960. In Biology of Weeds (J.L. Harper, ed.) pp. 153. Blackwell Scientific, Oxford.

Guenzi, W.D. and McCalla, T.M. (1962). Inhibition of germination and seedling development by crop residue. In: *Proceedings, Soil Science Society of America* **26:** 456-458.

Guenzi, W.D. and McCalla, T.M. (1966). Phenolic acids in oat, wheat, sorghum and corn residues and their phytotoxicity. *Agronomy Journal* **58:** 303-304.

Guenzi, W.D. and McCalla, T.M. (1966). Phenolic substance extracted from soil. In: *Proceedings Soil Science Society of America* **30:**214-216.

Guenzi, W.D. and McCalla, T.M. and Norstadt, F. (1967). Presence and persistence of phytotoxic substances in wheat, oat, corn and sorghum residues. *Agronomy Journal* **59**: 163-165.

Guenzi, W.D., Kehr, W.R. and McCalla, T.M. 1964. *Agron. J.* 56, 499-500.

Gulyaev, E.I., Nosko, V.K. and Ronsal, G.A. (1970). Productivity of mixed stands of maize and various varieties of soybean and possibility of using malva in agrocenoses with maize in south Ukraine. In: *Physiological and Biochemical Interactions Between the Plants in Phytocenosis* (Ed., A.M. Grodzinsky) Vol.**1:** 100-105. Naukova Dumka, Kiev, USSR.

Gupta, D.C. and Ram, K. 1981. *Indian J. Nematol* 11, 77-80.

Gupta, K., Narwal, S. S. and Wagle, D. S. (1987). Effect of aqueous root extract of soybean on rape and mustard. *Haryana Agricultural University Journal of Research* **17:** 208-215.

Hagin, R.D. (1989). Isolation and identification of 5-hydroxyindole-acetic acid and 5- hydroxytryptophan major allelopathic aglycones in quackgrass (*Agropyron repens* L. Beauv). *Journal of Agricultural and Food Chemistry* **37:** 1143-1149.

Hagin, R.D. (1989). Isolation and identification of 5-hydroxyindole-acetic acid and 5- hydroxytryptophan major allelopathic aglycones in quackgrass (*Agropyron repens* L. Beauv). *Journal of Agricultural and Food Chemistry* **37:** 1143-1149.

Hagwood Jr., E.S., Bauman, T.T., Williams Jr., J.L. and Schreiber, M.M. 1980. *Weed Sci.* 28, 729-734.

Hairston, J.E., Sanford, J.O., Pope, D.F. and Horneck, D.A. (1987). Soybean-wheat double cropping: Implication from straw management and supplemental nitrogen. *Agronomy Journal* **79:** 281-286.

Hall, A.B. 1980. *Diss. Abst. Int.* 41, 1220.

Hall, A.B., Blum, U. and Fites, K.C. (1982). Stress modification of allelopathy of *Helianthus annuus* L. debris on seed germination. *American Journal of Botany* **69:** 776-89.

Hall, A.B., Blum, U. and Fites, R.C. (1983). Stress modification of allelopathy of *Helianthus annuus* L. debris on seedling biomass production of *Amaranthus retroflexus* L. *Journal of Chemical Ecology* **9:** 1213- 1222.

Hamamura, Y. 1970. In Control of insect behaviour by natural products. (D.L. Wood, R.M. Silverstein and M. Nakajima eds). pp. 50-80. Academic Press, New York.

Hameed, S.F. 1970. *Indian J. Agri. Sci.* 40, 207-210.

Haq, I.U. and Hussain, F. 1979. *Pak. Tobacco.* 3(2), 17-19.

Harper, S.H.T. and Lynch, J.M. 1979. In Straw Decay and its Effects on Disposal and Utilization (E. Grosshard, ed.) pp. 289-291. John Wiley & Sons. Chichester, England.

Harris, J. 1984. *J. Aust. Inst. Agric. Sci.* 50, 180.

Harrison, H.F. and Peterson, J.K. (1991). Evidence that sweet potato (*Ipomea batatas*) is allelopathic to yellow nutsedge (*Cyperus esculentus*). *Weed Science* **39:** 308-312.

Hartung, A.C. and Stephens, C. T. (1983). Effect of allelopathic substances produced by *Asparagus* on incidence and severity of *Asparagus* decline due to *Fusarium* crown rot. *Journal of Chemical Ecology* **9:** 1163-1174.

Harvey, R.G. and Linscott, J.J. (1978). Ethylene production in soil containing quackgrass rhizomes and other plant materials. *Soil Science Society of American Journal* **42:** 721-724.

Harwood, R.R. 1976. In Nutrition and Agricultural Development (N.S. Serinnshaw and M. Behar, eds.) pp. 423-440. Plenum, Publishing Corp., New York.

Hatzios, K.K. (1987). Biotechnology applications in weed management: Now and in future. *Advances in Agronomy* **41:** 325-375.

Hawkins, R.S. 1925. *Agron. J.* 17, 91 (Abstract).

Hazebroek, J.P., Garrison, S.A. and Gianfagna, T. (1989). Allelopathic substances in asparagus roots: extraction, characterization and biological activity. *Journal of American Society of Horticultural Sciences* **114:** 152-158.

Hedin, P.A. 1985. Bioregulators for Pest Control ACS Symposium Series No. 276. American Chemical Society, Washington, DC.

Hedin, P.A., Maxwell, F.G. and Jenkins, J.N. 1974. In Proc. summer Inst. biol. control of Plant insects, diseases (F.G. Maxwell, F.A. Harris, eds). pp. 494-527. Univ. Press of Mississippi. Jackson, USA.

Hegelson, E.A. and Konzak, R. 1950. *North Dakota Agri. Exp. Stn. Bull.* 12, 71-76.

Heisey, R.M. and Delwiche, C.C. 1984. *Ann. J. Bot.* 71, 821-828.

Hepperly, P.R. and Diaz, M. (1983). The allelopathic potential of pigeonpea in Puerto Rico. *Journal of Agricultural University of Puerto Rico* **67:** 453-463.

Hicks, S.K. Wendt, C.W., Gannaway, J.R. and Baker, R.B. 1989. *Crop. Sci.* 29, 1057-1061.

Hicks, S.K., Wendt, C.W. and Gannaway, J.R. (1988). *Allelopathic Effects of Wheat Straw on Cotton Germination and Seedling Development.* The Texas Agricultural Experiment Station. The Texas A&M University, College Station. Texas. *Bulletin* pp. 1-5.

Hill, L.V. and Santelmann, P.W. (1969). Competitive .effects of annual weeds on Spanish peanuts. *Weed Science* **17:** 1-2.

Hodgson, J.M. 1958. *Weeds* 6, 1-11.

Holm, L., Pancho, J.V., Herberger, J.P. and Plucknett, D.L. 1979. A Geographical Atlas of World Weeds John Wiley, New York.

Hopen, H.J., Montazer-Zouhoor, A., Wu, S. and Rebeiz, C.A. 1985. *Weeds Today* 16(2), 4-5.

Horowitz, M. and Friedman, T. (1971). Biological activity of subterranean residues of *Cynodon dactylon, Sorghum halepense* and *Cyperus rotundus. Weed Research* **11:** 88-93.

Horricks, J.S. (1969). Influence of rape residue on cereal production. *Canadian Journal of Plant Science* **49:** 632-634.

Horsley, S.B. 1977. Proc. 4th *North American Forest Biol. Workshop.* (H.E. Wilcox and A.F. Hamer, eds.) pp. 93. School of Continuing Education, College of Environmental Sciences & Forestry, Syracause, New York.

Hosmani, M.M. 1978. *Striga : (A noxious root parasitic weed.* University of Agricultural Science, Dharwad, India. 165 pp.

Hoult, A.H.C. and Lovett, J.V. (1993). Biologically active secondary metabolites of barley. III. A method for identification and quantification of hordenine and gramine in barley by high performance liquid chromatography. *Journal of Chemical Ecology* **19**: 2245-2254.

Howell, C.R. and Stipanovic, J.B., Acker, R.D. Hadicke, E. and Schirmen, L. 1986. Abstract, Inst. Congress Pestic. Chem. 6th IUPAC, Ottawa. Abstract No. 2F-10.

Hozumi, Y., Nakayama, K. and Yoshida, K. 1974. *J. Central Agric. Exp. Stn.* (Konosu) 20, 87-102.

Hsiao, A.I., Worsham, A.D. and Moreland, D.E. 1981. *Weed Sci.* 29, 101-104.

Huber, D.M. and Abney, T.S. 1986. *J. Agron. Crop. Sci.* 157, 73-78.

Hussain F., Gaddon, M.A. and Akram, M. 1981. *Bangladesh J. Sci. Ind. Res.* 16, 105-110.

Hussain, F, 1980.*Oceologia* (Berlin). 45, 267-279.

Hussain, F. (1980). Allelopathic effects of Pakistani weeds: *Euphorbia granulata* Forsk. *Oecologia* (Berlin) **45:** 267-269.

Hussain, F. and Gaddon, M.A. 1981. *Oceologia* (Berlin) 51, 284-288.

Hussain, F., Mubarak, B. and Ilahi, I. (1987). Allelopathic effects of pakistani weed *Bothriochloa pertusa. Pakistan Journal of Agricultural Research* **8** (I) : 71-83.

Hussain, F., Nasrin, R. and Yasmin, 1988. *Pakistan J. Sci. Ind. Res.* 31, 293-296.

Isleib, D.R. 1960. *Weeds* 8, 631-635.

Isman, M.B. and Duffey, S.S. 1982. *J. Amer. Soc. Hort. Sci.* 107, 167-170.

Isogai, A., Chang, C., Murakoshi, S. and Suzuki, A. 1973. *J. Agric. Chem. Soc. Japan 47,* 443-447.

Itulya, F.M. 1987. *East Afr. Agric. For. J.* 52, 33-36.

Ivanov, V.P., 1962. *Fiziol. Rast.* **11:** 630-637.

Iverson, L.R. and Wali, M.K. (1982). Reclamation of coal mine lands: the role of *Kochia scoparia* and other pioneers in early succession. *Reclamation and Revegetation Research* **1:** 123-160.

Iwamoto, T., Hirota, A., Shima, S., Sakai, H., Isogai, A. 1985. *Agric. Biol. Chem.* **49:** 3323-3325.

Jain, R., Singh, M. and Dezman, D.J. (1989). Qualitative and quantitative characterization of phenolic compounds from lantana (*Lantana camara*) leaves. *Weed Science* **37:** 302-307.

Jain, R., Singh, M. and Dezman, D.J. 1989. *Weed Sci.* **37:** 302-307.

James, K.L., Banks, P.A. and Karnok, K.J. (1988). Interference of soybean (*Glycine max*) cultivars with sicklepod (*Cassia obtusifolia*). *Weed Technology* **2:** 404-409.

James, K.L., Banks, P.A., Kornok, K.J. 1988. *Weed Technol.* **2:** 404-409.

James, W.O. 1946. *Nature* **158:** 377.

Janzen, D.H., Juster, H.B. and Liener, I.E. 1976. *Science* **192:** 795-796.

Jarvis, R.B., Pena, N.B., Rao, M.M., Comezoglu, N.S., Comezoglu, T.F. and Mandava, N.B. (1985). Allelopathic agents from *Parthenium hysterophorus* and *Baccharis megapotamica.* In: *Chemistry of Allelopathy: Biochemical Interactions Among Plants* (Ed., A.C. Thompson). ACS Symposium Series No. **268:** 149-160. American Chemical Society, Washington DC.

Jayakumar, R. 1985. Studies on the effect of *Cassia sericea* Sw. on seed germination and seedling vigour of *Parthenium hysterophorus.* M.Sc. Thesis, Univ. Agri. Sci. Bangalore, India.

Jessop, R.S. and Lovett, J.V. (1982). The effect of stubble residues on the germination and early growth of wheat. In: *Proceedings of 2nd Australian Agronomy Conference "Agronomy Australia* 1982". Australian Society of Agronomy. Pp. 245.

Jessop, R.S. and Stewart, L.W. (1983). Effect of crop residues, soil type and temperature on emergence and early growth of wheat. *Plant and Soil* **74:** 101-109.

Jha, P.K. and Sen, D.N. (1981). Allelopathic potential of *Echinops echinatus* Roxb. *Indian Weed Science Society Conference* **2:** 197-201.

Jimenez, J.J., Schultz, K., Anaya, A.L., Hernandez, J. and Espejo, O. (1983). Allelopathic potential of corn pollen. *Journal of Chemical Ecology* **9:** 1011-10265.

Jimenez-Osorino, J.J. and Gliessman, S.R (1987). Allelopathic interactions in a wild mustard *(Brassica compestris* L.) and Broccoli *(Brassica oleracea* L. var. *italica)* in intercrop agroecosystem. In *Allelochemicals: Role in Agriculture and Forestry* (Ed., G.R. Waller) **330:** 262-274, American Chemical Society, Washington DC.

Jimenez-Osornio, J.J. and Schultz, C.K. 1981. Relaciones cultivo- arvenses en una chinampa. Teis delicenciatura Facultad de Ciencias. UNAM, Mexico.

Jobidon, R. (1986). Allelopathic potential of coniferous species to old-field weeds in eastern Quebec. *Forest Science* **32:** 112-119.

Jones, C.E. (1992). Crop rotation for the control of wild oats in wheat. In: *Proceeding 6th Australian Society of Agronomy Conference,* Armidale pp. 438-441. Australian Society of Agronomy.

Joshi, S. (1991). Biological control of *Parthenium hysterophorus* L. (Asteraceae) by *Cassia uniflora* Mill (Legumiosae) in Bangalore, India. *Tropical Pest Manag^ment* **37:** 182-184.

Joshi, S. (1991). Interference effects of *Cassia uniflora* Mill on *Parthenium hysterophorus* L. *Plant and Soil* **132:** 213-218.

Juntilla, O. 1975. *Physiol. Plant 33,* 22-27.

Juntilla, O. 1976. *Physiol. Plant.* 36, 374-378.

Juvik, I.A. and Stevens, M.A. (1982). Inheritance of foliar a-tomatine content in tomatoes. *Journal of American Society of Horticultural Sciences* **107:** 1061-1065.

Kahl, H. 1987. *J. Agron. Crop Sci.* 158, 56-64.

Kalantari, I. 1982. *Diss. Abst. Int.* B. 42, 4267-4268.

Kalmykova, N.A. 1972. In Physiological-Biochemical Basis of Plant Interactions in Phytocenosis (A.M. Grodzinsky, ed.) 3, 124-126. Naukova Dumka, Kiev.

Kamboj, H.B. 1991. Influence of isogenic lines of cotton `H-777' on biology, behaviour and control of major insect-pests of cotton M.Sc. Thesis, Haryana Agricultural University, Hisar, India.

Kanchan, S.D. (1975). Growth inhibitors from *Parthenium hysterophorus* L. *Current Science* **44:** 358- 359.

Kanchan, S.D. 1977. Agrophysiological impact of allelopathic effects of *Parthenium hysterophorus* L. Ph.D. thesis, Bangalore University, Bangalore.

Kanchan, S.D. and Jayachandra (1980). Pollen allelopathy-a new phenomenon. *New Physiologist* **84:** 739-742.

Kanchan, S.D. and Jayachandra 1980b. *Plant Soil* 55, 61-66.

Kanchan, S.D. and Jayachandra. (1979). Allelopathic effects of *Parthenium hysterophorus* L. Exudation of inhibitors through roots. *Plant and Soil* **53:** 27-35.

Kanchan, S.D. and Jayachandra. (1979). Allelopathic effects of *Parthenium hysterophorus* L. III. Inhibitory effects of the weed residue. *Plant and Soil* **53:** 37-47.

Kapoor, A., Mahor, R. Vaishampayan, W. and Gautam, N. 1981. *Geobios* 8, 66-67.

Karl, R., Lorez, G., Speakman, J.B., Acker, R.D. Hadicke, E. and Schirmen, L. 1986. Abstract, Int. Congress Pestic. Chem. 6th IUPAC, Ottawa. Abstract No. 2F-10.

Katekar, G.F. and Geissler, A.E. 1980. *Plant Physiol* 66, 1190-1195.

Kaurov, I.A. (1970). Interactions of birdsfoot and yellow lupine in pure and mixed cultures. In: *Physiological and Biochemical Interactions Between the Plants in Phytocenosis* (Ed., A.M. Grodzinsky) **1**: 67-71. Naukova Dumka, Kiev, USSR.

Kehr, W.R., Watkins, J.E. and Ogden, R.L. (1983). Alfalfa establishment and production with continuos alfalfa and following soybeans. *Agronomy Journal* **75:** 735- 738.

Kelsey, R.G. and Locken, L.J. (1987). Phytotoxity properties of cnicin, a sesquiterpene lactone from *Centaurea maculosa* (spotted kanpweed). *Journal of Chemical Ecology* **13:** 19-33.

Kenfield, D., Bunkers, G., Strobel, G.A. and Sugawara, F. (1988). Potential new herbicides - phytotoxins from plant pathogens. *Weed Technology* **2:** 519-524.

Khan, A.H. and Vaishya, R.D. 1992. In Proc. First Nat. Sym. Allelopathy in Agroecosystems (P. Tauro, S.S. Narwal, eds). Pp. 59-60. Indian Society of Allelopathy, Haryana Agricultural University, Hisar, India.

Khare, L.J. (1980). Phytotoxicity of the weed *Urgenia indica* Kunth on the seed gennination of associated crops. *Indian Journal of Botany* **3:** 87-91.

Khosla, S.N. and Sobti, S.N. 1979. *Pesticides* 13(7), 21- 26.

Kim, K.U., Lee, J., Jeong, H.J. and Kim, D.S. 1987. In Proc. 11th Asian-Pacific Weed Science Society Conference pp. 303-310.

Kim, Y.S. and Kil, B.S. (1987). A bioassay on susceptibility of selected species to phytotoxic substances from tomato plants. *Korean Journal of Botany* **30:** 59-67.

Kimber, R.W.L. (1973). Phytotoxicity from plant residue II. The effects of time of rotting straw from grasses and legumes on the growth of wheat seedling. *Plant and Soil* **38:** 347-361.

Kimber, R.W.L. (1967). Phytotoxicity from plant residues. I. The influence of rotted wheat straw on seedling growth. *Australian Journal of Agricultural Research* **18:** 361-374.

Kimber, R.W.L. (1973). Phytotoxicity from plant residues. II. The effect of time of rooting of straw from some grasses and legumes on the growth of wheat seedlings. *Plant and Soil* **38:** 347-61.

Kitahara, Y., Yanagawa, H., Kato, T. and Takahashi, N. (1972). Asparagusic acid, a new plant growth inhibitor in etiolated young asparagus shoots. *Plant and Cell Physiology* **13:** 923-925.

Kjaer, A. 1976. In "The Biology and Chemistry of Cruciferae" (J.Vaughan and A. Macleod, eds.) pp. 207-217. Academic Press, New York.

Kleifeld, Y. 1970. *Weed Sci.* 18, 16-18.

Kleifeld, Y., Goldwasser, Y., Herzlinger, G., Joel, D.M., Golan, S. and Khanna, D. (1994). The effects of flax (*Linum usitatissimum* L.) on Egyptian broomrape (*Orobanche aegyptica* Pers). *Weed Research* **34:** 37-44.

Klein, R. R. and Miller, D. A. (1980). Allelopathy and its role in agriculture. *Soil Science Plant Annals* **11:** 43-56.

Kloot, P.M. and Boyce, V.K. 1982. *Australian Weeds* 1, 11- 14.

Klun, J.A., Tipton, C.L. and Brindley, T.A. 1967. *J. Econ. Entomol.* 60, 1529-1523.

Knake, E.L. and Slife, F.W. (1962). Competition of *Setaria faberii* with corn and soybeans. *Weeds* **10:** 26-29.

Knake, E.L. and Slife, F.W. (1965). Giant foxtail seeded at various times in corn and soybeans. *Weeds* **13:** 331-334.

Kogan, M. 1986. *Iowa State J. Res.* 60, 501-527.

Kohli, R.K., Kumari, A. and Saxena, D.B. (1985). Auto and teletoxicity of *Parthenium hysterophorus* L. *Acta Universitatis Agriculturae* **33:** 253-264.

Kohlmuenzer, S. 1965. *Diss. Pharm.* 17, 357-379.

Komai, K. and Tang, C.H. (1989). Chemical constituents and inhibitory activities of essential oils from *Cyperus brevifolius* and *C. kyllingia. Journal of Chemical Ecology* **15:** 2171-2176.

Komai, K. and Ueki, K. 1975. *Weed Res.* (Japan) 20, 66-71.

Komai, K. and Ueki, K. 1980. *Weed Res.* (Japan) 25, 42-47.

Komai, K., Iwamura, J., Hemada, M. and Ueki, K. 1986. *Weed Res.* (Japan) 31, 280-286

Komai, K., Iwamura, J., Ueki, K. 1977. *Weed Res.* (Japan) 22, 14-18.

Komai, K., Sugiwaka, Y. and Sato, S. 1981. *Mem. Fac. Agri. Kinki Univ.* 14, 57-61.

Komai, K., Sugiwaka, Y. and Sato, S. 1982. *Mem. Fac. Agri. Kinki Univ.* 15, 33-41.

Kommedahl, T., Kotheimer, J.R. and Rernardini, J.V. (1959). The effects of quackgrass on germination and seedling development of certain crop plants. *Weeds* **7:** 1-12.

Kommedahl, T., Old, K.M. Ohman, J.H. and Ryan, E.W. 1970. *Weed Sci.* 18, 29-32.

Kovacs Jr., M.F. 1972. Dhurrin: An Allelopath Identified in Johnson grass Root Exudates Ph. D. Dissertation, University of Maryland, College Park.

Kovacs, L., Mikulas, J. and Polos, E. 1988. *Acta Agron.* (Hungary) 37, 65-70.

Kovacs, L., Mikulas, J. and Polos, F. (1988). Allelopathic activity of *Cirsium arvense* (L) Scop in Hungary. *Acta Agronomica Hungarica* **37 (1-2):** 65-69.

Kozel, P.C. and Tukey, H.B. Jr. 1968. *American J. Bot.* 55, 1184-1189.

Kraemer, M.E., Rangappa, M., Gade, W. and Benepal, P.S. 1987. *J. Econ. Entomol.* 80, 237-241.

Krasilnikova, N.A. 1958. Soil Microorganisms and Higher Plants. Academy of Sciences of USSR, Moscow.

Krasilnikova, N.A. and Garkina, N.I. 1946. *Microbiologiya* 15(2), 109-141.

Krupa, L.I. (1981). Phytotoxicity of soil under winter wheat. In: *Allelopathy in Natural and Artificial Phytocenosis: Collection of Scientific Papers* (Ed., A.M. Grodzinsky). Pp. 33-36. Naukova Dumka, Kiev, USSR.

Krupa, L.I. and Dzyubenko, N.H. (1981). On the problem of phenol carboxylic acids in soil under winter wheat. In: *Allelopathy in Natural and Artificial Phytocenosis: Collection of Scientific Papers* (Ed., A.M. Grodzinsky). pp. 37-42. Naukova Dumka, Kiev, USSR.

Kumari, A. Kohli, R.K. and Saxena, D.B. 1985. *Ann. Biol.* 1, 186-196.

Kuo, C.G., Cho, M.H. and Park, H.G. (1981). Effect of Chinese cabbage residue on mungbean. *Plant and Soil* **61:** 473-477.

Kuo, C.G., Cho, M.H. and Park, H.G. (1981). Effect of Chinese cabbage residue on mungbean. *Plant and Soil* **61:** 473-477.

Kurz, W.G.W. and Constabel, F. 1985. CRC *Crit. Rev. Biotechnol.* 2, 105-118.

Lad, S.K. 1983. *Sci. Cult.* 49, 187-188.

Lall, M. and Savongdy, H.O. 1981. *Proc. 8th Asian Pacific Weed Sci. Soc. Conf.* 1, 317-320.

Lambert, R.G. 1959. *Plant Dis. Reptr.* 43, 1117-1119.

Langston-Unkefer, P.L. Macy, P.A. and Durbin, R.D. 1984. *Plant* Physiol. 76, 71-74.

Lastuvka, Z. (1955). The effect of quackgrass on growth of wheat and rye. *Csekolovenoska Biologie* **4:** 165-175.

Lastuvka, Z. (1960). The effect of quackgrass on the anatomy of wheat. Publication Faculty of Science, University of Brno. *Tchecoslovaquie* **401:** 81-94.

Lastuvka, Z. (1970). Allelopathy and the processes of ion absorption and accumulation. In: *Physiological and Biochemical Interactions Between the Plants in Phytocenosis* (Ed., A.M. Grodzinsky) Vol. **1:** 37-40. Naukova Dumka, Kiev, USSR.

Lastuvka, Z. and Minarez, I. 1970. In "Physiological-Biochemical Basis of Plant Interactions in Phytocenosis" (A.M. Grodzinsky, ed.) 1, 55-59. Naukova Dumka, Kiev.

Lastuvka,Z. 1958. *Folia Biol.* 4 (2), 119-126.

Lauffer, G.A. and Garrison, S.A. 1977. *Hort. Sci.* 12, 385 (Abst.).

Laughlin, R.G., Munyon, R.L., Ries, S.K. and Wert, V.F. 1983. *Science* 219,1219-1220.

Law, J.H. and Regnier, P.E. 1971. *Annual Rev. Biochem.* 40, 533-553.

Lawson, D.R., Danehower, D.A., Shilling, D.G., Menetrez, M.L. and Spurr Jr., H.W. (1988). Allelochemical properties of *Nicotiana tabacum* leaf surface compounds. In: *Biologically Active Natural Products: Potential Use in Agriculture.* (Ed., H.G. Cutler). ACS Symposium Series No. **380:** 363- 377. American Chemical Society, Washington, D.C.

Lax, A.R., Shepherd, H.S. and Edwards, J.V. (1988). Tentoxin, a chlorosis-inducing toxin from *Alternaria* as a potential herbicide. *Weed Technology* **2:** 540-544.

Lazuakas, P. and Balnevichinte, Z. 1972. In "Physiological Biochemical Basis of Plant Interactions in Phytocenosis" (A.M. Grodzinsky, ed.) 3, 76-79. Naukova Dumka, Kiev.

Lea, P.J., Joy, K.W., Ramos, J.L. and Guerrero. 1984. *Phytochemistry* 23, 1-6.

Leather, G.R. (1983). Sunflowers (*Helianthus annuus* L.) are allelopathic to weeds. *Weed Science* **31:** 37-42.

Leather, G.R. (1983). Weed control using allelopathic crop plants. *Journal of Chemical Ecology* **9**: 983-1011.

Leather, G.R. (1987). Weed control using allelopathic sunflowers and herbicide. *Plant and Soil* **98:** 17-23.

Leather, G.R. and Forrence, L.E. 1979. *Proc. Weed Sci. Soc. Am.* pp. 79-80.

Leela, D. 1981. *Proc.* 8th *Asian Pacific Weed Sci. Soc. Conf.* pp. 401-404.

Lehle, F.R. and Putnam, A.R. (1982). Quantification of allelopathic potential of sorghum residues by novel indexing of Richards function fitted to cumulative cress seed germination curves. *Plant Physiology* **69:** 1212-1216.

Lehle, F.R., Frans, R. and McClell and, M. (1983). Allelopathic potential of white lupine (*Lupinus alba*) herbage and herbage extracts. *Weed Science* **31:** 513-519.

LeTourneau, D., Failes, G.D. and Heggeness, H.G. 1956. *Weeds* 4, 363-368.

LeTourneau., D. and Haggeness, H.G. 1957. *Weeds* 5, 12-19.

Levin, D.A. 1976. *Annu. Rev. Ecol. Syst.* 7, 121-159.

Levitt, J. and Lovett, J.V. (1984). Allelochemicals of *Datura stramonium* L. (Thorn-apple) in contrasting soil types. *Plant and Soil* **79:** 181-189.

Levitt, J., Lovett, J.V. and Garlick, P.R. (1984). *Datura stramonium* allelochemicals: Longevity in soil and ultrastructural effects on root tip cells of *Helianthus annuus* L. *The New Physiologist* **97:** 213-218.

Lewis, J.A. and Papavizas, G.C. 1975. In Biology and Control of Soil-borne Plant Pathogens (G.N. Burehl, ed.) p. 216. American Phytopathol. Soc. St. Paul. Minnesota.

Lichtenstein, E.P., Strong, F.M. and Morgan, D.G. 1962. *J. Agric Food Chem.* 10, 30-33.

Liebl, R.A. and Worsham, A.D. (1983). Inhibition of pitted morning glory (*Ipomoea iacurnosa* L.) and certain other weed species by phytotoxic components of wheat (*Triticum aestivum* L.) straw. *Journal of Chemical Ecology* **9**: 1027-1043.

Liebl, R.A. and Worsham, A.D. 1983b. *Weed Sci.* 36, 405- 414.

Liebman, M. (1988). Ecological suppression of weeds in intercropping systems: A review. In: *Weed Management in Agroecosystems: Ecological Approaches* (Eds., M.A. Altieri and M. Liebman). pp. 197- 212. CRC Press, Boca Raton, FL, USA.

Lindermann, R.G. and Toussoun, T.A. 1968. *Phytopathol* 58, 1571-1574.

Liu, D.L. and Lovett, J.V. (1993). Biologically active secondary metabolites of barley. I. Developing technique and allelopathy in barley. *Journal of Chemical Ecology* **19**: 2217-2230.

Liu, D.L. and Lovett, J.V. (1993). Biologically active secondary metabolites of barley. II. Phytotoxicity of barley allelochemicals. *Journal of Chemical Ecology* **19**: 2231-2244.

Liu, D.L. and Lovett, J.V. 1993c. *J. Chem. Ecol.* 19, 2245- 2254.

Livingston, B.E., Pritton, J.C. and Reid, F.R. 1905. Studies on the properties of an unproductive soil USDA Bureau of Soils Bull. 28.

Lockerman, R.H. and Putnam, A.R. (1979). Evaluation of allelopathic cucumbers (*Cucumis sativus*) as an aid to weed control. *Weed Science* **27:** 54-57.

Lockerman, R.H. and Putnam, A.R. (1981a). Growth inhibitors in cucumber plants and seeds. *Journal of the American Society of Horticultural Sciences* **106:** 418-422.

Lockerman, R.H. and Putnam, A.R. (198Ib). Mechanisms for different interferences among cucumber (*Cucumis sativus* L.) accessions. *Botanical Gazette* **142:** 427-430.

Lodhi, M.A.K. (1981). Accelerated soil mineralization, nitrification and revegetation of abandoned field due to the removal of crop-soil phytotoxicity. *Journal of Chemical Ecology* **7:** 685-694.

Lodhi, M.A.K., Bilal, R. and Malik, K.A. (1987). Allelopathy in agroecosystems: Wheat phytotoxicity and its possible roles in crop protection. *Journal of Chemical Ecology* **13:** 1881-1891.

Lodhi, M.A.K. (1979). Germination and decreased growth of *Kochia scoparia* in relation to its auto-allelopathy. *Canadian Journal of Botany* **57:** 1083-1088.

Lodhi, M.A.K. (1979). Allelopathic potential of *Salsola kali* L. and its possible role in rapid disappearance of weedy stage during revegetation. *Journal of Chemical Ecology* **5:** 429-437.

Lodhi, M.A.K. and Rice, E.L. (1971). Allelopathic effect of *Celtis laevigata. Bulletin of the Torrey Botanical Club* **98:** 83-89.

Lolas, P.C. 1986. *Weed Res.* 26, 1-8.

Lolas, P.C. and Coble, H.D. (1982). Non-competitive effects of johnsongrass (*Sorghum halepense*) on soybeans (*Glycine max*). *Weed Science* **30:** 589-593.

Lolas, P.H. 1981. Panhellenic Cong. Geotechnical Res. pp. 24 (Abstract).

Lorenzi, H. 1984. Plant Direto no Brasil. Pp. 24-35. Fundacao Cargill, Campinas.

Lovett, J.V. (1982). Allelopathy and self-defence in plants. *Australian Weeds* **2:** 33-36.

Lovett, J.V. (1983). Allelopathy and weed management in cropping systems. In: *Proceedings, IX Asian-Pacific Weed Science Conference.* pp. 31-46. Manila. Philippines: Asia Pacific Weed Science Society.

Lovett, J.V. (1983). Self-defence chemicals of plants. *10th International Congress on Plant Protection*. Pp. 838. Brighton, U.K.

Lovett, J.V. (1986). Allelopathy : The Australian experience. In: *The Science of Allelopathy* (Eds., A.R. Putnam and C.S. Tang). pp. 75-99. John Wiley & Sons Inc., New York.

Lovett, J.V. (1987). Allelopathy in Australia: Bacterial mediation. In: *Allelochemicals : Role in Agriculture and Forestry* (Ed., G.R. Waller). ACS Symposium Series no. **330:** 156-175. American Chemical Society, Washington DC.

Lovett, J.V. 1980. Proc. Aust. Agron. Conf. *Pathways to Productivity,* pp. 60. Qld. Agri. College, Lawes, Qld. Australia.

Lovett, J.V. 1982b. Proc. 2nd *Aust. Agron. Conf.* pp. 101-115. Australian Society of Agronomy, Parkville, Victoria.

Lovett, J.V. and Duffeild, A.M. (1981). Allelochemicals of *Camelina sativa. Journal of Applied Ecology* **18:** 283-290.

Lovett, J.V. and Jackson, H.F. (1980). Allelopathic activity of *Camelina sativa* L. Crantz in relation to its phyllosphere bacteria. *New Physiologist* **86:** 273-277.

Lovett, J.V. and Jessop, R.S. (1982). Effects of residues of crop plants on germination and early growth of wheat. *Australian Journal of Agricultural Research* **33:** 909-916.

Lovett, J.V. and Jokinen, K. (1984). A modified stairstep apparatus for studies of allelopathy and other phytotoxic effects. *Journal of Agricultural Science in Finland* **56:** 1- 7.

Lovett, J.V., Levitt, J., Duffeild, A.M. and Smith, N.G. (1981). Allelopathic potential of *Datura stramonium* L. *Weed Research* **21:** 165-170.

Lucena, J.M. 1974. *Revista Comalfi* 1, 40-57.

Lucena, J.M. and Doll, J. 1975. Sociedad Columbiana de Control de Malezas Y Figiologia Vegeta `COMALFI'. Resumenes de los Trabajos en el VII Seminaria p. 33.

Lucena, J.M. and Doll, J. 1976. *Revista Comalfi* 3, 241- 256.

Lukefahr, M.J. and Houghtaling, J.E. 1969. *J. Econ. Entomol.* 62, 588-591.

Lukefahr, M.J., Shaver, T.N., Cruhm, D.E. and Houghtaling, J.E. 1974. *Proc. Beltwide Cotton Prod. Res. Conf.* Pp. 93-95.

Lukefahr, M.J., Stipanovie, R.D., Bell, A.A. and Gray, J.R. 1977. Proc. *Beltwide Cotton Prod. Res. Conf.* Pp. 97-100.

Lydon, J. and Duke, S.O. (1989). The potential of pesticides from plants. In: *Herbs, Spices and Medicinal Plants: Recent Advances in Botany, Horticulture and Pharmacology* (Eds., E.L. Craker and J.E. Simon) Vol. **4:** 1-41. Oryx: Press, Phoenix, Arizona, U.S.A.

Lykhvar, D.F. and Nazarova, N.S. (1970). On importance of legume varieties in mixed cultures with maize. In: *Physiological and Biochemical Interactions Between the Plants in Phytocenosis* (Ed., A.M. Grodzinsky) Vol. **1:** 158-164. Naukova Dumka, Kiev, USSR.

Lynch, J.M. 1977. *J. Appl. Bacteriol.* 42, 81-87.

Lynch, J.M. and Peon, D.J. (1980). Damage to cereals caused by decaying weed residues. *Journal of the Science of Food and Agriculture* **31:** 321-324.

Lynch, J.M., Ellis, F.B., Harper, S.H.T. and Christian, D.G. 1981a. *Agric. Environ.* 5, 321-328.

Lynch, J.M., Penn, D.J. and Gussin, E.J. 1981b. *J. Sci. Food Agric.* 32, 628-629.

Macias, F.A., Varela, R.M., Toress, A. and Molinillo, J.M.G. (1993). Potential allelopathic guaianolides from cultivar sunflower leaves var. SH. 222. *Phytochemistry* **34:** 669-674.

Mainx, H.G. and Tauscher, B. (1983). Allelopathic effects-chemistry and bioassay. *Ber. Fachg. Herbologie* **24:** 221-227. (In German).

Mall, L.P. and Dagar, J.C. (1979). Effect of *Parthenium hysterophorus* L. extract on the germination and early seedling growth of three crops. *Indian Journal of Botanical Science* **58:** 40-43.

Mallik, M.A.B. and Tesfai, K. (1988). Allelopathic effect of common weeds on soybean growth and soybean bradyrhizobium symbiosis. *Plant and Soil* **112:** 177-182.

Mandava, B.N. 1979. In Plant Growth Substances (B.N. Mandava, ed.) p. 135, ACS Symposium Series No. 111, American Chemical Society, Washington, D.C.

Mandava, N.B. (1985). Chemistry and biology of allelopathic agents. In: *Chemistry of Allelopathy: Biochemical Interactions Among Plants* (Ed.,

A.C. Thompson). ACS Symposium Series No. **268:** 33-54. American Chemical Society, Washington DC.

Mankau, R. 1975. *J. Invert. Pathol.* 26, 333-339.

Mankau, R. and Minteer, R.J. 1962. *Plant Dis. Reptr.* 46, 376-378.

Mann, H.H. and Barnes, T.W. (1945). The competition between barley and certain weeds under controlled conditions. *Annals of Applied Biology* **32:** 15-22.

Mann, H.H. and Barnes, T.W. (1947). The competition between barley and certain weeds under controlled conditions. II. Competition with *Holcus mollis. Annals of Applied Biology* **34:** 252-267.

Mann, J .D., Steinhart, C.E. and Mudd, S.H. (1963). Alkaloids and plant metabolism. V. The distribution and formation of tyramine methylpherase during germination of barley. *Journal of Biological Chemistry* **238:** 678-681.

Manners, G.D. and Galitz, D.S. (1986). Allelopathic effect of small everlasting (*Antennaria microphylla*) identification of constituents to leafy spurge (*Euphorbia esula*). *Weed Science* **34:** 8-12.

Manske, R.H.F. 1950. Alkaloids Chemistry and Physiology. (R.H.F. Manske and H.R. Holmes, eds.) vol. 1, Academic Press, New York.

Marchaim, U., Brik, Y., Dovart, A. and Berman, T. (1972). Lucerne saponins as inhibitors of cotton seed germination. *Journal of Experimental Botany* **23:** 302-309.

Martin, P. 1957. *Z. Bot.* 45, 475-506.

Martin, P. and Rademacher, B. (1960). Studies on the mutual influences of weeds and crops. *Symposium of British Ecological Society* **1:** 143-152.

Mase, S. (1984). Meiji Herbiace (MW 801, SF-1293) (Common name: bialaphos) - A new herbicide. *Japan Pesticide Information* **45:** 27-30.

Mason-Sedun, W., Jessop, R.S. and Lovett, J.V. (1986). Differential phytotoxicity among species and cultivars of the genus *Brassica* to wheat. *Plant and Soil* **93:** 3-16.

Massantini, F., Caporali, F. and Zellini, G. (1977). EWRS Symp. on different methods of weed control and their integration vol. I, 23-30.

Matsui, S.H.O. and Nishiiri, K. (1989). *Japanese J. Crop. Sci.* 58, 337-341.

Maugh, T.H. (1981). *Science* 212, 33-34.

Maun, M.A. (1977). Suppressing effect of soybean on barnyardgrass. *Canadian Journal of Plant Sciences* **57:** 405-409.

Mavlanov, Kh. (1982). *Uzbekistan Biol. Zh.* 5, 73-74.

McCahon, C.B., Kelsey, R.G., Sheridan, R.P. and Shafizadeh, F. (1973). Physiological effects of compounds extracted from sagebrush. *Bulletin of the Torrey Botanical Club* **100:** 23-28.

McCalla, T.M. (1971). Biochemical Interactions Among Plants pp 39-43. National Academy of Sciences, Washington, D.C.

McCalla, T.M. (1976). Studies on phytotoxic residues from soil, organic matter and crop residues. *Nebraska Agric. Exp. Stn. Bull.* No. 225, pp. 1-8.

McCalla, T.M. and Army, T.J. (1961). *Adv. Agron.* 13, 125-196.

McCalla, T.M. and Duley, F.L. (1948). *Science* 108, 163.

McCalla, T.M. and Haskins, F.A. (1964). Phytotoxic substances from soil microorganisms and crop residues. *Bacteriological Review* **28:** 181-207.

McCalla, T.M. and Norstadt, F.A. (1974). Toxicity problems in mulch tillage. *Agriculture and Environment* **1:** 153-174.

McElgunn, I.D. and Heinrichs, D.H. (1970). Effect of root temperature and a suspected phytotoxic substance on the growth of alfalfa. *Canadian Journal of Plant Science* **50:** 307 -311.

McKinley, A.D. (1931). *Agron. J.* 23, 844-849.

McLaren, J.S. (1986). *Pestici. Sci.* 17, 559-578.

McLaughlin, J.P. (1977). Influence of yellow nutsedge on corn and sorghum growth and mineral nutrition, M.S. Thesis, Pennyylvania State University.

McWhorter, C. G. and Hartwig, E. E. (1972). Competition of johnsongrass and cocklebur with six soybean varieties. *Weed Science* **20:** 56-59.

Mecklenberg, R.A. and Tukey, H.B. (1964). *Plant Physiology* **39:** 533-536.

Megie, C.A. Pearson, R.W. and Hiltbold, A.E. (1967). Toxicity of decomposing crop residues to cotton germination and seedling growth. *Agronomy Journal* **59:** 197 -199.

Meissner, R., Nel, P.C. and Smith, N.S.H. (1979). Influence of red nutgrass (*Cyperus rotundus*) on growth and development of some crop plants. *Proceedings III National Weeds Conference of South Africa.* Pp. 39-52.

Melnichenko, A.N., Grechkanov, O.M. and Lodionov (1972). Allelopathic and ecological influences of white mustard on growth and development of nectar feeding mixture components. In: *Physiological and Biochemical Interactions Between the Plants in Phytocenosis* (Ed., A.M. Grodzinsky) Vol. **3:** 39-42. Naukova Dumka, Kiev, USSR.

Menges, R. M. (1987). Allelopathic effects of palmer amaranth (*Amaranthus palmerl*) and other plant residues in soil. *Weed Science* **35:** 339-347.

Menges, R.M. (1988). Allelopathic effects of palmer amaranth (*Amaranthus palmeri*) on seedling growth. *Weed Science* **36:** 325-328.

Mercado, B.L. and Manuel, J.S. (1978). *Weed Sci. Rep.* 1976-77 College of Agri., Univ. Philippines, Los Banos, pp. 64- 75.

Mersie, W. and Singh, M. (1987). Allelopathic effect of lantana on some agronomic crops and weeds. *Plant and Soil* **98:** 25-30.

Mersie, W. and Singh, M. (1987a). *J. Chem. Ecol.* 13, 1739- 1748.

Mikulas, J. (1984). *Scientarium Hungaricae*, 33, 423-427.

Miller, D.A. (1983). Allelopathic effects of alfalfa. *Journal of Chemical Ecology* **9:** 1056-1072.

Miller, E.C. (1931). Plant Physiology McGraw Hill Co. New York.

Millhollon, R.W. (1970). *Weed Sci.* 18, 333-336.

Minar, J. (1974). Folia Fac. Sci. Nat. Univ. Purkynianae Brun. 15, 1-84.

Miranda, R.P. and Mata, R. (1993). Tricolorin a major phytogrowth inhibitor from *Ipomoea tricolor. Journal of Natural Products* **56:** 571-82.

Misato, T. and Yamaguchi, I. (1984). Pesticides of microbial origin. *Outlook on Agriculture* **13:** 136-139.

Mishra, S.D. and Mojumdar, V. (1994). Toxic behaviour of neem products on soil and plant nematodes. In: *Abstracts* International Symposium *Allelopathy in Sustainable Agriculture, Forestry and Environment* (Eds., S. S. Narwal, P. Tauro, G.S. Dhaliwal and J. Prakash). pp. 130. Indian Society of Allelopathy, Haryana Agricultural University, Hisar, India.

Mishra, S.D. and Prasad, S.K. (1978). Indian J. Ento. 40, 42-53.

Mishushtin, E.N. and Naumova, A.N. (1955). Isv. Akad. Nauk. USSR Ser. Biol. 6, 3-9.

Mohmed-Saleem, M.A. and Fawusi, M.O.A. (1983). Agri. Ecosyst. Environ. 10, 347-352.

Mohnot, K. and Choudhary, B.L. (1981). Proc. Soc. Plant Physiol. Biochem. pp. 44-45. (Abstract).

Mohnot, K. and Soni, S. (1977). Ecophysiological studies of desert plants. II. Growth retarding factor in air-dried stem of *Solanum surattense* Burro. F. *Comparative Physiology and Ecology* **2:** 97-100.

Mohnot, K. and Soni, S.R. (1979). Presence of allelopathic principle in weeds of arid zone in relation to crop plants: Effect of aqueous extract of air dried leaves of *Amaranthus tricolor* on seed germination and seedling growth of sorghum (*Sorghum vulgaris*). In: *Recent Researches in Plant Sciences* (Ed., S.S. Bir). Pp. 449-454. Kalyani Publishers, New Delhi.

Mohnot, K. and Soni, S.R. (1977). Trans. Indian Soc. Desert Technol. 2, 47-52.

Molisch, H. (1937). *De Einflux einer Pflanze auf die Andere: Allelopathie.* Gustav Fischer, Jena, Germany.

Molisch, H. (1937). Dcr Enfusslinear Pflanze aufdie andere- Allelopathie Fischer, Jena, GDR.

Moore, D.R.E. and Waid, J.S. (1971). Soil Biol. Biochem. 3, 69-83.

Moroz, P.A. and Leshenko, S.G. (1972). In Physiological- Biochemical Basis of Plant Interactions in Phytocenosis (A.M. Grodzinsky, ed.) 3, 76-82. Naukova Dumka, Kiev.

Mosca, G., Merlo, D. and Toniolo, L. (1988). Riv. Agron. 22, 266-272.

Mothes, K. (1955). Annu. Rev. Plant Physiol. 6, 393-442.

Mountain, W.B. and Elliott, J.M. (1962). Can. J. Pl. Sci. 42, 642-645.

Mudd, S.H. (1961). 3-Amino methylindole and 3-methylaminomethylindole: a new constituent of barley. *Nature* **189:** 489.

Mudge, L.L., Gossett, B.J. and Murphy, J.R. (1984). Weed Sci. **32:** 591-594.

Muller, C. H. (1966). The role of chemical inhibition (allelopathy) in vegetational composition. *Bulletin of the Torrey Botanical Club* **93:** 332-351.

Muller, C.H. (1969). Allelopathy is factor in Ecological process. *Vegetation* **18:** 348-357.

Muller, C.H. (1971). Phytotoxins as plant habitat variables. In: *Biochemical Interactions Among Plants.* Pp. 64-72. Environmental Physiology Subcommittee for the International Biological Programme, Division of Biology and Agriculture, National Research Council, National Academy of Sciences, Washington DC.

Muller, W. H. and Muller, C. H. (1964). Volatile growth inhibitors produced by *Salvia* species. *Bulletin of the Torrey Botanical Club* **91:** 327-330.

Muller-Wilmes, V. and Zoschke, M. (1980). Ange. Bot. 54, 119-123.

Muminovic, S. (1991). Allelopathic effect of straw of crops on growth of weeds. *Savremeva polijoprturedia* **39**: 27-30.

Munakata, K. Yamada, S., Ishida, S., Futatsuya, F., Ito, K. and Yamamoto, H. (1973). Proc. Asian-Pacific Weed Sci. Soc. Conf. 4, 215-219.

Munniyapa, T.V. and Krishnamurthy, K. (1981). Proc. 8th Asian-Pacific Weed Sci. Soc. Conf. Vol. 2

Myers, H.E. and Hallsted, A.L. (1942). Proc. Soil Sci. Soc. Am. 7, 316-321.

Myras, H. and Juntilla, O. (1981). Interaction between *Heracleum laciniatum* and some other plants. *Holarctic Ecology* 4: 43-48.

Myras, H. and Juntilla, O. (1981). Hol. Arct. Ecol. 4(1), 43-48.

Nacional de Pesruise de Trigo, Passo Fundo, Brazil.

Nadkernichnyi, S.P. (1974). On the problem of distribution of microscopic toxins formed in soddy medium podzolic soil under some farm crops. In: *Physiological and Biochemical Interactions Between the Plants in Phytocenosis* (Ed., A.M. Grodzinsky) Vol. **5:** 97-100. Naukova Dumka, Kiev, USSR.

Nair, E.M., Whitenack, C.J. and Putnam, A.R. (1990). 2,2'-Oxo-1,1'-azobenzene a microbially transformed allelochemical from 2,3-benzoxazolinone. *Journal of Chemical Ecology* **16:** 353-364.

Nandal, S.N. and Bhatti, D.S. (1983). In III All India Nematol. Symp. Solan. Pp. 31.

Naqvi, H.H. and Muller, C.H. (1975). Pakistan J. Bot. 7, 139-147.

Narwal, S.S. and Gupta, K. (1991). Allelopathic effects of aqueous stubble extract of pearlmillet on the seed germination and seedling growth of oil seed crops. *Indian Journal of Ecology* **18:** 70-73.

Narwal, S.S. and Gupta, K. (1992). Allelopathic effects of aqueous stubble extract of sorghum and root extract of *Pennisetum glaucum* on seed germination and seedling growth of *Brassica juncea, Brassica napus,* var. *napus, Brassica napus* var. *brown sarson* and *Brassica napus.* In:

Proceedings of the First National Symposium on Allelopathy in Agroecosystems (Eds., P. Tauro and S.S. Narwal). Pp. 25-28. Indian Society of Allelopathy, Haryana Agricultural University, Hisar, India.

Narwal, S.S. and Gupta, K. (1991a). Haryana Agric. Univ. J. Res. 21, 118-125.

Narwal, S.S. and Gupta, K. (1992a). In Proc. I. Nat. Sym. Allelopathy in Agroecosystem (P. Tauro and S.S. Narwal, eds) pp. 22-25 Indian Society of Allelopathy, Hisar India.

Narwal, S.S., Gupta, K. and Pahuja, S.S. (1990). Allelopathic effects of aqueous stubble extract of sorghum on the seed germination and seedling growth of winter crops. *Indian Journal of Ecology* **17:** 171-174.

Narwal, S.S., Sarmah, M.K., Dahiya, D.S. and Kapoor, R.L. (1992). Smothering effect of pearlmillet genotypes on weed species. In: *Proceedings National Symposium Allelopathy in Agroecosystems* (Eds., P. Tauro and S.S. Narwal). Pp. 48-50. Indian Society of Allelopathy, Hisar, India.

Narwal, S.S., Singh, A., Singh, I. and Gupta, K. (1989). Allelopathic effects of stubble extract on pearlmillet on the seed germination and seedling growth of wheat, barley, lentil, chickpea and *taramira*. *Indian Journal of Ecology* **16:** 127-131.

Narwal, S.S., Singh, I., Singh, A. and Gupta, K. (1989). Allelopathic effects of pearlmillet extracts on the seed germination and seedling growth of Indian Colza. *Indian Journal of Ecology* **16:** 84-87.

Nath, R. (1981). Note on the effect of *Parthenium* extract on seed germination and seedling growth in crops. *Indian Journal of Agricultural Sciences* **51:** 601-603.

Nath, S.C. and Bordoloi, D.N. (1986). J. Res. Assam Agri. Univ. 4, 34-39.

Neill, R.L. and Rice, E.L. (1971). American Midl. Nat. 86, 344-357.

Netzy, D.H., Riopel, J.L., Ejeta, G. and Butler, L.G. (1988). Weed Sci. 36, 441-446.

Neustrueva, S.N. and Terekhina, T.A. (1975). In Physiological- Biochemical Basis of Plant Interactions in Phytocenosis (A.M. Grodzinsky, ed.) 6, 49-54. Naukova Dumka, Kiev.

Newman, E.I. (1982). The possible relevance of allelopathy in agriculture. *Pesticide Science* **13:** 575-582.

Nicollier, G.F. and Thompson, A.C. (1982). Phytotoxic compounds from *Melilotus alba* (white sweet clover) and isolation and identification of two new flavonoids. *Journal of Agricultural and Food Chemistry* **30:** 760- 764.

Nicollier, G.F., Pope, D.F. and Thompson, A.C. (1983). Biological activity of dhurrin and other compounds from Johnson grass (*Sorghum halepense*). *Journal of Agricultural and Food Chemistry* **31:** 744- 748.

Nicollier, G.F., Pope, D.F. and Thompson, C. (1985). Phytotoxic compounds isolated and identified from weeds. In: *Chemistry of Allelopathy: Biochemical Interactions Among Plants* (Ed., A.C. Thompson). ACS Symposium Series No. **268:** 207-218. American Chemical Society, Washington DC.

Nielsen, K.F., Cuddy, T.F. and Woods, W.B. (1960). The influence of the extract of some crops and soil residues on germination and growth. *Canadian Journal of Plant Science* **40:** 188-197.

Nielsen, R.L., Stuthman, D.D. and Barnes, D.K. (1981). Interference between oats and alfalfa in mixed seedlings. *Agronomy Journal* **73:** 635-638.

Nishimura, H., Nakamura, T. and Mizutani, J. (1984). Phytochemistry 23, 2777-2779.

Nishimura, N., Kaku, K., Nakamura, T., Fukuzawa, Y. and Mizutani, J. (1982). Agric. Biol. Chem. 43, 319-320.

Norman, A.G. (1959). Proc. Soil Sci. Soc. Am. 23, 368-370.

Norris, D.M. (1977). In The Chemical basis for plant resistance to pests. American Chemical Society, Washington, DC.

Norstadt, F.A. and McCalla, T.M. (1963). Phytotoxic substances from a species of *Penicillium. Science* **140:** 410-411.

Norstadt, F.A. and McCalla, T.M. (1968). Proc. Soil Sci. Soc. Am. 32. 241-245.

Norstadt, F.A. and McCalla, T.M. (1969). Soil Sci. 107, 188-193.

Norstadt, F.A. and McCalla, T.M. (1971). Soil Sci. 111, 236-243.

Ohman, J.H. (1961). Diss. Abst. Int. B. 22, 1769.

Ohman, J.H. and Kommedahl, T. (1960). Relative toxicity of extracts from vegetative organs of quackgrass to alfalfa. *Weeds* **8:** 66-670.

Ohman, J.H. and Kommedahl, T. (1964). Plant extracts, residues and soil minerals in relation to competition of quackgrass with oats and alfalfa. *Weeds* **12:** 222-231.

Ohman, J.H. and Kommedahl, T. (1969). Phytopatholgy 49, 548 (Abst.).

Okafor, L.I. and DeDatta, S.K. (1976). Competition between upland rice and purple nutsedge for nitrogen, moisture and light. *Weed Science* **24:**43-46.

Oleszek, W. (1987). Allelopathic effects of volatiles from some *Cruciferae* species on lettuce, barnyard grass and wheat growth. *Plant and Soil* **102:** 271-273.

Oleszek, W., Jurzysta, M. and Gorski, P.M. (1992). Alfalfa saponins -the allelopathic agents. In: *Allelopathy: Basic and Applied Aspects* (Eds., S.J.H. Rizvi and V. Rizvi). pp. 151-167. Chapman and Hall, London.

Oliver, L.R. (1979). Weed Sci. 24, 43-46.

Olmstead, C. (1981). War on Weeds - helping plants themselves - Michigan Science in action. Michigan Agric. Exp. Sta. Bull. 44.

Omura, S., Hinotozawa, K. Iwamura, N. and Murata, M. (1984). J. Antibiot. 37, 939-940.

Osvald, H. (1947). Vaxtodling 2, 288-290.

Osvald, H. (1948). J. Ecol. 36, 192-193.

Ota, T. (1983). Japanese J. Crop Sci. 52, 447-451.

Overland, L. (1966). The role of allelopathic substances in the barley crop. *American Journal of Botany* **53**: 423-432.

Panasuik, O., Bills, D.D. and Leather, G.R. (1986). Allelopathic influence of *Sorghum bicolor* on weeds during germination and early development of seedlings. *Journal of Chemical Ecology* **12:**1533-44.

Pandya, B.M. and Sidha, V.K. (1987). Competition between *Celosia argentea* L. and sorghum with reference to weed position. *Journal of Indian Botanical Society* **66:** 15-21.

Pandya, S.M. (1975). Effects of *Celosia argentea* extracts on root and shoot growth of pearlmillet seedlings. *Geobios* **2:** 175-178.

Pandya, S.M. (1976). Effect of *Celosia argentea* extract on dry weight of pearlmillet seedlings. *Geobios* **3:** 137-138.

Pandya, S.M. (1977). On the relative nature of the inhibiting effects of *Celosia argentea* L. of different ages. *Science and Culture* **43:** 343-344.

Pandya, S.M. and Dash, D.K. (1982). Competition between the weed, *Celosia argentea* L. and maize with reference to the time of weed introduction. *International Journal of Ecology and Environment Science* **8:** 53-62.

Pandya, S.M. and Pota, K.B. (1978). On the allelopathic potentials of root exudates from different ages of *Celosia argentea* L. weed. *National Academy of Science Letters* **1:** 56-58.

Pandya, S.M., Dave, V.R. and Vyas, K.G. (1984). Effect of *Celosia argentea* L. on root nodules and nitrogen content of three legume crops. *Science and Culture* **50:** 161-162.

Pardales Jr., J.R. and Dingal, A.G. (1988). An allelopathic factor in taro residues. *Tropical Agriculture* (Trinidad) **65:** 21-24.

Pareek, R.P. and Gaur, A.C. (1973). Plant Soil 39, 441-444.

Parenti, R.L. and Rice, E.L. (1969). Inhibitional effects of *Digitaria sanguinalis* and possible role in old field succession. *Bulletin of the Torrey Botanical Club* **96:** 70- 78.

Parkinson, V., Efron, Y., Bello, L. and Dashiell, R. (1987). Plant Prot. Bull. FAO 35(2), 51-54.

Parr, J.F. and Renszer, H.W. (1962). Soil Sci. Soc. Am. Proc. 26, 552-556.

Paszkowski, W.L. and Kremer, R.J. (1988). Biological activity and tentative identification of flavonoid components in vefvetleaf (*Abutilon theophrasti* Medik.) seed coats. *Journal of Chemical Ecology* **14:** 1573-1582.

Patel, R.M. and Desai, M.V. (1964). Plant Dis. Reptr. 48, 167-168.

Pathak, M.D. and Dale, D. (1983). In Chemistry and world food supplies : The new frontiers (L.W. Scheinit, ed). PP. 129-141. New York.

Patrick, Z.A. and Koch, L.W. (1958). Inhibition of respiration, germination & growth by substances arising during the decomposition of certain plant residues in soil. *Canadian Journal of Botany* **36:** 621-647.

Patrick, Z.A. (1955). The peach replant problem in Ontario. II. Toxic substances from microbial decomposition products of peach root residues. *Canadian Journal of Botany* **33:** 461-486.

Patrick, Z.A. (1971). Phytotoxic substances associated with the decomposition of plant residues in soil. *Soil Science* **111:** 13-18.

Patrick, Z.A. and Koch, L.W. (1963). The adverse influence of phytotoxic substances from decomposing plant residues on resistance of tobacco to black root rot. *Canadian Journal of Botany* **41:** 747-758.

Patrick, Z.A. and Toussoun, T.A. (1965). *Ecology of Soil Borne Plant Pathogens* (Eds., K.C. Baker and W.C Snyder). Pp. 440-459. University of California Press, California.

Patrick, Z.A., Toussoun, T.A. and Koch, L.W. (1964). Effect of crop residue decomposition products on plant roots. *Annual Review of Phytopathology* **2:** 267-292.

Patrick, Z.A., Toussoun, T.A. and Snyder, W.C. (1963). Phytopathology 53, 152-161.

Patterson, D.T. (1984). "Research Methods in Weed Science" Weed Science Society of American, Champaign, Illinois.

Pedersen, M.W. (1965). Agron. J. 57, 516-517.

Penn, D.J. and Lynch, J. M. (1981). Effect of decaying couch grass rhizomes on the growth of barley. *Journal of Applied Ecology* **18:** 669-674.

Penn, D.J. and Lynch, J.M. (1981). In Proc. Conf. Grass Weeds in Cereals in the United Kingdom. Association of Applied Biologists, pp. 59-64.

Pepperman A.B. and Bradow, J.M. (1988). Weed Sci 36, 719-725.

Perez, F.J. (1990). Allelopathic effect of hydroxamic acids from cereals on *Avena sativa* and *A. fatua. Phytochemistry* **29:** 773- 776.

Perez, F.J. and Ormeno-Nunez, J. (1991). Difference in hydroxamic acid content in roots and root exudates of wheat (*Triticum aestivum* L.) and rye (*Secale cereale* L.): Possible role in allelopathy. *Journal of Chemical Ecology* **17:** 1037-1043.

Perez, F.J. and Ormeno-Nunez, J. (1993). Weed growth interference from temperate cereals: The effect of a hydroxamic acid exuding rye (*Secale ceoreale* L) cultivars. *Weed Research* **33:**115-119.

Perez-Pico, J.E. and Ashton, F.M. (1984). 81st Ann. Meet. Amer. Soc. Hort. Sci.

Periturin, F.T. (1913). Izv. Mosk. S-Kh. Inst. Kn. 4.

Peters, E.J. and Luu, K.T. (1985). Allelopathy in tall fescue. In: *Chemistry of Allelopathy: Biochemical Interactions Among Plants* (Ed., A.C. Thompson). ACS Symposium Series No. **268:** 273-283. American Chemical Society, Washington DC.

Peters, J.A., Gastal, M.F.C. and Finger, F.L. (1982). In Proc. 4th Brazilian Congress on herbicides and herbaceous Weeds and 64 th Congress of Latin American Weed Association. pp. 14-15.

Peterson, J.K. and Harrison, H.F. (1991). Differential inhibition of seed germination by sweet potato (*Ipomoea batatas*) root periderm extracts. *Weed Science* **39:** 119-123.

Peterson, J.K. and Harrison, H.F. (1991). Isolation of substances from sweet-potato (*Ipomoea batatas*) periderm tissue that inhibits seed germination. *Journal of Chemical Ecology* **17:** 943-951.

Peterson, J.K. and Harrison, H.F. (1990). Weed Sci. 39, 119-123.

Petrova, A.G. (1977). In "Interaction of Plants and Microorganisms in Phytocenosis" (A.M. Grodzinsky, ed.) pp. 91-97. Naukova Dumka, Kiev.

Pickerign, S.V. (1914). J. Agric. Sci. Camb. 6, 136-151.

Picman, A.K. (1987). Biochem. Syst. Ecol. 15, 361-364.

Piedrahita, W., Mesio, R. and Doll, J. (1975). Revista Comalfi. 2, 89-103.

Plhak, F. (1967). The effect of quackgrass on succeeding plants. *Plant and Soil* **27:** 273-284.

Plinius Secundus, C. 1 AD. Natural History vols. 1-20 (Eng. transl. by H. Rackam, WHS Jones, and D.E. Eichholz) Harvard University Press, Cambridge, Massachusetts, 1938-63.

Pope, D.F., Thompson, A.C. and Cole, A.W. (1985). The effect of root exudates on soybean germination, root growth, nodulation and dry matter production. In: *Chemistry of Allelopathy: Biochemical Interactions Among Plants* (Ed., A.C. Thompson). ACS Symposium Series No. **268:** 235-241. American Chemical Society, Washington DC.

Porwal, M.K. and Gupta, O.P. (1986). Allelopathic influence of winter weeds on germination and growth of wheat. *International Journal of Tropical Agriculture* **5:** 276-279.

Porwal, M.K. and Gupta, O.P. (1987). Trop. Agric. 4, 276-279.

Prasad, S.K., Mishra, S.D. and Gour, A.C. (1972). Indian J. Ento. 34, 307-311.

Prihar, S.S., Sandhu, K.S. and Khera, K.L. (1975). Indian J. Ecol. 2, 13-22.

Pronin, V. A., Voronkova, F. V. Adakovlev, A. A. (1972). Plant relations in mixed stand as dependent on biological and biochemical process in soil. In: *Physiological Biochemical Basis of Plant Interactions in Phytocenosis* (Ed., A.M. Grodzinsky) **3**:121-124. Naukova Dumka, Kiev, USSR.

Pronin, V.A. (1970). Methods for studying interrelations between plants under sterile conditions. In: *Physiological and Biochemical Interactions Between the Plants in Phytocenosis* (Ed., A.M. Grodzinsky) Vol. **1**: 16-24. Naukova Dumka, Kiev, USSR.

Pronin, V.A. and Yakovlev, A.A. (1970). Influence of nutrition conditions and rhizospheric microorganisms on the interrelations of maize and fodder beans in mixed cultures. In: *Physiological and Biochemical Interactions Between the Plants in Phytocenosis* (Ed., A.M. Grodzinsky) Vol. **1**: 93-100. Naukova Dumka, Kiev, USSR.

Prutenskaya, N.I., Yurchak, Z.D. and Soroka, M.A. (1970). Physiology of active substance of micro organism and decomposing plant residue. In: *Physiological Biochemical Basis of Plant Interactions in Phytocenosis* (Ed., A.M. Grodzinsky) Vol. **1**: 218-222. Naukova Dumka, Kiev.

Prutenskaya, N.I. (1972). Presence of inhibitors and stimulators of *Sinapsis arvensis* L. in germinating seeds of cultivated plants. In: *Physiological and Biochemical Interactions Between the Plants in Phytocenosis* (Ed., A.M. Grodzinsky) Vol. **3:** 73-75. Naukova Dumka, Kiev, USSR.

Prutenskaya, N.I. (1974). Effect of barley leachates and germinating seeds of crops on wild mustard. In: *Physiological and Biochemical Basis of Plant Interactions in Phytocenosis* (Ed., A.M. Grodzinsky) **3:** 73-75. Naukova Dumka, Kiev, USSR.

Prutenskaya, N.I., Yurchak, L.D. and Soroka, M.A. (1970). Physiologically active substances of microorganisms and decomposing plant residue. In: *Physiological and Biochemical Interactions Between the Plants in Phytocenosis* (Ed., A.M. Grodzinsky) Vol. **1**: 218-22. Naukova Dumka, Kiev, USSR.

Purvis, C.E., Jessop, R.S. and Lovett, J. V. (1985). Selective regulation of germination and growth of annual weeds by crop residues. *Weed Research* **25:** 415-421.

Purvis, C.E., Jessop, R.S. and Lovett, J.V. (1985). Weed response to crop stubbles. In: *Proceedings of 3rd Australian Agronomy Conference*. Pp. 375. University of Tasmania, Hobart.

Putman, A.R. (1988). Allelochemicals from plants as herbicides. *Weed Technology* **2:** 510-518.

Putnam, A.R. (1985). Weed allelopathy. In: *Weed Physiology, Reproduction and Ecophysiology* (Ed., S.O. Duke) **1:** 131-155. CRC Press, Boca Raton, Florida, USA.

Putnam, A.R. and DeFrank, J. (1983). Use of phytotoxic plant residues for selective weed control. *Crop Protection* **2**: 173-181

Putnam, A.R. and Duke, W.B. (1974). Biological suppression of weeds: Evidence for allelopathy in accessions of cucumber. *Science* **185**: 370-71.

Putnam, A.R. and Tang, C.S. (1986). Allelopathy : State of the science. *The Science of Allelopathy* (Eds., A. R. Putnam and C. S. Tang). Pp. 1-19. John Wiley & Sons Inc., New York.

Putnam, A.R. (1978). Allelopathy in agroecosystems. *Annual Review of Phytopathology* **16:** 431-451.

Putnam, A.R. (1983). Allelopathic chemicals: Nature's herbicide in action. *Chemical and Engineering News* **61:** 34-45.

Putnam, A.R. (1986). Allelopathy : Can it be managed to benefit horticulture ? *Hort Science* **21:** 411 - 413.

Putnam, A.R. (1983b). Fruit Grower 103, 10.

Putnam, A.R. (1988). In Weed management in agroecosystems : Ecological approaches (M.A. Altieri and M. Liebman, eds.) pp. 77- 88. CRC Press Inc., Boca Raton, FL. USA.

Putnam, A.R. and DeFrank, J. (1979). Proc. 9th Int. Cong. Plant Prot. pp. 580-582.

Putnam, A.R. and Weston, L.A. (1986). Adverse impacts of allelopathy on agricultural systems. In: *The Science of Allelopathy* (Eds., A.R. Putnam and C.S. Tang). pp. 43-56. John Wiley & Sons Inc., New York.

Putnam, A.R., De Frank, J. and Barnes, J.P. (1983). Exploitation of allelopathy for weed control in annual and perennial cropping systems. *Journal of Chemical Ecology* **9:** 1001-1010.

Qasem, J.R. and Abu-Irmaileh, B.E. (1985). Allelopathic effects of *Salvia syriaca* L. (Syrian sage) in wheat. *Weed Research* **25:** 47-52.

Qasem, J.R. and Hill, T.R. (1989). Possible role of allelopathy in the competition between tomato, *Senecio vulgaris* L. and *Chenopodium album* L. *Weed Research* **29:** 349-356.

Quinn, J.A. (1974). *Convolvulus sepium* in old field succession on the New Jersey Piedmont. *Bulletin of the Torrey Botanical Club* **101:** 89-95.

Rademacher, B. (1940). Pflanzenbau 17, 131-145.

Rajan, L. (1973). Curr. Sci. 42, 729-730.

Rakhteenko, I.B., Kaurov, I.A. and Minko, I.F. (1973). On the problem of exchange with root excretions in some agricultural plants in agrophytocenosis. In: *Physiological and Biochemical Interactions Between the Plants in Phytocenosis* (Ed., A.M. Grodzinsky) Vol. **4:** 16-19. Naukova Dumka, Kiev, USSR.

Rakhteenko, I.N. and Egorova, R.N. (1971). Physiological peculiarities of mutual influence of lupine and oat in mixed and pure cultures. In: *Physiological and Biochemical Interactions Between the Plants in Phytocenosis* (Ed., A.M. Grodzinsky) Vol **2:** 79-84. Naukova Dumka, Kiev, USSR.

Rakhteenko, I.N., Kaurov, I.A. and Minko, I.F. (1973). Effect of water soluble metabolites of a series of crops on some physiological processes. In: *Physiological and Biochemical Interactions Between the Plants in Phytocenosis* (Ed., A.M. Grodzinsky) Vol. **4:** 23-26. Naukova Dumka, Kiev, USSR.

Ramakudzibga, A. M. (1991). Allelopathic effect of aqueous wheat (*Triticum aestivum* L.) straw extract on the germination of eight arable weeds found in Zimbabwe. *Zimbabwe Journal of Agricultural Research* **29:** 77-79.

Rands, R.D. and Dopp, E. (1938). J. Agric. Res. 56, 53-67.

Rao, A.N., Dagar, J.C. and Dubey, P.S. (1977). Indian J. Weed Sci. 9, 24-30.

Rao, J.V.S. Rao, K.R.M. and Murthy, S.S. (1979). Trop. Ecol. 20, 5-8.

Rasmuseen, J.A. and Einhellig, F.A. (1975). Amercian Midl. Nat. 94, 478-483.

Rasmussen, J.A. and Einhellig, F.A. (1979). South West Nat. 24, 637-644.

Read, J. J. and Jensen, E. H. (1989). Phytotoxicity of water-soluble substances from alfalfa and barley soil extracts on four crop species. *Journal of Chemical Ecology* **15:** 619-628.

Reinhardt, C.F., Meissner, H. and Nel, P.C. (1993). Allelopathic effect of sweet potato (*Ipomoea batatas*) cultivars on certain wood and vegetable species. *South African Journal of Plant and Soil* **10:** 41-44.

Rempe, E. Kh. (1967). Sel-Khoz. Biol. (Russia), 2, 592-597.

Rempe, E. Kh. and Gryunberg, T.M. (1970). Sel-Khoz. Biol. (Russia) 5, 703-708.

Retiveld, W.J. (1983). J. Chem. Ecol. 9, 259-268.

Rhode, R.A. (1972). Ann. Rev. Phytopath 10, 233-252.

Rice, E.L. (1974). *Allelopathy.* Academic Press, New York.

Rice, E.L. (1984). *Allelopathy*, 2nd ed. New York Academic Press.

Rice, E.L. (1964). Inhibition of nitrogen-fixing and nitrifying bacteria by seed plants. *Ecology* **45:** 824- 837.

Rice, E.L. (1971). Inhibition of nodulation of inoculated legumes by leaf leachates of pioneer plant species from abandoned fields. *American Journal of Botany* **58:** 368-371.

Rice, E.L. (1977). Some roles of allelopathic compounds in plant communities. *Biochemical Systematic and Ecology* **5:** 201-206.

Rice, E.L. (1979). Allelopathy- an update. *Botanical Review* **45:** 105-109.

Rice, E.L. (1980). Effects of decaying rice straw on growth and nitrogen fixation of a blue green alga. *Botanical Bulletin of Academia Sinica* **21:** 111-117.

Rice, E.L. (1985). Allelopathy - an overview. In: *Chemically Mediated Interactions Between Plants and Other Organisms* (Eds. G.A. Cooper-Driver, S. Tony and E.E. Coon). pp. 81-105. Plenum Publishing Corporation, New York.

Rice, E.L. (1983). "Pest control with Nature's Chemicals" Univ. Oklahoma Press, Norman, pp. 224.

Rice, E.L., Lin, C.Y. and Hirano, C.Y. (1981). Effects of decomposing rice straw on growth and nitrogen fixation by rhizobium. *Journal of Chemical Ecology* **7:** 333-344.

Ries, S.K., Wert, V., Sweely, C.C. and Leavitt, R.A. (1977). Science 195, 1339-1341.

Rizvi, S.J.H. and Rizvi, V. (1987). Improving crop productivity in India: Role of allelochemicals. In: *Allelochemicals : Role in Agriculture and Forestry* (Ed. G.R. Waller). ACS Symposium Series **330:** 69-75, American Chemical Society, Washington.

Rizvi, S.J.H., Mishra, G.P. and Rizvi, V. (1989). Allelopathic effects of nicotine on maize. I. Its possible importance in crop rotation. *Plant and Soil* **116:** 289-291.

Rizvi, S.J.H., Mishra, G.P. and Rizvi, V. (1989). Allelopathic effects of nicotine on maize. II. Some aspects of its mechanism of action. *Plant and Soil* **116:** 292-293.

Rizvi, S.J.H., Mukerji, D. and Mathur, S.N. (1980). A new report on a possible source of natural herbicide. *Indian Journal of Experimental Biology* **18:** 777-779.

Rizvi, S.J.H., Mukherji, D. and Mathur, S.N. (1981). Selective phytotoxicity of 1,3,7-trimethylxanthine between *Phaseolus mungo* and some weeds. *Agricultural and Biological Chemistry* **45:** 1255- 1256.

Rizvi, S.J.H., Singh, V.K., Rizvi, V. and Waller, G.R. (1988). Geraniol, an allelochemical of possible use in integrated pest management. *Plant Protection* **3:** 112-114.

Robbins, W., Crafts, A. S. and Raynor, R. N. (1982). *Weed control.* McGraw Hill, New York.

Robertson, W.K., Lundy, H.W., Prine, G.M. and Currey, W.L. (1976). Agron. J. 68, 271-281.

Robeson, D.J. and Strobel, G.A. (1984). Phytochemistry 23, 1597- 1599.

Robeson, D.J., Strobel, G.A. Matusumoto, C.F., Fischer, E.L., Chen, M.H. and Clardy, J. (1984). Experientia 40, 1248-1250.

Robinson, E.L. (1976). Effect of weed species and placement on seed cotton yield. *Weed Science* **24:** 353-355.

Robinson, T. (1983). *The Organic Constituents of Higher Plants.* 5th edition. Cordus Press, North Amherst, Massachusetts.

Robinson, T. (1967). The Organic Constituents of Higher Plants. Second Edition, pp. 319. Burgess, Minneapolis, Minnesota, USA.

Roeth, F.W. (1973). Weed Sci. 21, 474-476.

Rose, S.J., Burnside, O.C. Specht. J.E. and Swisher, B.A. (1984). Agron. J. 76, 523-528.

Rose, S.J., Burnside, O.C., Specht, J.E., Williams, J.H. and Swisher, B.A. (1983). Weed Sci. Soc. Am. 16, Abstrect 7.

Rovira, A. D. (1969). Plant root exudates. *Botanical Review* **35**: 35-37.

Rovira, A.D. (1959). Plant Soil 11, 53-64.

Rudiger, W. and Lohaus, E. (1987). Germination and growth inhibitors as allelochemicals. In: *Allelochemicals: Role in Agriculture and Forestry* (Ed., G.R. Waller). ACS Symposium Series No. **330:** 118- 128. American Chemical Society, Washington DC.

Ruiz-Sifre, G.V. and Ries, S.K. (1983). Response of crops to sorghum residues. *Journal of American Society of Horticultural Science* **108:** 262-266.

Ruth, M., Nel, P.C. and Beyers, E.A. (1986). South Afr. J. Plant Soil 3, 176-180.

Sadhu, M.K. and Das, T.M. (1971). Root exudates of rice seedlings. The Influence of one variety on another. *Plant and Soil* **34:** 541-546.

Sadhu, M.K. (1975). Nature of inhibitory substances in the root exudates of rice seedlings. *Indian Journal of Experimental Biology* **13:** 577-579.

Sahid, I. B., Tasrif, A., Sastroutomo, S.S. and Latiff, A. (1993). Allelopathic potential of legume cover crops on selected weed species. *Plant Protection Quarterly* **8(2):** 49-53.

Sanford, J.O. (1982). Straw and tillage management practices in soybean-wheat double cropping. *Agronomy Journal* **74:** 1032-1035.

Sanford, J.O. (1982). Agron. J. 74, 1032-1035.

Santos, H.P., Lhamby, J.C.B. and Pereira, L.R. (1983). In "Soja Resultados de Pesruisa, 1982-83". pp. 55-65. Centro

Sapankevitch, P.V., Maksimova, N.A. and Khomenko, B.P. (1965). Agribiol. No. 6, 915-916.

Sarma, K.K.V. (1974). Geobios. 1, 137.

Sarma, K.K.V. (1983). Allelopathic potential of the phytoextracts of *Aristida adscensionis* L. *Tropical Ecology* **24:** 19-21.

Sarma, K.K.V., Giri, G.S. and Subramanyan, K. (1976). *Trop. Ecol.* 17, 76-77.

Sarmah, M.K. (1992). *Allelopathic Effect of Wheat Residues on the Succeeding Crops and Weeds.* Doctoral Dissertation, Haryana Agricultural University, Hisar, India.

Sarmah, M.K., Narwal, S.S. and Yadava, J.S. (1992). Smothering effect of *Brassica* species on weeds. In: *Proceedings, First National Symposium Allelopathy in Agroecosystems* (Eds., P. Tauro and S.S. Narwal). Pp. 51-55. Indian Society of Allelopathy, Haryana Agricultural University, Hisar, India.

Scheffer, R.P. (1983). In "Toxins and plant pathogenesis". (J.M. Daly and B.J. Deverall, eds.) Pp. 1-40. Academic Press, New York.

Schmidt, S.K. (1988). *J. Chem. Ecol.* 14, 1561-1571.

Schon, M.K. and Einhellig, F.A. (1982). Allelopathic effects of cultivated sunflower on grain sorghum. *Botanical Gazette* **143:** 505-510.

Schreiber, M.M. (1992). Influence of tillage, crop rotation and weed management on giant foxtail (*Setaria faberi*) population dynamics and corn yield. *Weed Science* **40:** 645-653.

Schreiber, M.M. and Williams Jr., J.L. (1967). Toxicity of root residues of weed grass species. *Weeds* **15:** 80-81.

Schreiner, O. and Reed, H.S. (1907). The Production of deleterious excretions by roots. *Bulletin of the Torrey Botanical Club* **34:** 279-303.

Schreiner, O. and Reed, H.S. (1908). The toxic action of certain organic plant constituents. *Botanical Gazette* (Chicago) **45:** 73-102.

Schreiner, O. and Reed, H.S. (1907a). Certain organic constituents of Soil in relation to soil fertility. *U S D A Bur. Soils Bull.* No. 47.

Schreiner, O. and Shorey, E.C. (1909). The isolation of harmful organic substances from soils. *USDA Bureau of Soils, Bulletin No.* **53:** 34-53.

Schreiner, O. and Sullivan, M.X. (1909). *J. Biol. Chem.* 6, 39-50.

Schumacher, W.J., Thill, D.C. and Lee, G.A. (1982). Allelopathic potential of wild oat (*Avena fatua*) on spring wheat (*Triticum aestivum*) growth. *Journal of Chemical Ecology* **9:** 1235-1245.

Schumacher, W.J., Thill, D.C. and Lee, G.A. (1982). *Proc. North Am. Symp. Allelopathy,* pp. 14 (Abst.). Urbana, Champaign Illinois.

Sekizawa, Y. and Takematsu, T. (1983). In Pesticide Chemistry, Human Welfare and Environment, vol. 2. Natural Products (N. Takahashi, H. Yoshioka, T. Misato and S. Matusunaka, eds.) pp. 261- 268. Pergamon Press, Oxford.

Sen, D.N. (1987). Ecological studies on Weed Crop Association in Indian arid zone. pp. 97-116. Botany Deptt. Jodhpur Univesity, Jodhpur.

Sen, D.N. Chawan, D.D. and Sharma, K.D. (1969). *Proc. Indian Acad. Sci.* B. 70, 111-117.

Seredyuk, L.S. (1982). On the role of organic substances in crop production. In: *Role of Allelopathy in Plants, Science Colloquim* (Ed., A.M. Grodzinsky) pp.155-159. Naukova Dumka, Kiev, USSR.

Sethi Mathur, A. and Mohnot, K. (1988). Allelopathic influence of leaf extract of *Trianthema portulacastrum* on germination and growth of mothbean (*Vigna aconitifolius*). *Journal of Current Biosciences* **5:** 61-63.

Sewell, M.C. (1933). *Bot. Gaz.* 75, 1-26.

Sharma, K.D. and Sen, D.N. (1971). Growth regulators in the fruit pulp of *Solanum surattense. Zeitschrift fuer Pflanzenphysiologie* **65:** 458-460.

Sharma, K.D. and Singhvi, N.R. (1981). *Geobios* 8, 223-225.

Sharma, K.D., Sidana, K.L. and Singhvi, N.R. (1982). Allelochemic effect of *Peganum harmala* L. on *Pennisetum typhoides* L. *Indian Journal of Botany* **5:** 115-119.

Sharma, V. and Nathawat, G.S. (1987). Allelopathic effect of *Argemone mexicana* L. on species of *Triticum, Brassica, Raphanus* and *Pennisetum. Current Science* **56:** 427-428.

Shaukat, S.S., Gazala, P., Khan, D. and Ahmed, M. (1985). *Pakistan J. Bot.* 17, 235-246.

Shettel, N.L. and Balke, N.W. (1983). Plant growth response to several allelopathic chemicals. *Weed Science* **31:** 293-298.

Shilling, D.G. (1984). *Diss. Abst. Int.* B. 45, 28.

Shilling, D.G., Jones, L.A., Worsham, A.D., Parker, C.F. and Wilson, R.F. (1986). Isolation and identification of some phytotoxic compounds from aqueous extracts of rye (*Secale cereale* L.). *Journal of Agricultural and Food Chemistry* **34:** 633-38.

Shilling, D.G., Liebl, R.A. and Worsham, A.D. (1985). Rye (*Secale cereale* L.) and wheat (*Triticum aestivum* L.) mulch: The suppression of certain broad-leaved weeds and the isolation and identification of phytotoxins.

In: *Chemistry of Allelopathy* (Ed., A.C. Thompson). ACS Symposium Series **268**: 243-271. American Chemical Society, Washington DC.

Shilling, D.G., Worsham, A.D. and Danehower, D.A. (1986). Influence of mulch, tillage and diphenamid on diphenamid on weed control, yield and quantity on no till flue cured tobacco (*Nicotiana tabacum*). *Weed Science* **34:** 738- 744.

Shukle, R.H. and Murdock, L.I. (1983). *Environ. Entomol.* 12 : 787-791.

Siegel, S.M. (1950). *Science* 112, 754-755.

Sigareva, D.D. (1982). Susceptibility of winter wheat, maize and sugarbeet to parasitic nematodes under continuous cropping and in crop rotation with organic and inorganic fertilizers. In: *Role of Allelopathy in Plants, Science Colloquim* (Ed. A.M. Grodzinsky) pp.129-137. Naukova Dumka, Kiev, USSR.

Singh, C.M., Angiras, N.N. and Singh, S.D. (1988). *Indian J. Weed Sci.* 20, 63-66.

Singh, K., Malik, A.S. and Narwal, S.S. (1983). *Haryana Agric. Univ. J. Res.* 13, 587-592.

Singh, R. and Agrawal, R.A. (1988). *J. Cotton Res. Dev.* 2, 62-65.

Singh, S.C. and Sen, D.N. (1982). Observations on seedling allelopathy in weeds. *Current Science* **51:** 45-46.

Singh, S.C. and Sen, D.N. (1981). *Annu. Conf. Indian Weed Sci.* pp. 2 (Abst.)

Singh, S.P. (1967). Presence of a growth inhibitor in the tubers of nutgrass (*Cyperus rotundus* L.). *Proceedings, Indian Academy of Science* **67** (Part II): 19-23.

Singh, S.P., Pal, U.R. and Luka, K. (1989). *J. Agron. Crop. Sci.* 162, 236-240.

Singhal, B.K. and Sen, D.N. (1985). Allelopathic potential of pigweed (*Amaranthus viridis* L.) on rainfed crops of Indian desert. (Eds., S.D. Mishra, D.N. Sen and I. Ahmed). *Geobios* **(Special volume)**. pp. 213-221.

Singhal, B.K. and Sen, D.N. (1982). *Proc.* 69th *Indian Sci. Cong. Sec.* X. *Agric. Sci.* pp. 88-89 (Abstract). Indian Science Congress Association, Calcutta.

Singhvi, N.R. and Sharma, K.D. (1984). *Indian Soc. Desert Technol.* 9, 95-100.

Singhvi, N.R. and Sharma, K.D. (1985). *Indian Soc. Desert Technol.* 10, 53-55.

Skinner, J.J. (1913). *Plant World* 16, 342-348.

Smeda, R. and Putnam, A.R. (1984). *Proc. North Central Weed Control Conf.* 39, 145.

Smeda, R.J. and Putnam, A.R. (1988). Cover crops suppression of weeds and influence on strawberry yields. *Hort Science* **23:** 132-134.

Smith, A.M. and Secoy, D.M. (1977). *Furrow* 13, 27.

Smith, C.M. (1986). *Agric. Ecosyst. Environ.* 18, 1-7.

Smith, G.L. and Secoy, D. (1975). *J. Argic. Food Chem.* 23, 1050-1055.

Smith, Jr., R.J. (1974). Competition of barnyardgrass with rice cultivars. *Weed Science* **22:** 423-426.

Sokolova, E.A. (1973). Effect of mustard root excretions on absorption of phosphorus and synthesis of nutrients in pea plants. In: *Physiological and Biochemical Interactions Between the Plants in Phytocenosis* (Ed., A.M. Grodzinsky) Vol. **4:** 20-23. Naukova Dumka, Kiev, USSR.

Sokolova, E.A. and Mikryukov, G.I. (1970). On the influence of the white mustard and barley on the morphological, anatomical structure and yield of pea. In: *Physiological and Biochemical Interactions Between the Plants in Phytocenosis* (Ed., A.M. Grodzinsky) Vol. **1**: 72-74. Naukova Dumka, Kiev, USSR.

Sokolova, E.A. and Mikryukov, G.I. (1971). Change in ratio of components in pea-mustard and pea- barley agrophytocenosis. In: *Physiological and Biochemical Interactions Between the Plants in Phytocenosis* (Ed., A.M. Grodzinsky) Vol. **2:** 84-88. Naukova Dumka, Kiev, USSR.

Solomon, M.J. and Bhandari, D.C. (1981). Allelopathic potential of *Gomphrena decumbens* Jacq on two ram-fed crops. *Geobios* **8** : 9-12.

Song, N.H., Jin, Y.M. and Kuo, G.C. (1985). Research Report, Rural Development Administration, Plant Environment, Mycology and Farm Prodcuts Utilization, Korea Republic 27(1), 17-22.

Soni, S.R. and Mohnot, K. (1988). Chlorophyll synthesis in sorghum seedlings as affected by allelopathic complex of the leaves of *Celosia argentea. Comparative Physiology and Ecology* **13:** 152-154.

Soni, S.R. and Mohnot, K. (1978). *Proc. Symp. Recent Advances in Plant Physiology in India and Seminar on Plant Growth* and *Metabolism.* Abstract No. 48. Indian Soc. Plant Physiol., New Delhi.

Soni, S.R. and Mohnot, K. (1981). *Proc. Nat. Symp. Seed Physiology* and *Biochemistry* Abstract No. 55, Indian Society of Plant Physiology and Biochemistry, Lucknow.

Spelce, D.L. and Muselman, L.J. (1981). *Z. Pflanzenphysiol.* 104, 281-283.

Spencer, G.F., Tjarks, L.W., England, R.E. and Seest, E.P. (1986). The effect of naturally occurring naphthoquinones on velvetleaf *(Abutilon theophrasti)* germination. *Journal of Natural Products* **49:** 530-533.

Spring, O., Albert, K. and Hagar, A. (1982). Three biologically active *Heliangolides* from *Helianthus annuus. Phytochemistry* **21:** 2552-2553.

Springer, J.P., Cutler, H.G., Crumely, H.G., Cox, R.H. Davis, F.E. and Thean, J.E. (1981). *J. Agri. Food Chem.* 29, 853-855.

Srivastava, A.K. (1969). Effect of seed extract of *Heliotropium eichwaldi* Steud, on seed germination of *Phaseolus aureus* Roxb. *Current Science* **38:** 440-441.

Srivastava, J.N., Shukla, J.P. and Srivastava, R.C. (1985). Effect of *Parthenium hysterophorus* L. extract on the seed germination and seedling growth of barley, pea and wheat. *Acta Botanica Indica* **13:** 194- 197.

Stachon, W.J. and Zimdahl, R.L. (1980). Allelopathic activity of Canada thistle (*Cirsium arvense*) in Colorado. *Weed Science* **28:** 83-86.

Staniforth, D.W. (1957). Effects of annual grass weeds on the yield of corn. *Agronomy Journal* **49:** 551- 555.

Staniforth, D.W. (1958). Soybean-foxtail competition under varying soil moisture conditions. *Agronomy Journal* **50:** 13-15.

Staniforth, D.W. (1961). Response of corn hybrids to yellow foxtail competition. *Weeds* **9:** 132-136.

Staniforth, D.W. (1965). Competitive effects of three foxtail species on soybeans. *Weeds* **13:** 191-193.

Steenhagen, D.A. and Zimdahl, R.H. (1979). Allelopathy of leafy spurge (*Euphorbia esula*). *Weed Science* **27:** 1-3.

Stefureac and Fratilescu-Sesan. (1979). *Biol. Veg.* 31, 55-61.

Steinsiek, J.W., Oliver, L.R. and Collins, F.C. (1980). The effect of phytotoxic substances from wheat straw on selected weeds. In: *Proceedings of the 33rd Annual Meeting of the Southern Weed Science Society*. Pp. 223.

Steinsiek, J.W., Oliver, L.R. and Fred, F.C. (1982). Allelopathic potential of wheat (*Triticum aestivum*) straw on selected weed species. *Weed Science* **30:** 495-487.

Stepanova, L.P. and Prutenskaya, N.I. (1975). On allelopathic properties of lucerne. In: *Physiological and Biochemical Interactions Between the Plants in Phytocenosis* (Ed., A.M. Grodzinsky) Vol. **6:** 69-72. Naukova Dumka, Kiev, USSR

Stevens Jr., G.A. and Tang, C.H. (1985). Inhibition of seedling growth of crop species by recirculating root exudates of *Bidens pilosa* L. *Journal of Chemical Ecology* **11:** 1411-1425.

Stevens, G.A. and Tang, C.H. (1987). Inhibition of crop seedlings growth by hydrophobic root exudates of the weed *Bidens pilosa*. *Journal of Tropical Ecology* **3:** 91-94.

Stevens, K.L. (1986). Allelopathic polyacetylenes from *Centaurea repens* (Russian knapweed). *Journal of Chemical Ecology* **12:** 1205-1210.

Stretton, A.O., Campbell, W.C. and Babu, J.R. (1987). In "Vistas in nematology" (J.A. Veech and D.W. Dickson, eds.) Pp. 136-146.

Su, S.M. and Fang, X.L. (1981. *Acta Agron. Sinica* 7, 123-128.

Sugawara, F., Strobel, G. Fisher, L.E., Van Duyne, G.D. and Clardy, J. (1985). *Proc. Natl. Acad. Sci.* USA 82, 8291-8294.

Sugha, S.K. (1978). Allelopathic potential of superb lily (*Gloriosa superba* L.). *Science and Culture* **44:** 461-462.

Sugha, S.K. (1979). Effect of weed extracts on wheat germination. *Science and Culture* **45:** 65-66.

Sugha, S.K. (1980). An assessment of allelopathic potentials of three common weeds. *Food Farming and Agriculture* **11:** 261-262.

Sukul, N.C. (1970). *Proc. Zool. Soc.* Calcutta 23, 139-146.

Swain, T. (1977). *Annu. Rev. Plant Physiol.* 28, 479-501.

Szazepanski, A.J. (1977). Allelopathy as a means of biological control of winter weeds. *Aquatic Botany* **3:** 193-197.

Takatori, F. and Souther, F. (1978). Proc. Workshop on Asparagus Deptt. Plant Sci., Univ. California, Riverside, California.

Takiguchi, Y., Yoshikawa, H., Terahara, A. Torikata, A. and Terao, M. (1979). *J. Antibiot.* 32, 857-867.

Takijama, Y. (1963). *Bull. Natl. Inst. Agri. Sci. Tokyo* 13, 217.

Tames, S.R., Gasto, M.D.V. and Vietiz, E. (1973). *Physiol. Plant* 28, 195-200.

Tang, C.S. and Waiss Jr., A.C. (1978). *J. Chem. Ecol.* 4, 225-232.

Teasdale, J.R. (1988). *Proc. N.E. Weed Sci. Soc.* 42, 73.

Teasdale, J.R. (1993). *Weed Science* 41, 46-51.

Teasdale, J.R. and Daughtery, C.S.T. (1993). Weed suppression by live desiccated hairy vetch (*Vicia villosa*). *Weed Science* **41:** 207-210.

Teasdale, J.R. and Mohler, C.L. (1992). In Proc. I. International Weed Control Congress, Vol. 2, 516-518. Weed Science Society of Victoris Melbourne, Australia.

Ten Eych, A.M. and Shoesmith, A.M. (1906). "Indian Corn" pp. 256- 262. *Kansas Exp. Ste. Bull.* No. 147.

Theophrastus, Ca 300 B.C. Enquiring into plants and in minor work on odour and weather signs vol. 2 (Eng. transl. by Hort, A. and Heineman, W.) London 1916.

Thompson, A.C. (1985). Chemistry of Allelopathy ACS Symposium Series No. 268. American Chemical Society, Washington, D.C.

Tinnin, R.O. and Muller, C.H. (1971). The allelopathic potential of *Avena fatua*: Influence on herb distribution. *Bulletin of the Torrey Botanical Club* **98:** 243-250.

Tischler, W. (1975). Worlerbucher der Biologie-Okologie VEB Gustav Fischer, Jena.

Toai, T.V. and Linscott, D.L. (1979). Phytotoxic effect of decaying quackgrass (*Agrophyron repens*) residues. *Weed Science* **27:** 595-598.

Tomerlin, A.H. and Smart G.C. Jr. (1969). *J. Nematol* 1, 29-30.

Tomshevskaya, E.G. and Lugovskaya, E.Ya. (1970). In: Physiological- Biochemical Basis of Plant Interactions in Phytocenosis (A.M. Grodzinsky, ed.) 1, 75-82. Naukova Dumka, Kiev.

Touchette, R., Leroux, G.D. and Deschenes, I.M. (1988). Allelopathic activity of quackgrass (*Agropyron repens*) extracts and residues on alfalfa (*Medicago sativa*). *Canadian Journal of Plant Science* **68:** 785-792.

Toussoun, T.A. and Patrick, Z.A. (1963). *Phytopathology* 53, 265- 270.

Toussoun, T.A., Weinhold, A.R., Linderman, R.G. and Patrick, Z.A. (1968). *Phytopathology* 58, 41-45.

Towers, G.H.N. and Arnason, J.T. (1988). Photodynamic herbicides. *Weed Technology* **2:** 545-549.

Trenbath, B.R. (1976). Plant interactions in mixed crop communities. In: *Multiple Cropping* (Ed., M. Stelly). ASA Special Publication No.**27:** 129-169. American Society of Agronomy, Madison, Wisconsin, USA.

Triffit, K.J. (1930). *J. Helminthol* 8, 19-48.

Tripathi, R.S., Singh, R.S. and Rai, J.P.N. (1981). Allelopathic potential of *Eupatorium adenophorum*-a dominant ruderal weed of Meghalaya. In: *Proceedings of Indian Natural Science Academy* **847 (3) :** 458-465.

Trivedi, R.N. and Trivedi, M.P. (1987). *Acta Bot. Indica* 15, 314- 315.

Tsuzuki, E. (1980). Studies on the allelopathy in buckwheat plants. *Buckwheat Symposium,* Ljublajana, Yugoslavia. *Supplementary Volume*: Pp. 13-23.

Tsuzuki, E. and Araki, M. (1984). Studies on allelopathy among higher plants. III. Allelopathic substances from *Spergula arvensis* L. *Weed Research* (Japan) **29:** 208-213 (In Japanese).

Tsuzuki, E. and Kawagoe, H. (1984). Studies on allelopathy among higher plants. IV. On allelopathy in leguminous crops. *Bulletin, Faculty of Agriculture, Miyazaki University* **31:** 189-195 (In Japanese).

Tsuzuki, E. and Vamamoto, V. (1987). Studies on allelopathy among higher plants. Isolation and identification of phenolic substances from wild perennial buckwheat (*Fagopyrum cymosum* M.). *Bulletin of the Faculty of Agriculture, Miyazaki University* **34:** 289-295.

Tsuzuki, E., Ando, H. and Nishiyama, H. (1984). Studies on allelopathy among higher plants. II. Effects of root exudates from crops and weeds on the growth of plants in identical or different species. *Weed Research* (Japan) **29:** 203-207 (In Japanese).

Tsuzuki, E., Katsuki, A., Shida S. and Nagatomo, T. (1977). On the growth inhibitors contained in buckwheat plants. *Bulletin of Faculty Agriculture Miyazaki University* **24:** 41-46.

Tsuzuki, E., Vamamoto, V. and Shimizu, T. (1987). Fatty acids in buckwheat are growth inhibitors. *Annals of Botany* **60:** 69-70.

Tsuzuki, E., Watanabe, J. and Shida, S. (1978). Studies on allelopathy among higher plants. I. Effects of residues and aqueous and various organic solvent extracts from lawn pennywort (*Hydrocotyle sibthorpioides* Lam.) on the seed germination and growth of some vegetables and crop plants. *Bulletin Faculty of Agriculture Miyazaki University* **25:** 403-409.

Tukey Jr., H.B. (1969). Implications of allelopathy in agricultural plant science. *Botanical Review* **35:** 1-17.

Tukey Jr., H.B. and Mecklenburg, R.A. (1964). Leaching of metabolites from foliage and subsequent reabsorption and redistribution of the leachate in plants. *American Journal of Botany* **5:** 737- 742.

Tukey, Jr., H.B. (1970). *Annu. Rev. Plant Physiol.* 21, 305-324.

Turner, W.B. (1971). Fungal Metabolities. Academic Press, New York.

Uhlenbrock, J.H. and Bijloo, J.D. (1969). *Reuceil dex Travux* Chimiques des Pays-Bas 78, 382-390.

USDA, (1977). Report of the research planning conference on the role of secondary compounds in plant interactions (allelopathy), USDA, Washington, D.C.

Uteush, Y.A. (1979). Rape in forage production. Naukova, Dumka, Kiev.

Uygur, F.N., Cinar, F. and Koseli (1990). The allelopathic effect of *Raphanus sativus. Journal of Plant Diseases and Plant Protection* **12:** 259-264.

Varela, R.M. (1982). Allelopathic studies on cultivar of sunflower. M.S. Disseration University of Cadiz, Spain.

Vaughan, M.A. and Vaughan, K.C. (1988). Mitotic disrupters from higher plants and their potential uses as herbicides. *Weed Technology* **2:** 533-539.

Ventura, W., Watanabe, I., Komada, H., Nishio, M., Cruz, A.D. and Castillo, M. (1984). *IRRI Research Paper* Series No. 99 Int. Rice Res. Inst. Los Banos, Philippines.

Vesonder, R.F., Stodola, F.H., Wickerham, L.J., Ellis, J.J. and Rohwedder, (1991). *Can. J. Chem.* 49, 2029-2032.

Vezina, L. and Doyon, D. (1983). *Phytoprotection* 64, 77-81.

Vietez, B.A. and Vietze, A.M. (1979). *Bot. Gaz.* 140, 433-436.

Vilhordo, B.W., Burin, M.E., Gandolfi, V.H., Barni, N.A., Gomes, J.E. and Goncalves, J.C. (1985). *Agronomi a Sulriogr.* 21, 55-64.

Vinall, N.H. and Getty, R.E. (1921). Growing and Utilizing Sorghum. USDA Farmers Bull. No. 1158.

Vir, D. and Sharma, R.K. (1985). *Asian Farm Chem.* 1, 23-24.

Voichenko, T.I., Romanova, G.A., Kirilyuk, R.V. and Mazorchuk, L.I. (1982). Allelopathic activity of non-traditional and low distributed crops. In: *Role of Allelopathy in Plants, Science Colloquim* (Ed. A.M. Grodzinsky) pp. 84-87. Naukova Dumka, Kiev, USSR.

Vyas, K.G. (1982). Nitrogen fixation by leguminous crop plants and interference thereof by dominant weeds. Ph.D. thesis, Saurashtra Univ., Rajkot.

Vyas, K.G., Bhaskar, A. and Pandya, S.M. (1983). Inhibition of nodulation and nitrogen fixing bacteria of *Vigna mungo* L. by root exudates from *Celosia argentea* L. *Flora* **174:** 489-495.

Walker, D.W. and Jenkins, D.D. (1986). Influence of sweet potato plant residue on growth of sweet potato vine cuttings and cowpea plants. *Hort Science* **21:** 426-428.

Walker, D.W., Hubbel, T.J. and Sedberry, J.E. (1989). *Agric. Ecosyst. Environ.* 26, 45 52.

Waller, G.R. (1987). Allelochemicals : Role in Agriculture and Forestry. ACS Symposium Series No. 330. American Chemical Society, Washington, D.C.

Waller, G.R. and Dermer, O.C. (1981). Enzymology of alkaloid metabolism in plants and microorganisms. *The Biochemistry of Plants* **7:** 317-402.

Waller, G.R. and Nowacki, E.K. (1978). The role of alkaloids in plants. In: *Alkaloid Biology and Metabolism in Plants* (Eds., G.R. Waller and E.K. Nowacki). Pp. 143-181. Plenum Press, New York.

Waller, G.R., Jurzysta, M and Thorne, R.L.Z. (1993). *Bot. Bull. Acad. Sin. (Taipei)*

Waller, G.R., Kenzer, F.G., McPherson, J.K. and McGoan, S.R. (1987). *Plant Soil* 98, 5-15.

Wang, T.S.C., Cheng, S.Y. and Tung, H. (1967a). *Soil Sci.* 103, 360-366.

Wang, T.S.C., Cheng, S.Y. and Tung, H. (1967b). *Soil Sci.* 104, 138-144.

Wang, T.S.C., Li, S.W. and Ferng, Y.L. (1978). *Soil Sci.* 126, 16- 21.

Wang, T.S.C., Yang, T.K. and Chuang, T.T. (1967). Soil phenolic acids as plant growth inhibitors. *Soil Science* **103:** 239-246.

Wang. T.S.C., Yeh, K.L., Cheng, S.Y. and Yang, T.K. (1971). "Biochemical Interactions Among Plants" pp. 113-120. National Academy of Sciences, Washington, D.C.

Warnes, D.D. (1985). *Proc. North Central Weed Control Conf.* 40, 73.

Weaver, T.W. and Klarich, D. (1977). Allelopathic effects of volatile substances from *Artemisia tridentata* Nutt. *The American Midland Naturalist* **97:** 508-512.

Webster, G.R., Khan, S.U. and Moore, A.W. (1967). Poor growth of alfalfa (*Medicago sativa*) on some Alberta soils. *Agronomy Journal* **59:** 37-41.

Welbank, P.J. (1963). Toxin production during decay of *Agropyron repens* (couch grass) and other species. *Weed Research* **3:** 205-214.

Welbank, P.J. (1960). In "The Biology of Weeds" (J.L. Harper, ed.) pp. 158-164. Blackwell, Oxford.

Went, F.W., Juhren, G. and Juhren, M.C. (1952). *Ecology* 33, 351- 364.

Weston, L.A. (1986). *Diss. Abst. Inst.* B. 47, 1366.

Weston, L.A. and Putnam, A.R. (1985). Inhibition of growth, nodulation and nitrogen fixation of legumes by quackgrass. *Crop Science* **25:** 561-565.

Weston, L.A. and Putnam, A.R. (1986). Inhibition of legume seedling growth by residues and extracts of quackgrass (*Agropyron repens*). *Weed Science* **34:** 366-372.

Weston, L.A., Burke, B.A. and Putnam, A.R. (1987). Isolation, characterisation and activity of phytotoxic compounds from quackgrass (*Agropyron repens* L. Beauv.). *Journal of Chemical Ecology* **13:** 403- 421.

Weston, L.A., Harmon, R. and Muller, S. (1989). Allelopathic potential of sorghum-sudangrass hybrid (Sudan). *Journal of Chemical Ecology* **15:** 1855-1865.

White, R.H., Worsham, A.D. and Blum, O. (1989). Allelopathic potential of legume debris and aqueous extracts. *Weed Science* **37:** 674-679.

Whitehead, D.C. (1963). *Soil Fert.* 26, 217-229.

Whitehead, D.C. (1964). *Nature* 202, 417-418.

Whittaker, R.H. (1971). The chemistry of communities. In: *Biochemical Interactions Among Plants.* Pp. 10-18. Environmental Physiology Subcommittee for the International Biological Programme, Division of Biology and Agriculture, National Research Council, National Academy of Sciences, Washington DC.

Whittaker, R.H. (1970). In Chemical Ecology (E. Sondheimer and J.B. Simeone, eds.) pp. 43-70. Academic Press, New York.

Whittaker, R.H. and Feeny, P.P. (1971). Allelochemics: Chemical interactions between species. *Science* **171:** 757-770.

Wild, A. and Mandersheid, R. (1984). *Z. Naturforsch* 39C, 500-504.

Wild, G. and Rovira, A. (1977). *Soil Biol. Biochem.* 9, 203-206.

Williams, J.T. (1964). A study of the competitive ability of *Chenopodium album* L. *Weed Research* **4:** 283-295.

Williams, R.D. and Hoagland, R.E. (1982). The effect of naturally occurring phenolic compounds, on seed germination. *Weed Science* **30**: 206-212.

Williams, R.D. and Warren, G.F. (1975). *Weed Sci.* 23, 317-323.

Wilson, Jr. R.G. (1981). *Weed Sci.* 29, 159-164.

Wilson, R.E. and Ricc, E.L. (1968). Allelopathy as expressed by *Helianthus annuus* and its role in old-field succession. *Bulletin of the Torrey Botanical Club* **95:** 432-448.

Wink, M. 1983). *Planta* 158, 365-368.

Winter, A.G. and Schonbeck, F. (1953a). *Naturvissons-Chaften* 40, 168-169.

Winter, A.G. and Schontack, F. (1953b). *Naturvissonehaften* 41, 145-146.

Wolf, R.B., Spencer, G.P. and Kwolek, W.E. 1984). *Weed Sci.* 32, 612-615.

Woods, F.W. (1960). Biological antagonisms due to phytotoxic root exudates. *Botanical Review* **26:** 546-569.

Woodward, M.D. and Glenn, S. (1983). *Proc.* 37th *Ann. Meet North-eastern Weed Sci. Soc.* p. 114.

Worsham, A.D. (1984). Crop residues kill weeds. Allelopathy at work with wheat and-rye. *Crop and Soil Magazine* **37(1):** 18-20.

Worsham, A.D. (1989). Current and potential techniques using allelopathy as an aid in weed management. In: *Phytochemical Ecology: Allelochemicals, Mycotoxins and Insect Pheromones and Allomones* (Eds., C.H. Chou and G.R. Waller). Institute of Botany, Academia Sinica Monograph Series No. **9:** 275-291. Taipei, China.

Worsham, A.D. (1991). Allelopathic cover crops to reduce herbicide input. In: *Proceedings Southern Weed Science Society,* USA **44:** 58-64.

Worsham, A.D. and Blum, U. (1992). In Proc. Ist International Weed congress Vol. 2, 577-579. Weed Sci. Soc. Victoria, Melbourne, Australia.

Wozniak, K.L., Sullivan, W.M., Flowerday, A.D. and Walden, R.P. (1981). *Proc.* 73rd *Ann. Meet Am. Soc. Agron.* p. 116 (Abst.)

Wu, W.M.H., Liu, C.L. and Chao, C.C. (1976a). *J. Chinese Agri. Chem. Soc.* 96, 16-37.

Wu, W.M.H., Liu, C.L., Chao, C.C., Shieh, S.W. and Lin, M.S. (1976b). *J. Agric. Assoc. China, Taipei* 96, 16-37.

Yamada, O., Kaise, Y., Futatsunga, F., Issida, S., Ito, K., Yamamoto, H and Munakata, H. (1972). *Agric. Biol. Chem.* 36, 2013-2015.

Yang, S.X., Chen, R.Y. and Wu, Z.L. (1987). *Plant Physiol. Commun.* 1, 18-21.

Young, A. (1804). The Farmers Calender London.

Young, C.C. (1984). Autointoxication in root exudates of *Asparagus officinalis* L. *Plant and Soil* **82:** 247-253.

Young, C.C., Thorne R.L.Z. and Waller, G.R. (1989). Phytotoxic potential of soils and wheat straw in rice rotation cropping systems of subtropical Taiwan. *Plant and Soil* **120:** 95-101.

Yuhara, I. (1971). *Bull. Sugarbeet Res.* 13, 201-205.

Zabyalyendzick, S.F. (1973). *Vyestsi Akad. Navuk, BSSR Syer Biyal Navuk* 5, 31-34 (In Byelorussian) Russian Summary.

Zimdahl, R.L. (1987). Allelopathy. *FAO Plant Protection Bulletin* **35:** 70- 74.

Zimdahl, R.L. (1980). Weed Crop Competition - A Review. Int. Plant Prot. Centre, Oregon State Univ. Coravallis, OR.

Zink, J. and Hurle, K. (1990). Effect of cover crops on weed infestation in maize. *Zeitschriftfuer Pflanzenkrankheiten und Pflanzenshutz* **12:** 237-247.